Recycling and Development of New Building Materials or Products

Recycling and Development of New Building Materials or Products

Guest Editor

Yuri Ribakov

MDPI

Basel • Beijing • Wuhan • Barcelona • Belgrade • Novi Sad • Cluj • Manchester

Guest Editor
Yuri Ribakov
Department of Civil
Engineering
Ariel University
Ariel
Israel

Editorial Office
MDPI AG
Grosspeteranlage 5
4052 Basel, Switzerland

This is a reprint of the Special Issue, published open access by the journal *Materials* (ISSN 1996-1944), freely accessible at: https://www.mdpi.com/journal/materials/special_issues/recycling_building_materials.

For citation purposes, cite each article independently as indicated on the article page online and as indicated below:

Lastname, A.A.; Lastname, B.B. Article Title. *Journal Name* **Year**, *Volume Number*, Page Range.

ISBN 978-3-7258-3056-5 (Hbk)
ISBN 978-3-7258-3055-8 (PDF)
https://doi.org/10.3390/books978-3-7258-3055-8

Contents

materials

MDPI

Article

Designing Efficient Flash-Calcined Sediment-Based Ecobinders

Mouhamadou Amar [1,2,*], Mahfoud Benzerzour [1,2] and Nor-Edine Abriak [1,2]

1 IMT Nord Europe, Institut Mines-Télécom, Centre for Materials and Processes, F-59000 Lille, France
2 Univ. Lille, Institut Mines-Télécom, Univ. Artois, Junia, ULR 4515—LGCgE—Laboratoire de Génie Civil et Géo-Environnement, F-59000 Lille, France
* Correspondence: mouhamadou.amar@imt-nord-europe.fr

Abstract: To ensure the optimum navigation of boats and protection against flooding, waterways and ports are regularly dredged. The volume of dredged materials represents 56 million m^3 in France and 300 million m^3 in Europe. These materials show a high potential for a use as supplementary cementitious material (SCM). In this paper, sediments treated by the flash calcination method (STFC), which is based on a low-energy consumption process, are utilized as a mineral admixture in a cementitious matrix. The results of the physical, chemical, and mineralogical characterization prove that this heat treatment has an interesting impact on the final properties of the sediments. Mortars based on the flash-calcined product have comparable mechanical properties to control mortar. For a substitution rate below 10%, the performances are even equivalent to a metakaolin (MK80)-based mortar. Calorimetry testing demonstrated that calcined materials also improve hydration processes in the cement matrixes by generating additional heat release due to sediment pozzolanic activity. Across this study, it is shown that waste material including sediment can be transformed after optimized heat treatment into a valuable resource for the building and infrastructure sector.

Keywords: sediment; characterization; flash calcination; substitution; pozzolanic activity

Citation: Amar, M.; Benzerzour, M.; Abriak, N.-E. Designing Efficient Flash-Calcined Sediment-Based Ecobinders. *Materials* **2022**, *15*, 7107. https://doi.org/10.3390/ma15207107

Academic Editor: Yuri Ribakov

Received: 11 June 2022
Accepted: 10 October 2022
Published: 13 October 2022

Publisher's Note: MDPI stays neutral with regard to jurisdictional claims in published maps and institutional affiliations.

1. Introduction

In the early 21st century, the environment has become one of the biggest issues. Climate change and environmental degradation pose an existential threat to Europe (EU) and the rest of the world. To implement a new growth strategy regarding environmental issues, several countries, notably in the EU, are conducting the "Green Deal" approach leading to sustainable development. The principle is to dissociate economic growth from resource use and also to reach zero greenhouse gas emissions by 2050. Recently, the International Conference on Climate Change (COP 26, 2022) was held in Scotland and endeavored to regulate the global carbon market, climate change, worldwide temperature rise, water supply problems, etc. Several government policies nowadays aim to optimize productivity through better efficiency in order to reduce the use of natural resources, energy consumption, and waste production.

Ordinary Portland Cement (OPC), which is the essential component in concretes and mortars, is responsible for at least 7% of global CO_2 emissions resulting from the annual worldwide manufacture of 4.6 billion tons of this product [1]. According to recent predictions, the share of global CO_2 emissions attributed to cement production is likely to reach 25% in 2050, hence the urgent need to improve its sustainability [2]. In China, the cement industry, which is one of the most energy-intensive sectors emitting 1.2 to 2 billion tons of carbon per year [3], is developing technologies to reduce its impact. Through Carbon Capture and Storage (CCS), alternative raw material use, and energy efficiency, it is targeted to cut down 30% CO_2 emissions by 2050 [4,5]. The adopted strategy in Europe and the USA is also similar through the use of Limestone Calcined Clay Cement (LC3) and nanotechnologies [1,6].

In the construction sector, the aggregate need was 445 million tons in France in 2021. More than 96% of these aggregates are of natural origin. In 2019, the turnover in the aggregates sector increased by up to 3.66 billion euros and the ready-mixed concrete (RMC) market production was 3.77 billion euros [7]. Meanwhile, cement consumption in France was approximately 18 million tons. This constitutes huge financial profits for industrial companies but is also alarming for the environment and resource use. For all these reasons, sediments can potentially be used as suitable materials (as SCMs) in the construction sector [8–12]. The clinker content can be reduced significantly by replacing cement with SCMs, which have multiple effects: pozzolanic, hydraulic, and filler properties. Cement's high level of pollution is mainly due to limestone decarbonization, which is fundamental in order to achieve the correct chemical chemistry of Portland cement [13].

Saving natural resources is crucial, and environmental regulations are increasing. It is therefore sensible to develop the use of alternative materials in the construction sector. Several projects (ECOSED chair, FISP (US), USAR (EU), SETARMS, SURICATES, etc.) have already studied and considered using sediments as a resource for sustainable development purposes [8–12,14]. Every year in France, around 56 million m^3 [15] of sediments are dredged, stored, and treated. In Europe, the total volume is 300 million m^3, while in Brazil, it has reached 80.3 million m^3 [16,17]. Due to the high and growing demand for cement, the availability of conventional SCMs, such as blast furnace slag, limestone filler, silica fume, and natural pozzolans will not be enough to fully cover it. However, the beneficial use of dredged material relies strongly on numerous factors: mineral composition, grain size distribution, and physicochemical characteristics [18]. Furthermore, for this purpose, numerous scientific obstacles need to be resolved, including the inorganic contaminant content of sediments, such as heavy metals (lead, copper, chromium, etc.), salts, cyanides, and organic hydrocarbons (PAHs, PCBs, and TBT) [19].

Transforming raw sediment, considered waste, into a new material resource still faces technical challenges. One of the options is to process the sediments into an SCM by calcination, leading to the activation of mineral fractions. Heat treatment through "direct" or traditional calcination using a laboratory kiln has already been studied in sediments to improve some properties [20–28], but this method is expensive with high energy consumption and environmental impact. However, another innovative technique, called "flash calcination", has been applied in some studies for the activation of clays and to transform kaolin into metakaolin [29–34]. Cremona et al. [35] investigated the properties of fresh and hardened concrete and the durability of several metakaolin-based binders. It was concluded that flash-calcined clays improve workability, porosity, and creeping. Similar conclusions were stated by Sullivan et al. [36], showing that MK contributes to prolonging the service life of infrastructures. Danilevich et al. [37] activated thermally aluminum hydroxide in gibbsite using a reactor called TSEFLAR. Examining the characteristics of the final product, the study has shown that process parameters such as temperature, the residence time of the particles, and raw material consumption play a major role in activation processes. This new calcination technique has the advantage of lowering energy consumption and gas release due to the quickness of the process. As a matter of fact, this process includes dehydroxylation that activates the material with thermal excitation between 700 °C and 850 °C. For products containing undesirable non-active materials, complex calcium-based minerals, sulfur-rich phases [38], or phosphates [39], flash calcination can be considered to be more advantageous than traditional treatment methods such as rotary calcination. Flash calcination is also able to activate Muscovite, illite, phlogopite, and kaolinite, which are among the clay minerals that show pozzolanic activity after appropriate heat treatment and are generally present in sediment minerals. Such results were previously obtained by Ambroise [40] after calcination at 750 °C. This triggers and activates amorphous phases [30,41]. The activation of the clay fractions present in sediments has recently been studied [2,26,42–44]. The final product is also usable as SCM for traditional Portland cement replacement [45] and in geopolymer applications [46]. Amar et al. [47] studied the reactivity of flash-calcined sediment. By using various pozzolanic tests, it was demonstrated that the calcination process has a positive impact on the pozzolanic activity of sediments [48]. Snellings et al. [26,42] also used the flash calcination technique on the same sediment studied by Van Bunderen et al. [25]. This last study investigated the influence of the temperature of treatment and the final reactivity of sediment. The authors tested three calcination temperatures: 820 °C, 865 °C, 905 °C, and showed the influence on the final product properties. Van Bunderen et al. [34] also studied the effect of flash-calcined sediment on mechanical properties and the shrinkage of concretes. After testing cement replacement at 20, 30, and 40 wt%, the results clearly demonstrated that flash-calcined sediments are suitable for the production of sustainable cement and concrete. Inocente et al. [33] recently used this activation technique to produce highly reactive metakaolin. The chosen calcination temperatures ranged from 900 °C to 1100 °C. It was also shown that flash calcination is suitable for transforming unreactive material into a product with a high degree of amorphization.

The flash calcination treatment technique based on a low-energy process is studied in the following. This innovative method is not widely used in the waste-reuse sector. This method was shown to be capable of enhancing the pozzolanic activity of sediments. As a result, these materials that were initially considered waste can be transformed into a suitable SCM for the construction sector.

First, this paper will focus on the characteristics of sediments treated with the flash calcination method (STFC) and will then highlight the impact of the treatment. Next, mortars using STCF as eco-efficient SCM will be formulated and tested. Finally, a comparative study will be made of raw sediment (RS), metakaolin (MK), and limestone filler (LF).

2. Materials, Methods, and Experimental Work

2.1. Materials

Portland cement CEM I 52.5N (OPC), which complies with European standard EN 197-1 (2012), was used in this study. The mortar formulations were based on a siliceous standardized river-origin sand (ISO 679: 2009), with rounded grains.

The marine sediments were dredged from the Grand Port Maritime de Dunkerque (GPMD), situated in the North of France. Using an oven at 40 °C, all raw sediments were dried, then crushed, and sieved at 120 μm. In this study, the flash-treated sediments are designated by STFC, RS = raw sediment, NS = natural sand. MK80 is a metakaolin with Dmax = 80 μm, and LF80 is a limestone filler with Dmax = 80 μm.

2.2. Characterization Techniques

Different physical and chemical methods were used to identify the characteristics of the materials. Granulometry analysis was performed using a laser diffraction particle size analyzer COULTER LS12330. With this device equipped with a dry module, the particle distribution was determined from 0 to 2 mm. The BET (Brunauer–Emmett–Teller) tests, also called mass area and which are an estimation of fineness (NF EN ISO 18757 [49]), were performed on a Micromeritics ASAP2020 analyzer using nitrogen gas. The samples were outgassed for 16 h at 40 °C. Blaine surface was measured following the NF EN 196-6 [50] standard. The organic fraction (LOI) was also determined (XP P94-047 [51]), and the total organic carbon (TOC) content was evaluated. The specific density of the products was determined by a MICROMETRICS AccuPyc1330 helium pycnometer (NF EN 1097-7 [52]). Heat release investigations using a semi-adiabatic calorimeter CERILH (Langavant method) according to the NF EN 196-9 [50] standard were also carried out for the first seven days following initial cement hydration. The apparatus was 400 × 160 mm with a thermally sealed container equipped with heat sensors. Differential Scanning Calorimetry (DSC) and Thermogravimetric Analysis (TGA) were performed on a NETZSCH STA 449 using nitrogen gas. The analysis program was set to work between 40 °C and 1100 °C and CO, CO_2, and water detection were activated. The mineral composition was investigated using X-ray diffraction on a Bruker D8 device equipped with a Co anticathode. The Rietveld refinement method was also applied, especially for the quantitative measurements of the clay fractions. Analysis was conducted as follows: between 6–80° angle 2θ, 0.2° pitch, and acquisition time 0.5 s. Scanning Electron Microscope (SEM) analysis was performed using a Hitachi S-4300SE/N with a field emission gun; the speed voltage was adjusted to 5 kV.

2.3. Mortar Preparation

The composition of mortars used to study mechanical strength is given in Table 1. The consistency test was executed according to NF EN 196-3:(2009). This test guarantees to have an equivalent fresh state between the control mortar and those including SCMs to keep results comparable. Mechanical strength was evaluated using prismatic mortar specimens 4 cm × 4 cm × 16 cm (NF EN 196-1 [53]). RM is the control mortar. MRS and MSTFC are the respective raw-sediment-based and flash-calcined-based mortars. Formulations using MK and LF are designated by MMK80 and MLF80 in this order.

Table 1. Mortar composition of RM, MRS, MSTDC, and MSTFC for the heat of hydration test (NF EN 196-9) [50].

<table>
<tr><th>Constituent</th><th>RM</th><th colspan="3">MRS</th><th colspan="3">MSTFC</th><th colspan="3">MMK80</th><th colspan="3">MLF80</th></tr>
<tr><td>Cement CEM I 52.5 N (g)—5% | 10% | 15%</td><td>450.0</td><td colspan="12">427.5 | 405 | 382.5</td></tr>
<tr><td rowspan="2">Sediment (g)</td><td rowspan="2">---</td><td>5%</td><td>10%</td><td>15%</td><td>5%</td><td>10%</td><td>15%</td><td colspan="3" rowspan="2">---</td><td colspan="3" rowspan="2">---</td></tr>
<tr><td>28.6</td><td>57.2</td><td>85.8</td><td>26.8</td><td>53.5</td><td>80.3</td></tr>
<tr><td rowspan="2">Metakaolin (g)</td><td rowspan="2">---</td><td colspan="3" rowspan="2">---</td><td colspan="3" rowspan="2">---</td><td>5%</td><td>10%</td><td>15%</td><td colspan="3" rowspan="2">---</td></tr>
<tr><td>25.9</td><td>51.7</td><td>77.6</td></tr>
<tr><td rowspan="2">Limestone filler (g)</td><td rowspan="2">---</td><td colspan="3" rowspan="2">---</td><td colspan="3" rowspan="2">---</td><td colspan="3" rowspan="2">---</td><td>5%</td><td>10%</td><td>15%</td></tr>
<tr><td>26.2</td><td>52.5</td><td>78.7</td></tr>
<tr><td>Normalized sand (g)</td><td>1350</td><td colspan="3">1350</td><td colspan="3">1350</td><td colspan="3">1350</td><td colspan="3">1350</td></tr>
<tr><td>Water (g)</td><td>225</td><td colspan="3">225</td><td colspan="3">225</td><td colspan="3">225</td><td colspan="3">225</td></tr>
</table>

2.4. Flash Calcination Method

Flash calcination is a heat treatment technique consisting of the exposure of finely ground products to high temperatures in the presence of air. In flash calcination, gas–solid heat transfer is exploited effectively to calcinate the material quickly. This technique was developed in France by Salvador in 1992 [31]. Originally, the flash calcination technique was applied to activate clays such as kaolinite and to trigger their pozzolanic properties. This treatment method provokes a dihydroxylation process following dehydration [26,30,54]. This fits with the elimination of a hydroxyl bond (−OH) that occurs between 450 °C and 750 °C [55]. In addition, as the flash process is spontaneous, the flash calcination initiates amorphous phases and therefore boosts the reactivity of the product [2,56]. In fact, the temperature rise can be estimated from 0.5 to 1.5×10^4 °C/s [57]. Due to the rapid process, the material obtained has a lower density and higher porosity translating to increased reactivity. Direct calcination using a laboratory kiln at 750 °C for a residence time between 1 h to 3 h can lead to such results. However, in this case, energy costs and the CO_2 footprint, in particular, are usually high. In flash calcination, the product is heated rapidly (up to 1200 °C), held at this high temperature for a short period (0.1 to 1 s), and finally quenched rapidly with the counter flow of cold air before storing [44,54]. The visual aspect of the sediment throughout the treatment process is also shown in Figure 1.

Adjusting the flash calcination process requires the control of two parameters: exposure time and temperature, which can reach 1200 °C inside the calciner unit [30]. The energy used in this flash calcination process can be evaluated at 2 GJ/t according to San Nicolas [30,58]. This study aimed to apply this treatment process to the marine dredged sediments, which will be adapted to activate sediment phases (including clays). The calcining chamber unit, flash tower, and pilot installation are shown in Figure 2.

Figure 1. The visual aspect of sediment through treatment stages.

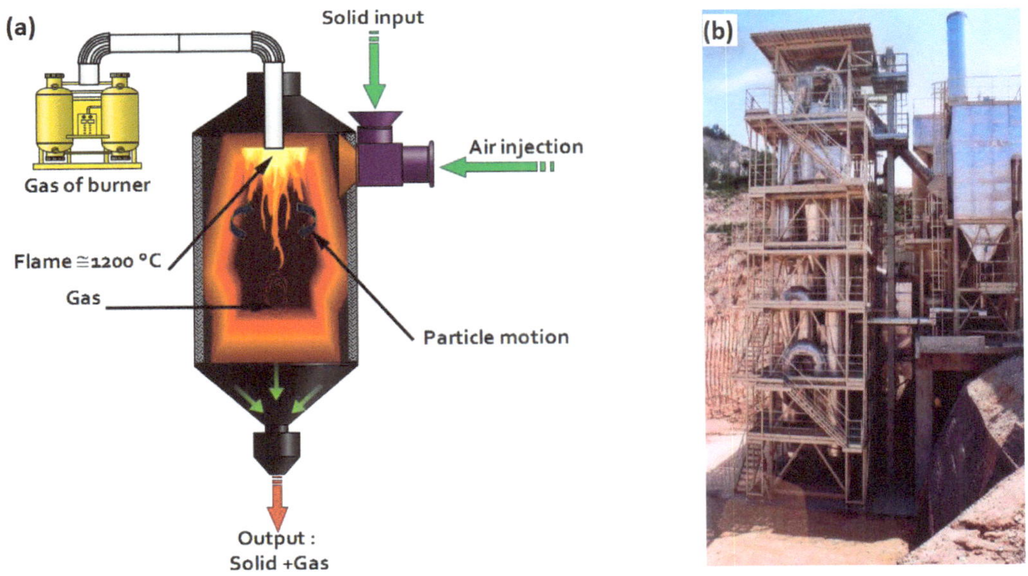

Figure 2. *Cont.*

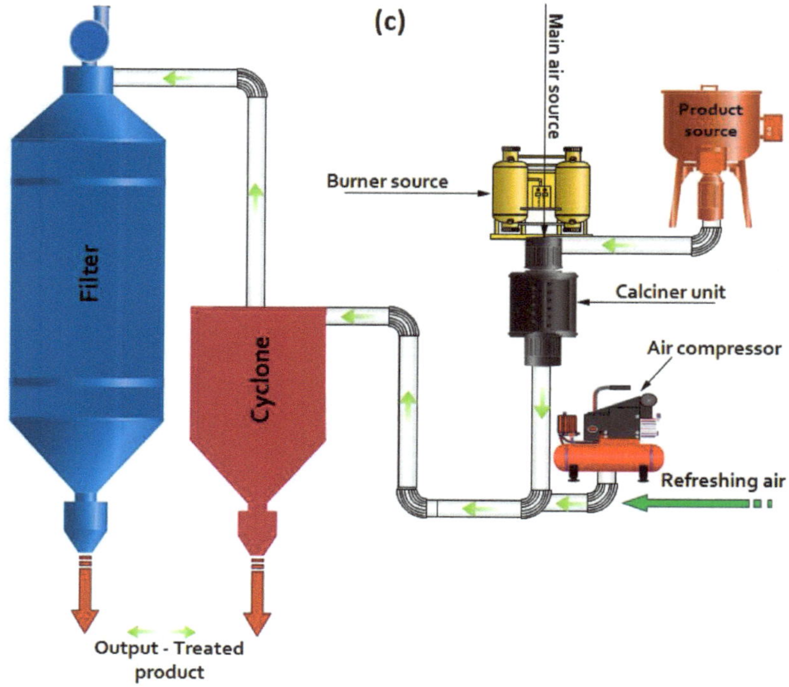

Figure 2. (**a**) Flash calciner chamber; (**b**) Semi-mobile flash tower [22]; (**c**) Pilot installation.

3. Experimental Results and Discussion

3.1. Physical Characterization

3.1.1. Granulometry

The results of the laser granulometry test are shown in Figure 3. The calcination process affects the particle size distribution (D_{50} = 2.46 µm for RS, and D_{50} = 5.93 µm for STFC). The sintering effect between particles may be the main reason RS is finer than STCF and Portland cement. This result agrees with SEM observations showing the presence of very fine particles in the RS sample. Particle size distribution plays a major role in pore size distribution. It was also shown through fractal analysis that it impacts hydration, porosity, durability, and mechanical resistance [59,60].

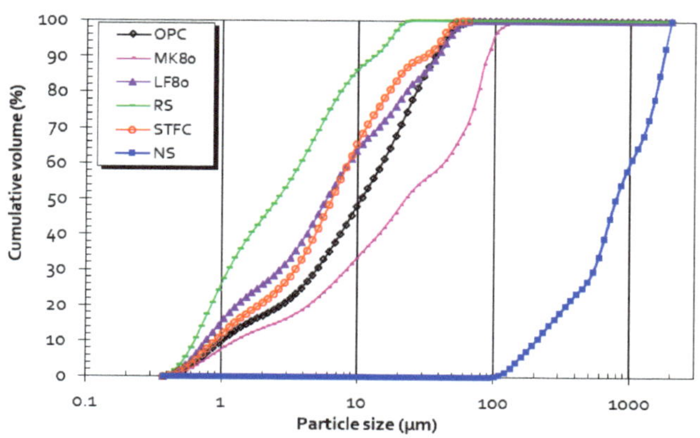

Figure 3. Particle size distribution for OPC, MK80, LF80, NS, RS, and STFC.

3.1.2. Physical Properties

The results of the BET and the Blaine tests are given in Table 2. The results demonstrate that the flash calcination process has an impact on the fineness of the materials. This effect is the outcome of the organic matter elimination, which wraps the sediment particles. Therefore, with the calcination, organic matter is eliminated, and the BET/Blaine value decreases. Fineness is a parameter connected to the reactivity of the material and its performance. For the flash-calcined sediments, the LOI (1.7%) of the final product is low and proves the efficiency of calcination.

Table 2. Determination of the physical properties of the materials.

Materials	Cement (CEM I 52.5 N)	Raw Sediment (RS)	Flash Calcined Sediment (STFC)	Metakaolin (MK80)	Limestone Filler (LF80)	Natural Sand (NS)
Density (g/cm^3)	3.15	2.48	2.65	2.74	2.70	2.65
Blaine (cm^2/g)	3669	10,093	4106	4820	7181	4548
BET (cm^2/g)	9194	86,207	59,930	94,600	9744	9507
TOC * (%)	---	5.76	0.99	---	---	---
LOI * (%)	1.90	9.92	1.70	---	---	---

TOC * = total organic carbon = 1.72 * LOI [61]; LOI * = loss of ignition.

Generally, a decrease in the specific surface area is observed after the consistent thermal process. This variation is inversely proportional to density, as shown in Figure 4. It may indeed be explained by physical and chemical modifications due to sintering [62]. In Figure 1, STDC is the sediment treated by the traditional calcination process. Snellings et al. [42] measured the effect of the flash calcination process on sediment and established that the higher the calcination temperature, the lower the specific surface area. According to [29], the calcination process induces the formation of spherical particles that are the product of multiple melting–sintering and destructuration processes.

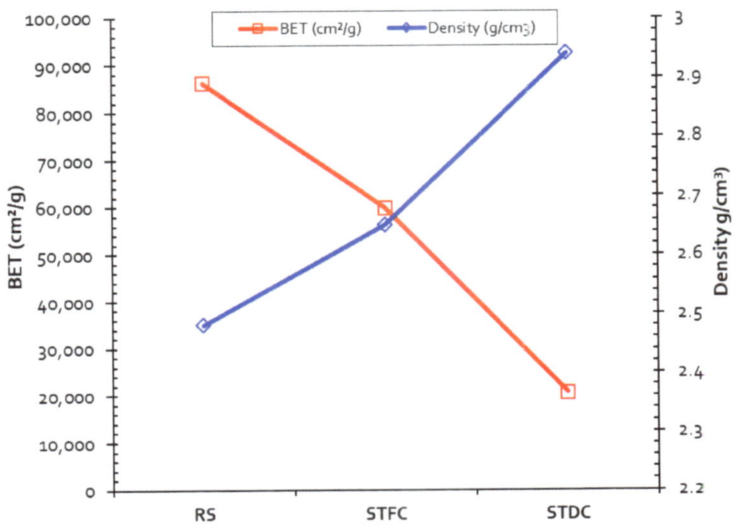

Figure 4. Evolution of density and BET surface area for RS, STFC, and STDC.

Finally, it is important to highlight that a heat treatment usually produces the following effects [63]: (a) Modification of the bulk density (increase and decrease from 600 °C);

(b) Modification of the specific area (increase and decrease from 400 °C); (c) Decrease of internal porosity by densification; (d) Change in grain size (D_{50} increases strongly at 700 °C).

3.1.3. TGA and DSC Results

The TGA results displayed in Figure 5a highlight a peak in the mass loss at 450 °C corresponding to the dehydration phase. The release of carbon monoxide (CO) and carbon dioxide (CO_2) at 420 °C and 480 °C, respectively, is due to the combustion reactions of organic matter and organic pollutants (PAHs, PCBs, TBTs) present in the sediment [64]. A second CO_2 peak appearing at 700 °C is caused by the decarbonation of calcite and dolomite [23]. These calcareous materials originate from the shells of marine species naturally present in sediments. According to these results, the temperature that guarantees organic matter elimination and also complete decarbonation must be greater than 700 °C.

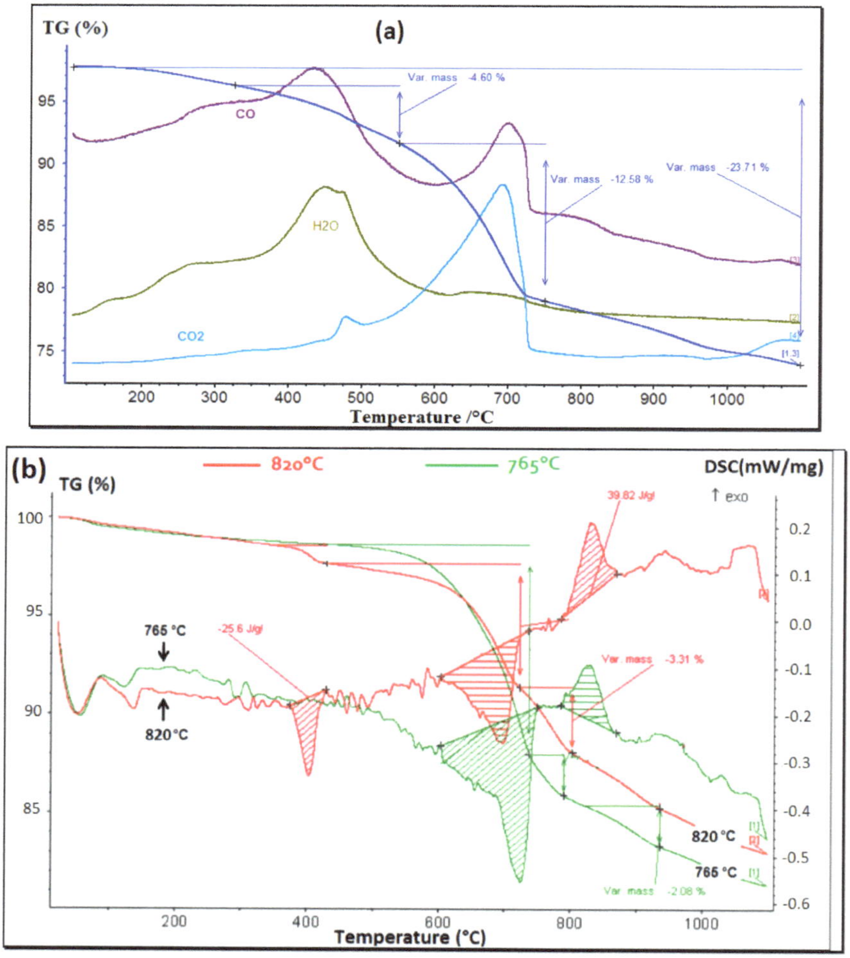

Figure 5. (**a**) TGA results for RS and (**b**) TGA on pre-flash-calcined sediment (STFC).

The choice of the optimal calcination temperature relies on the DSC results depicted in Figure 5b. This test was performed on sediments that were flash calcined at 765 °C and 820 °C. The first peak at 400 °C probably corresponds to the organic matter decomposition as stated in Figure 5a. The analysis reveals the recrystallization area at 810 °C is narrow, indicating that recrystallization was incomplete. These results prove that flash calcination at 765 °C or 820 °C leads to similar results, but is more energy costly at 820 °C. Due to its efficiency, the choice was made to adopt 765 °C as the optimized temperature of calcination.

3.2. Chemical and Mineralogical Analyses

The purpose of the chemical analysis was to determine the proportion of each chemical element using X-ray Fluorescence (XRF) analysis (see Table 3). These values are close to those identified in previous studies [65,66] conducted on GPMD sediments. As for the calcined sediments, the main oxide constituents SiO_2, Al_2O_3, and Fe_2O_3 were similar to the RS values. These results confirm the possible moderate impact of the calcination process on chemical composition. The composition of cement is shown in Table 4 and corresponds to commercial Portland cement regarding its chemical composition.

Table 3. Concentrations (%) of the major oxide elements for OPC, RS, and STFC.

	SiO_2	Al_2O_3	Fe_2O_3	CaO	SO_3	K_2O	TiO_2	Na_2O	MgO	P_2O_5	MnO	ZnO
OPC	20.0	5.1	3.4	63.5	3.1	0.8	0.3	0.3	0.8	0.4	0.1	0.2
RS	51.9	8.2	9.3	22.1	0.2	1.9	0.4	2.0	2.0	0.4	0.2	0.1
STFC	52.8	8.0	9.3	21.6	0.2	1.8	0.4	1.9	2.0	0.4	0.2	0.1

Table 4. Chemical composition of cement (%).

	Alite (C_3S)	Belite (C_2S)	Aluminate (C_3A)	Ferrite (C_4AF)	Anhydrite ($CaSO_4$)	Gypsum ($CaSO_4$, $2H_2O$)	Lime (CaO)
OPC	63.7	16.1	4.3	9.8	0.5	3.8	1.8

3.3. X-ray Diffraction Analysis (XRD)

This test revealed the high presence of quartz (SiO_2), calcite ($CaCO_3$), and other phases such as pyrite (FeS_2) or clays (kaolinite) in the RS. The results of the XRD analysis performed on the RS, STDC (traditional calcination) and STFC materials are given in Figure 6 and show major crystalline modifications. The heat probably initiated some physicochemical processes. The XRD clearly demonstrates a drop in the occurrence of crystalline phases such as calcite due to the decarbonization phase. Clay phases such as kaolinite must be transformed into reactive metakaolin [30,56]. It is also noted that there was an initiation of phases such as anhydrite ($CaSO_4$), which is the result of reactions between calcium carbonates and sulphate. Anhydrite can modify cement hydration reactions as stated in the investigation led by Snellings et al. [42]. Furthermore, mineral additions with a high calcite ($CaCO_3$) content must favor the hydration of alite (C_3S) [57,68]. This effect is particularly accentuated as the calcite content and granularity are relatively high. It seems that in the presence of water, calcium carbonate reacts with alite (C_3S) and aluminate (C_4AH_{13}) to form hydrated mono-carboaluminate calcium [C_3A, $CaCO_3$, $11H_2O$], which crystallizes in hexagonal platelets [69,70].

These analyses were performed on a single-oriented glass slide to identify clay minerals; the results are presented in Figure 7. They are based on the determination of lattice spacings (001) and relative intensities. The calcination process transforms the majority of the clay minerals present into illite (86%). According to Dang et al. [23], the occurrence of newly created phases can be attributed to the heat treatment effect.

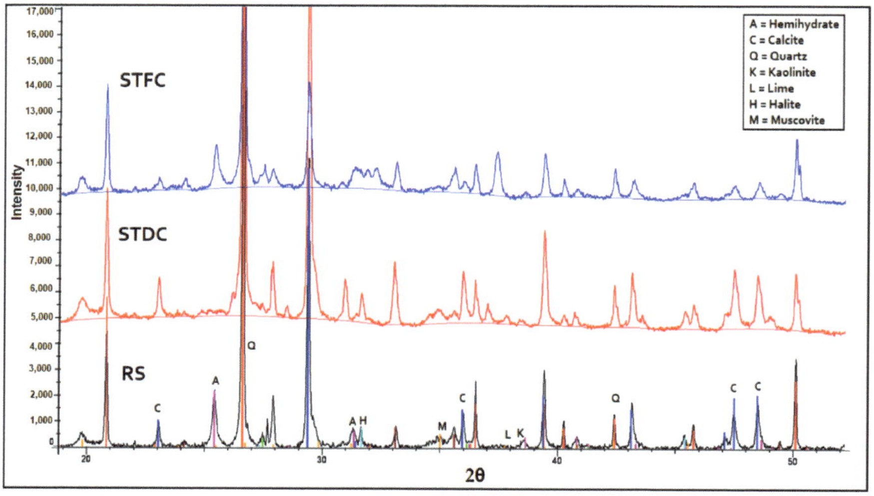

Figure 6. XRD analysis of RS, STDC, and STFC.

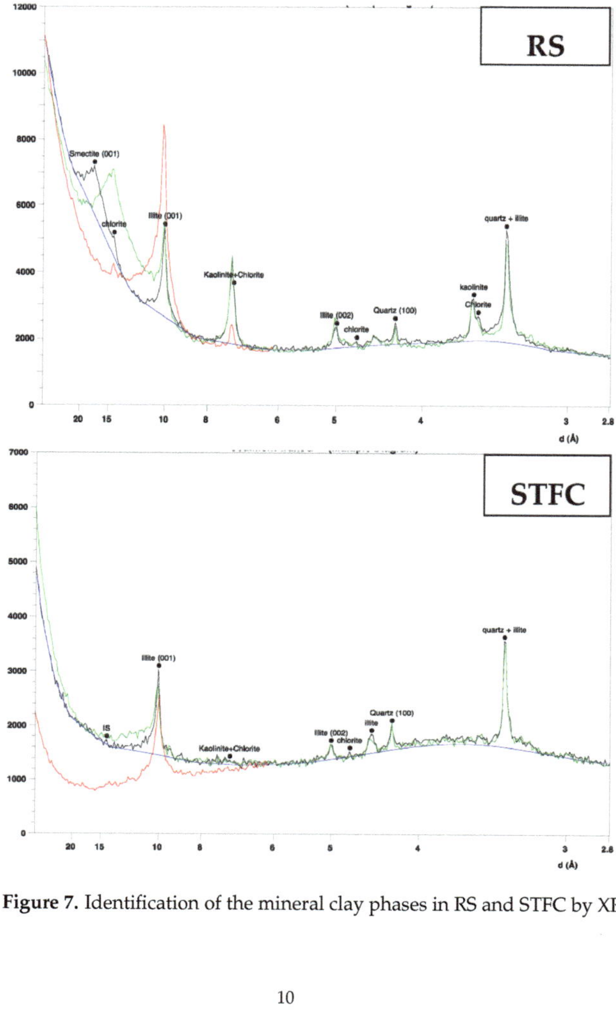

Figure 7. Identification of the mineral clay phases in RS and STFC by XRD.

3.4. XRD Analysis of Clays

The Rietveld semi-quantitative analysis is based on optimization techniques and refinement. The methodology applied relies on the difference between the measured and the calculated diagrams. The statistical criterion generally applied for the fitting quality is the weighted least square error value. The tests were performed on the sediment RS and STCF. Samples were prepared using a single-oriented glass slide to identify clay mineral proportions. This analysis is also based on the determination of lattice spacings (001) and relative intensities. Samples should be prepared carefully, and specific skills are required by the operator. The results are presented in Table 5 and Figure 7.

Table 5. Identification of clay minerals by the XRD method.

Samples	Smectite	Illite	Kaolinite	Chlorite	Interlayered 10-14S Swellings	Additional Minerals
RS	32%	36%	20%	12%	—	Quartz
STFC	—	86%	5%	—	9%	Quartz

The analysis reveals that the RS samples are composed of kaolinite, which represents 20% of the clay phases, and chlorite (12%). The main phases are illite (36%) and smectite (32%). The STFC sediment is composed mainly of 86% illite and the remaining 5% of kaolinite. The calcination procedure turns the majority of the clay minerals into illite (86%). These figures demonstrate that heat treatment initiates internal structural modification in clay minerals [42]. This can be partly the cause of chemical activity noticed previously by Amar et al. [47] in these sediments. The occurrence of these newly created phases can be explained by the effect of heat treatment [23,41].

3.5. Scanning Electron Microscopy (SEM)

The SEM analysis consists of examining the internal structure of the materials at a very high resolution. SEM analysis provides powerful magnification, up to 500,000 times. As a result, the majority of the micron constituents of mortar can be examined, scrutinized, and studied. In a cementitious matrix, these constituents can be CSH hydrates (≈100 Å) and strips of clay particles (≈10 Å) from SCM, ettringite crystals (≈50 μm), crystals of portlandite $Ca(OH)_2$ (≈20 μm), or the morphology of finely-ground material. The results in Figure 8d reveal different particle forms that are mineral or animal (seashell) in origin. These types of particles are commonly observed in sediments. As a result of sedimentation and the presence of minerals such as limestone ($CaCO_3$), the phenomenon of cementing between the particles can be observed (Figure 8b). For the RS material, the presence of pyrite particles and some plate-like particles can be clearly distinguished (Figure 8a). Pyrite is a mineral originally present in raw sediments and is visible in Figure 8a. The presence of organic constituents or pollutants that needed to be eliminated justifies the need for an appropriate treatment choice (using heat treatment, for instance).

For STFC, physical changes were confirmed by the SEM observations. The heating process modified the form of the particles with the development of regular spherical particles with a diameter of 20 μm (Figure 8c). In previous studies led by Nicolas et al. [30] and also by Teklay et al. [44], these spherical particles were identified on flash-calcined kaolin clay, representing almost 20% of the final product. Similarly, Claverie et al. [29] showed that these particles originated from an agglomeration process of submicron particles, as shown in Figure 8c. They are mainly composed of gases and nano-particles of aluminum silicates that cover the external surface (Figure 8c). Through later studies, Snellings et al. [26,42] highlighted the local melting process origin of this rounded effect.

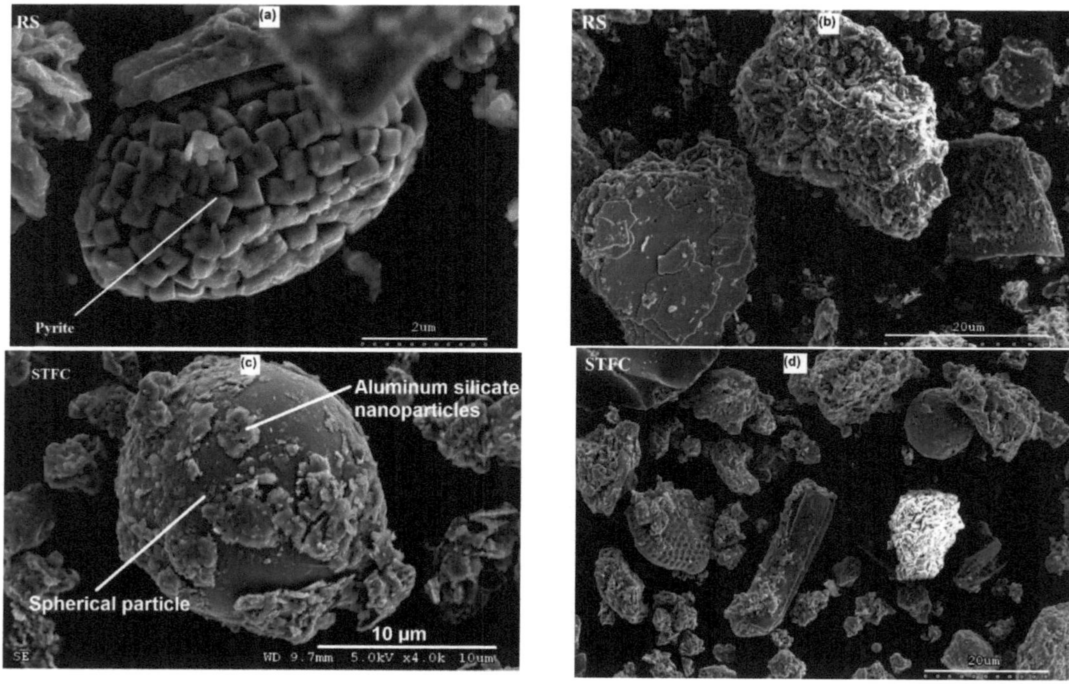

Figure 8. Electron microscopy of RS and treated STFC sediment. (**a**) Pyrite particle, (**b**) plate-like cemented particles, (**c**) Spherical aluminosilicate nanoparticle, (**d**) Organic particles and seashells.

3.6. Impact of the Presence of Sediments on Cementitious Matrix

Due to their physicochemical composition, sediments modify the cementitious matrix properties [23,41]. In fact, this is materialized by the presence of lime as a result of the thermal conversion of calcite. Depending on their fineness and shape, sediments are able to play a role in filling space and thus contribute to improving overall compactness by reducing the intergranular space. This contributes to enhancing the hydration reactions and ameliorating general properties. However, the presence of constituents resulting from pollution, such as lead (Pb) [71], zinc (Zn) [72], cadmium (Cd), or chromium (Cr) [71], could disturb the normal hydration processes and the setting of mortar [73,74]. The mechanisms that usually lead to the formation of hydrates (CSH) or the setup of the porous network can be disrupted by some of these minerals [71,73,75]. The effect of metal contaminants has been studied by Minocha et al. and Park et al. [71,76]. It has been shown that the presence of inorganic contaminants in the cementitious matrix generally has a detrimental effect on mechanical strength and also on durability properties [76].

The formulated mortars are designated as follows: RM = control mortar (with no SCM); MRS = mortars with raw sediment; MSTFC = mortars including flash-treated sediments; MMK = metakaolin-based mortar, and MLF = mortar with limestone filler. The number after these notations stands for the substitution rate (e.g., MRS5 = 5% cement replacement).

3.7. Mechanical Strength

The mechanical strength of prepared mortars was monitored following NF EN 196-1 [53]. Prismatic samples $4 \times 4 \times 16$ cm^3 were tested in compression for all formulated mortars. The results of the compression tests indicate that STCF-based mortars have significant resistance. After 28 days of maturation, the most resistant mortars were MSTFC5 and MRS5, with respective compressive strengths of 62.2 MPa and 59.2 MPa (Figures 9 and 10). The substituted mortars containing 10% sediment exhibited the same strength as RM at

28 days, while the 15%-substituted mortars had lower strength than RM, in the range of 6% and 12%. The composition of mortars used to assess mechanical strength is given in Table 1. These results suggest that the presence of sediment, appropriately used in the cementitious matrix, results in an improvement in strength overall. Benezet et al. [77] demonstrated in a previous study that finely crushed particles of quartz (less than 5 µm or BET > 10,000 cm^2/g) are highly active. In the present study, we must bear in mind that D50 ≤ 6 µm for all calcined material and BET > 20,000 cm^2/g (Table 2). Therefore, one could argue that the flash calcination triggered the pozzolanic activity and impacted the materials' fineness. This is correlated with the elimination of organic matter and densification.

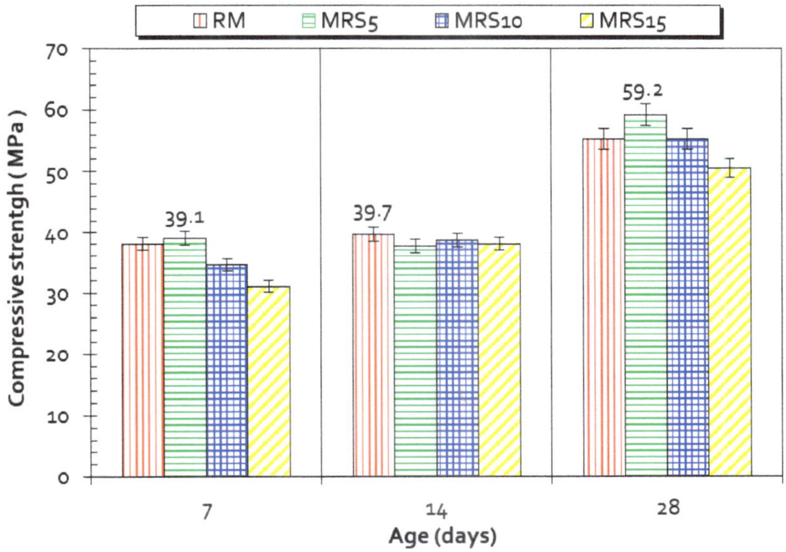

Figure 9. Compressive strength of RM and MRS, (5%, 10%, and 15%) mortars.

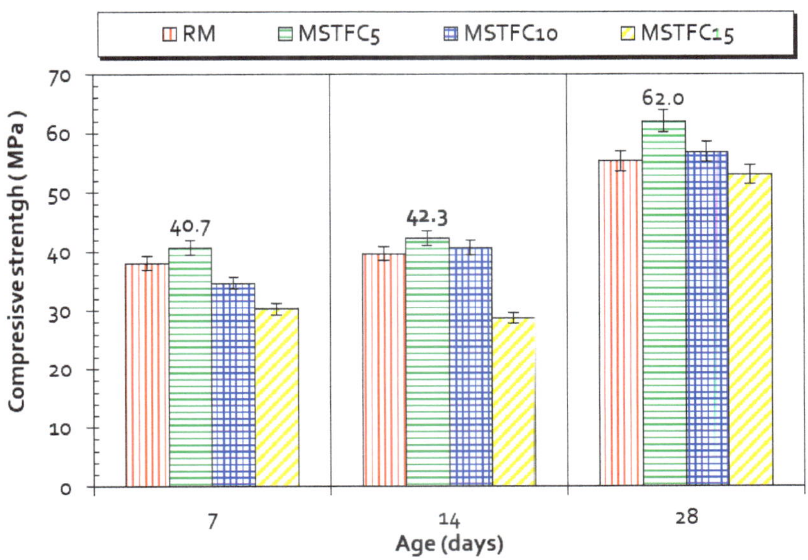

Figure 10. Compressive strength of RM and MSTFC (5%, 10%, and 15%) mortars.

Finally, a comparison between RM, MRS, MSTFC, MK-based mortars, and LF-based mortars was performed (Figure 11), clearly showing that MSTFC5 mortars are similar to the MMK5 and MLF5 mortars after 28 days of maturation. Nonetheless, there is slightly less resistance for MSTFC10 and MSTFC15. Their resistance values are in the range of 13% to 15% less than those for mortars containing MK and LF. This can be explained by the fact that MK is a highly pozzolanic material. Its activity coefficient could be considered to be χ = (0.9 to 1) [30,78].

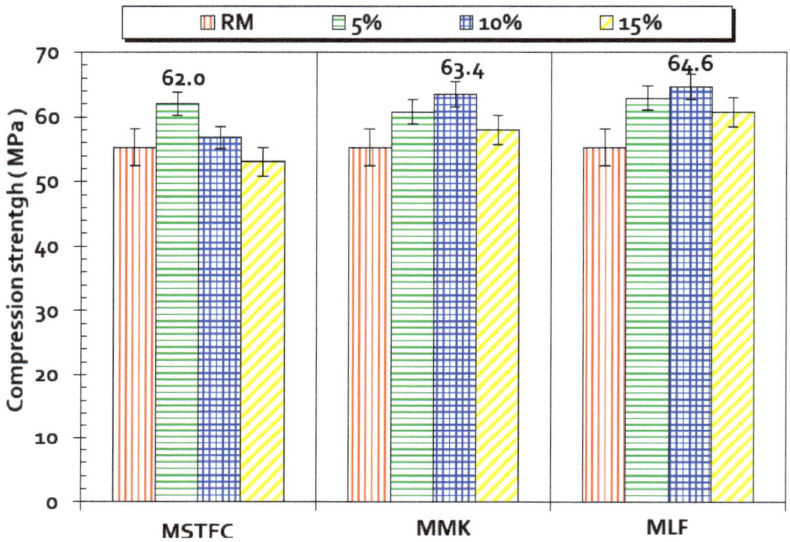

Figure 11. Compressive strength of MSTFC, MK and LF (5%, 10%, and 15%) mortars at 28 days maturation.

Studies conducted by Nicolas et al. [30] and Claverie et al. [29] on kaolin clays treated with flash calcination subsequently showed that the effect of the treatment should be recognized. It was demonstrated that physicochemical activity was noticed in the material after the clay phases had been activated. This suggests that calcined clays improve hydration reactions [2,26]. The decomposition of the physical and chemical effects indicates some strong connection between that physical effect and the phenomenon of heterogeneous nucleation. Nevertheless, the chemical effect seems to be controlled by complex chemical reactions, such as precipitation/dissolution, fineness [79], and amorphousness [26,80,81]. These effects probably triggered the enhancement observed in the mechanical performance [16,26,82–84]. Van Bunderen et al. [2,25] demonstrated that the addition of calcined dredging sediments reduces the autogenous shrinkage of cement and could potentially be used as an alternative SCM. According to Hentschel et al. [85] and Felekoglu [86], some factors such as grain shape and roughness can directly impact the rheology, fluidity, and porosity of mortars. The finely crushed minerals present can have some activity in the cementitious matrix [87].

3.8. Heat of Hydration—Langavant Calorimeter

The heat flow test can be considered a rapid screening test for supplementary cementitious materials (SCMs). The principle of this test is to monitor the variation in temperature and heat flow of mortars inside sealed measuring cells. The results are plotted in the graphs depicted in Figure 12. Mortar composition is reported in Table 6 (NF EN 196-9) [88], and the heat released can be deduced from the expression given in Equation (1). MK-based mortars are designated by MMK80, and MLF80 is an LF-based mortar. For the formulations, one substitution rate is tested (10%) because it seems to be the limit above which the compressive strength of MSTFC begins to decrease and becomes less than that of RM for a greater replacement rate.

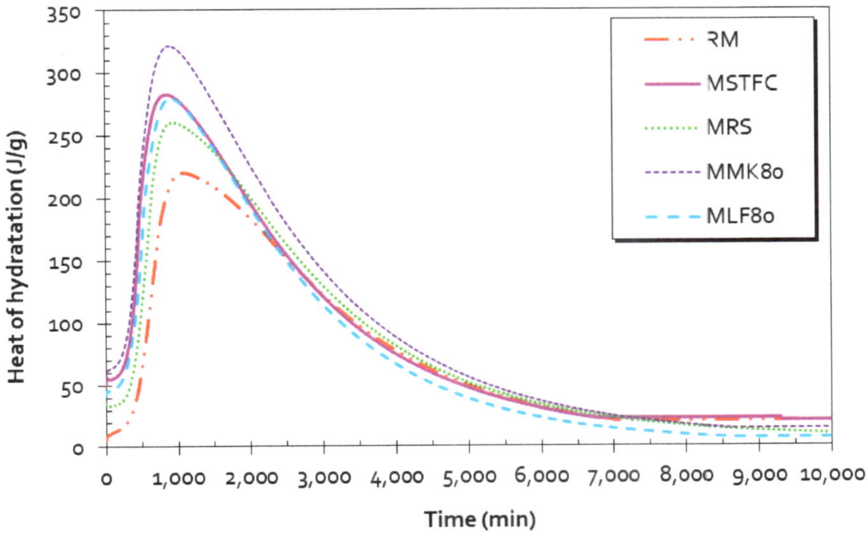

Figure 12. Relationship between the heat of hydration and time for RM, MRS, and MSTFC mortars.

Table 6. Mortar composition of RM, MRS, and MSTFC for the heat of hydration test (NF EN 196-9) [50].

Constituent	RM	MRS	MSTFC	MMK80	MLF80
Cement CEM I 52.5 N (g)	360.0	331.8	329.8	328.7	329.2
SCMs (g)	—	28.2	30.2	31.3	30.8
Normalized sand (g)	1080.0	1080.0	1080.0	1080.0	1080.0
Water (g)	180.0	180.0	180.0	180.0	180.0

$$Q = \frac{C}{m_c}\theta_t + \frac{\int_0^t \alpha * \theta_t * dt}{m_c} \quad (1)$$

where:

C = Total heat capacity (J/K);
θ_t = Temperature difference between the reference cell and measuring cell (K);
m_c = Mass of cement (g);
α = Heat dissipation coefficient (J/(g*K);
t = Time (s).

RM reached a maximal temperature of 51.9 °C after 17.80 h, which corresponds to a generated heat of Q = 240.08 J/g. For MSTFC mortar, the heat released at the end of 15.90 h of screening was Q = 285.63 J/g, and MRS had a lower value. MMK80 is the most active mortar due to the high pozzolanic activity of MK (Q = 324.5 J/g). An increase in the heat flow was noticed following the replacement ratios of cement by SCM. This is mainly attributed to the filler effect (physical effect) but also partly to the early pozzolanic reaction (chemical effect) of sediments, as discussed by Snellings et al. [26].

The cumulative heat patterns demonstrate that MRS and MSTFC show higher heat than RM and appear to react faster. The finding is that additional thermic or chemical activity should be generated by the activated sediment particles. It can also be related to the lime phase, which reacts with water following Equation (2), or the hydration of hemihydrate ($CaSO_4 \cdot 0.5H_2O$) in Equation (3) that produces additional heat output [89].

$$CaO + H_2O \rightarrow Ca(OH)_2 + Heat \quad (2)$$

$$\langle CaSO_4\ |0.5H_2O\rangle + 1.5H_2O \rightarrow \langle CaSO_4\ |2H_2O\rangle + Heat \quad (3)$$

The cumulative heat patterns indicate that both MRS and MSTFC show greater heat release than RM and appear to lead accelerate chemical reactions. The result is that additional thermic or chemical activity could have been generated by sediment particles, possibly related to the lime phase and pozzolanic reactions. One possibility to lower this effect if it is not desired could be to use cement with a low heat of hydration, such as Portland-limestone cement (PLC). Skibsted and Snellings [41] demonstrated that two mechanisms limit SCM reactions: water availability and space accessibility in the cementitious matrix. These parameters should therefore have a real influence if a higher replacement ratio of cement to SCM is considered. Snelson et al. [90] led a similar study using two pozzolanic materials: metakaolin and fly ash. They showed that MK generates an enhancement of the overall hydration and the heat release is higher due to the beginning of rapid pozzolanic reactions. In cementitious materials, pozzolanic reactions are controlled by the supply of Ca^{2+} ions, as discussed by Snelson [90]. This can explain the heat excess generated by the sediments due to their high water requirements (typically ≅ 50%) and calcium content (≅15%). This suggests that sediments change heat release [23,26] and can efficiently enhance hydration reactions [25,26,30].

3.9. Initial Setting Time

The initial setting time test (VICAT) is the determination of the time it takes for the standardized needle to sink into the mixture to a depth of 6 mm. The results and mortar formulations are displayed in Table 7. RM is the mortar with the most rapid initial setting time (285 min), and the mortar with the longest time is the MRS (321 min). The MSTFC showed an initial setting time of 326 min.

Table 7. Mortar composition of RM, MRS, MSTDC, and MSTFC for the initial setting time VICAT test (NF EN 196-3) [88].

Constituent	RM	MRS	MSTFC
Cement CEM I 52.5 N (g)	500.0	425.0	425.0
Calcined sediment (g)	0	75.0	75.0
Water (g)	150.0	165.0	165.0
Setting time	285 min (4 h:45 min)	321 min (5 h:21 min)	326 min (5 h:26 min)

Figure 13 shows the initial setting time test results. These patterns show that the initial setting time increases with the presence of sediment in the mortar formulation. At the beginning of the test, needle penetration was quite similar in the MRS and the MSTFC. We then argued that the presence of clayey phases or calcite in the sediments could explain this phenomenon by a contraction effect in the matrix. Hence, it can be concluded that there is no improvement in the setting process when sediments are used.

The initial setting time test (VICAT) results highlighted the fact that the presence of sediments leads to a delay in setting processes. This could be related to the presence of sulfates or contaminants (mineral or organic). Indeed, according to Arliguie and Grandet [72], the presence of zinc (Zn) has the most detrimental impact. Its presence leads strongly to lower compressive strength and decreases density. Zinc inhibits the hydration of C_3S and C_3A with a more pronounced effect on C_3S. This phenomenon is due to the formation of an amorphous impermeable film around anhydrous grains, called zinc hydroxide ($Zn(OH)_2$) (Equation (4)).

$$Zn^{2+} + 2(OH)^- \rightarrow Zn(OH)_2 \quad (4)$$

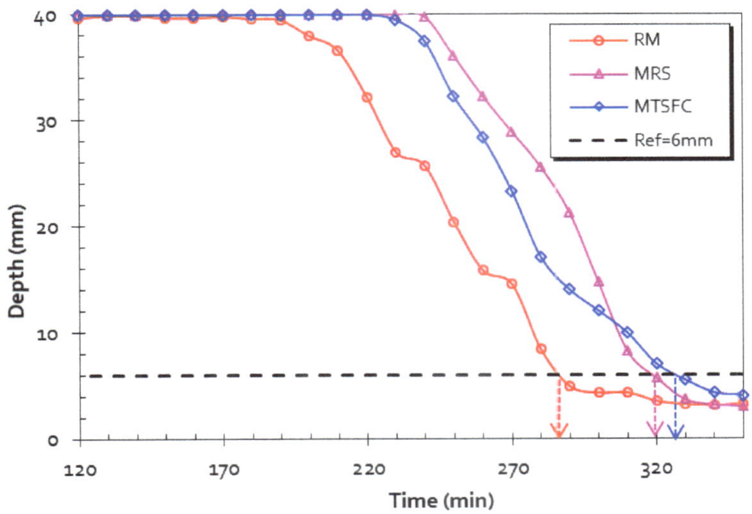

Figure 13. Evolution of the initial setting time for RM, MRS, and MSTFC.

However, the presence of Ca^{2+} and OH^- ions seems likely to allow the destruction of the seal by the formation of a calcium hydroxyzincate layer and therefore enables hydration reactions to continue (Equation (5)).

$$Zn(OH)_4^{2-} + Ca^{2+} + 2H_2O \rightarrow \langle CaZn(OH)_4 | \ 2H_2O \rangle \quad (5)$$

Similarly, the presence of nitrate (Ni) leads to a reduction in the hydration of cement by more than 50% [73,74]. Hemihydrate identified in XRD analysis (Equation (3)) leads into the production of gypsum ($CaSO_4$, $2H_2O$), which should have a retarder effect in a cementitious matrix.

The initial setting time delay could be related to the presence of mineral or organic contaminants. Likewise, the marine nature of sediment probably accounts for this behavior. In fact, the presence of marine salt such NaCl, KCl, and sulphates could have caused a delay in the setting process, as demonstrated by Minocha et al. [73]. The presence of organic matter may also influence the hydration process. It was identified in a previous study to affect setting and hardening due to the formation of a particular protective film around the cement grains [91,92].

4. Conclusions

The effect of heat treatment processes on sediment performances was investigated in this paper. The purpose was to determine and compare the characteristics of flash-calcined sediment and regular SCMs. Sediment-based mortars were formulated and tested after 7, 14, and 28 days. The main conclusions that can be drawn from this work are the following:

Heat treatment modifies sediment properties such as density (by removal of organic matter), densification, and dehydroxylation. However, fineness is lowered due to a sintering/melting process.

XRD analysis highlights the fact that after flash calcination, the clayey phases are transformed mainly into illite. In addition, SEM analysis identified spherical particles made up of gases and aluminum silicates that could enhance the hydration process.

From a mechanical analysis, the presence of finely ground and flash-treated sediments improves mechanical resistance. The compressive strength of MSTFC mortars was 11.3% higher than that of the RM (control) at 28 days for 5% cement substitution, whereas the mechanical performances of MSTFC10 were equivalent to those of RM at 28 days.

The investigation conducted on the heat of hydration using the Langavant calorimetry method (NF EN 196-9) demonstrates that the presence of sediment generally modifies hydration kinetics and processes. It also shows that the presence of sediment generates additional heat release.

A later study of the durability and environmental impact of sediment-based mortars will be carried out in the near future.

Author Contributions: Investigation, M.A.; Methodology, M.A.; Project administration, N.-E.A.; Resources, M.B.; Validation, M.B. and N.-E.A.; Writing—original draft, M.A. All authors have read and agreed to the published version of the manuscript.

Funding: This research received no external funding.

Institutional Review Board Statement: This study did not require ethical approval.

Informed Consent Statement: Not applicable.

Data Availability Statement: The data used in this study are available from the corresponding author on submission of a reasonable request.

Acknowledgments: This study is supported by "Region des Hauts de France" and all partners of the ECOSED chair. The authors express their sincere thanks for their support.

Conflicts of Interest: The authors declare that they have no conflict of interest.

References

1. Scrivener, K.L.; Gartner, E.M. Eco-efficient cements: Potential economically viable solutions for a low-CO2 cement-based materials industry. *Cem. Concr. Res.* **2018**, *114*, 2–26. [CrossRef]
2. Van Bunderen, C.; Snellings, R.; Horckmans, L.; Dockx, J.; Vandekeybus, J.; Vandewalle, L.; Cizer, Ö. Concrete with Flash-Calcined Dredging Sediments as a Novel Supplementary Cementitious Material. In *High Tech Concrete: Where Technology and Engineering Meet*; Springer International Publishing: Cham, Switzerland, 2018; pp. 109–116.
3. Lin, B.; Zhang, Z. Carbon emissions in China's cement industry: A sector and policy analysis. *Renew. Sustain. Energy Rev.* **2016**, *58*, 1387–1394. [CrossRef]
4. Zhang, C.Y.; Yu, B.; Chen, J.M.; Wei, Y.M. Green transition pathways for cement industry in China. *Resour. Conserv. Recycl.* **2021**, *166*, 105355. [CrossRef]
5. Dinga, C.D.; Wen, Z. China's green deal: Can China's cement industry achieve carbon neutral emissions by 2060? *Renew. Sustain. Energy Rev.* **2022**, *155*, 111931. [CrossRef]
6. Poudyal, L.; Adhikari, K. Environmental sustainability in cement industry: An integrated approach for green and economical cement production. *Resour. Environ. Sustain.* **2021**, *4*, 100024. [CrossRef]
7. Union Nationale des Industries de Carrières et Matériaux de Construction. *UNICEM Rapport et Statistiques sur les Granulats*; Rapport Technique (2015)—Technical Report (2015); Union Nationale des Industries de Carrières et Matériaux de Construction: Clichy, France, 2015.
8. Xu, Y.; Yan, C.; Xu, B.; Ruan, X.; Wei, Z. The use of urban river sediments as a primary raw material in the production of highly insulating brick. *Ceram. Int.* **2014**, *40*, 8833–8840. [CrossRef]
9. Aoual-Benslafa, F.K.; Kerdal, D.; Ameur, M.; Mekerta, B.; Semcha, A. Durability of Mortars Made with Dredged Sediments. *Procedia Eng.* **2015**, *118*, 240–250. [CrossRef]
10. Said, I.; Missaoui, A.; Lafhaj, Z. Reuse of Tunisian marine sediments in paving blocks: Factory scale experiment. *J. Clean. Prod.* **2015**, *102*, 66–77. [CrossRef]
11. Millrath, K.; Kozlova, S.; Shimanovich, S.; Meyer, C. *Beneficial Use of Dredged Material 2*; Echo Environmental Inc.: New York, NY, USA; Columbia University: New York, NY, USA, 2001.
12. Zhao, Z.; Benzerzour, M.; Abriak, N.E.; Damidot, D.; Courard, L.; Wang, D. Use of uncontaminated marine sediments in mortar and concrete by partial substitution of cement. *Cem. Concr. Compos.* **2018**, *93*, 155–162. [CrossRef]
13. Flatt, R.J.; Roussel, N.; Cheeseman, C.R. Concrete: An eco material that needs to be improved. *J. Eur. Ceram. Soc.* **2012**, *32*, 2787–2798. [CrossRef]
14. Asokan, P.; Saxena, M.; Asolekar, S.R. Recycling hazardous jarosite waste using coal combustion residues. *Mater. Charact.* **2010**, *61*, 1342–1355. [CrossRef]
15. Amar, M.; Benzerzour, M.; Safhi, A.E.M.; Abriak, N.-E. Durability of a cementitious matrix based on treated sediments. *Case Stud. Constr. Mater.* **2018**, *8*, 258–276. [CrossRef]
16. Mymrin, V.; Stella, J.C.; Scremim, C.B.; Pan, R.C.Y.; Sanches, F.G.; Alekseev, K.; Pedroso, D.E.; Molinetti, A.; Fortini, O.M. Utilization of sediments dredged from marine ports as a principal component of composite material. *J. Clean. Prod.* **2016**, *142*, 4041–4049. [CrossRef]

17. Snellings, R.; Scrivener, K.L. Rapid screening tests for supplementary cementitious materials: Past and future. *Mater. Struct.* **2016**, *49*, 3265–3279. [CrossRef]
18. Junakova, N.; Junak, J.; Balintova, M. Reservoir sediment as a secondary raw material in concrete production. *Clean Technol. Environ. Policy* **2015**, *17*, 1161–1169. [CrossRef]
19. Rozière, E.; Samara, M.; Loukili, A.; Damidot, D. Valorisation of sediments in self-consolidating concrete: Mix-design and microstructure. *Constr. Build. Mater.* **2015**, *81*, 1–10. [CrossRef]
20. Ramaroson, J. Calcination des Sédiments de Dragage Contaminés. Etudes des Propriétés Physico-Chimiques. Ph.D. Thesis, INSA, Lyon, France, 2008.
21. Agostini, F. Inertage et Valorisation des Sédiments de Dragage Marins. Doctoral Dissertation, Université des Sciences et Technologie de Lille-Lille I, Villeneuve-d'Ascq, France, 2006.
22. Tribout, C.; Husson, B.; Nzihou, A. Use of treated dredged sediments as road base materials: Environmental assessment. *Waste Biomass Valorization* **2011**, *2*, 337–346. [CrossRef]
23. Dang, T.A.; Kamali-Bernard, S.; Prince, W.A. Design of new blended cement based on marine dredged sediment. *Constr. Build. Mater.* **2013**, *41*, 602–611. [CrossRef]
24. Gastaldini, A.L.G.; Hengen, M.F.; Gastaldini, M.C.C.; do Amaral, F.D.; Antolini, M.B.; Coletto, T. The use of water treatment plant sludge ash as a mineral addition. *Constr. Build. Mater.* **2015**, *94*, 513–520. [CrossRef]
25. Van Bunderen, C.; Snellings, R.; Vandewalle, L.; Cizer, Ö. Early-age hydration and autogenous deformation of cement paste containing flash calcined dredging sediments. *Constr. Build. Mater.* **2019**, *200*, 104–115. [CrossRef]
26. Snellings, R.; Horckmans, L.; Van Bunderen, C.; Vandewalle, L.; Cizer, Ö. Flash-calcined dredging sediment blended cements: Effect on cement hydration and properties. *Mater. Struct. Constr.* **2017**, *50*, 1–12. [CrossRef]
27. Scrivener, K.; Snellings, R.; Lothenbach, B.; Group, F. *A Practical Guide to Microstructural Analysis of Cementitious Materials*; CRC Press: Boca Raton, FL, USA, 2017; ISBN 978-1-4987-3865-1.
28. Hadj Sadok, R.; Belas, N.; Tahlaiti, M.; Mazouzi, R. Reusing calcined sediments from Chorfa II dam as partial replacement of cement for sustainable mortar production. *J. Build. Eng.* **2021**, *40*, 102273. [CrossRef]
29. Claverie, M.; Martin, F.; Tardy, J.P.P.; Cyr, M.; De Parseval, P.; Grauby, O.; Le Roux, C. Structural and chemical changes in kaolinite caused by flash calcination: Formation of spherical particles. *Appl. Clay Sci.* **2015**, *114*, 247–255. [CrossRef]
30. Nicolas, R.S.; Cyr, M.; Escadeillas, G. Characteristics and applications of flash metakaolins. *Appl. Clay Sci.* **2013**, *83–84*, 253–262. [CrossRef]
31. Salvador, S.; Pons, O. Semi-mobile flash dryer/calciner unit to manufacture pozzolana from raw clay soils—Application to soil stabilisation. *Constr. Build. Mater.* **2000**, *14*, 109–117. [CrossRef]
32. Salvador, S. Pozzolanic properties of flash-calcined kaolinite: A comparative study with soak-calcined products. *Cem. Concr. Res.* **1995**, *25*, 102–112. [CrossRef]
33. Inocente, J.M.; Elyseu, F.; Jaramillo Nieves, L.J.; Jiusti, J.; Cargnin, M.; Peterson, M. Production and characterization of high-reactivity metakaolins calcined in flash reactor. *Appl. Clay Sci.* **2021**, *213*, 106247. [CrossRef]
34. Van Bunderen, C.; Benboudjema, F.; Snellings, R.; Vandewalle, L.; Cizer, O. Experimental analysis and modelling of mechanical properties and shrinkage of concrete recycling flash calcined dredging sediments. *Cem. Concr. Compos.* **2021**, *115*, 103787. [CrossRef]
35. Cremona, C.; Vildaer, S.; Cadillac, M. Experimental Study of Blended Binders with Metakaolin. *Sustainability* **2021**, *13*, 10548. [CrossRef]
36. Sullivan, M.S.; Chorzepa, M.G.; Hamid, H.; Durham, S.A.; Sonny Kim, S. Sustainable materials for transportation infrastructures: Comparison of three commercially-available metakaolin products in binary cementitious systems. *Infrastructures* **2018**, *3*, 17. [CrossRef]
37. Danilevich, V.; Isupova, L.; Parmon, V. The process for preparation of active aluminum hydroxyoxide via flash calcination of gibbsite in a new energy-efficient centrifugal drum-type reactor. *Clean. Eng. Technol.* **2021**, *3*, 100118. [CrossRef]
38. Hanein, T.; Thienel, K.C.; Zunino, F.; Marsh, A.T.M.; Maier, M.; Wang, B.; Canut, M.; Juenger, M.C.G.; Ben Haha, M.; Avet, F.; et al. Clay calcination technology: State-of-the-art review by the RILEM TC 282-CCL. *Mater. Struct. Constr.* **2022**, *55*, 3. [CrossRef]
39. Blazy, P.; Bouhaouss, A. Removal of organic matter in Moroccan Youssoufia phosphate by flash calcination. *Mining Metall. Explor.* **2005**, *22*, 107–115. [CrossRef]
40. Ambroise, J.; Murat, M.; Péra, J. Hydration reaction and hardening of calcined clays and related minerals V. Extension of the research and general conclusions. *Cem. Concr. Res.* **1985**, *15*, 261–268. [CrossRef]
41. Skibsted, J.; Snellings, R. Reactivity of supplementary cementitious materials (SCMs) in cement blends. *Cem. Concr. Res.* **2019**, *124*, 105799. [CrossRef]
42. Snellings, R.; Cizer, Ö.; Horckmans, L.; Durdziński, P.T.; Dierckx, P.; Nielsen, P.; Van Balen, K.; Vandewalle, L. Properties and pozzolanic reactivity of flash calcined dredging sediments. *Appl. Clay Sci.* **2016**, *129*, 35–39. [CrossRef]
43. Scrivener, K.L.; Crumbie, A.K.; Laugesen, P. The interfacial transition zone (ITZ) between cement paste and aggregate in concrete. *Interface Sci.* **2004**, *12*, 411–421. [CrossRef]
44. Teklay, A.; Yin, C.; Rosendahl, L. Flash calcination of kaolinite rich clay and impact of process conditions on the quality of the calcines: A way to reduce CO_2 footprint from cement industry. *Appl. Energy* **2016**, *162*, 1218–1224. [CrossRef]

45. Nguyen, Q.D.; Castel, A. Reinforcement corrosion in limestone flash calcined clay cement-based concrete. *Cem. Concr. Res.* **2020**, *132*, 106051. [CrossRef]
46. Pouhet, R.; Cyr, M.; Bucher, R. Influence of the initial water content in flash calcined metakaolin-based geopolymer. *Constr. Build. Mater.* **2019**, *201*, 421–429. [CrossRef]
47. Amar, M.; Benzerzour, M.; Abriak, N.-E.; Mamindy-Pajany, Y. Study of the pozzolanic activity of a dredged sediment from Dunkirk harbour. *Powder Technol.* **2017**, *320*, 748–764. [CrossRef]
48. Amar, M.; Benzerzour, M.; Kleib, J.; Abriak, N.-E. From dredged sediment to supplementary cementitious material: Characterization, treatment, and reuse. *Int. J. Sediment Res.* **2021**, *36*, 92–109. [CrossRef]
49. *NF EN ISO 18757*; Fine Ceramics (Advanced Ceramics, Advanced Technical Ceramics)—Determination of Specific Surface Area of Ceramic Powders by Gas Adsorption Using the BET Method. AFNOR: Ile-de-France, France, 2006.
50. *NF EN 196-9*; Methods of Testing Cement—Part 9: Heat of Hydration—Semi-Adiabatic Method. AFNOR: Ile-de-France, France, 2010.
51. *XP P94-047*; Soils: Investigation and Testing. Determination of the Organic Matter Content. Ignition Method. AFNOR: Ile-de-France, France, 1998.
52. *NF EN 1097-7*; Tests for Mechanical and Physical Properties of Aggregates—Part 7: Determination of the Particle Density of Filler—Pyknometer Method. AFNOR: Ile-de-France, France, 2008.
53. *NF EN 196-1*; Methods of Testing Cement—Part 1: Determination of Strength. AFNOR: Ile-de-France, France, 2006.
54. Teklay, A.; Yin, C.; Rosendahl, L.; Køhler, L.L. Experimental and modeling study of flash calcination of kaolinite rich clay particles in a gas suspension calciner. *Appl. Clay Sci.* **2015**, *103*, 10–19. [CrossRef]
55. Sperinck, S.; Raiteri, P.; Marks, N.; Wright, K. Dehydroxylation of kaolinite to metakaolin—A molecular dynamics study. *Mater. Chem.* **2011**, *21*, 2118–2125. [CrossRef]
56. Bich, C.; Ambroise, J.; Péra, J. Influence of degree of dehydroxylation on the pozzolanic activity of metakaolin. *Appl. Clay Sci.* **2009**, *44*, 194–200. [CrossRef]
57. Slade, R.C.T.; Davies, T.W.; Atakül, H.; Hooper, R.M.; Jones, D.J. Flash calcines of kaolinite: Effect of process variables on physical characteristics. *J. Mater. Sci.* **1992**, *27*, 2490–2500. [CrossRef]
58. Nicolas, S. Approche Performantielle des Bétons Avec Métakaolins Obtenus par Calcination Flash. Doctoral Dissertation, Université Paul Sabatier-Toulouse III, Toulouse, France, 2012.
59. Wang, L.; Yu, Z.; Liu, B.; Zhao, F.; Tang, S.; Jin, M. Effects of Fly Ash Dosage on Shrinkage, Crack Resistance and Fractal Characteristics of Face Slab Concrete. *Fractal Fract.* **2022**, *6*, 335. [CrossRef]
60. Peng, Y.; Tang, S.; Huang, J.; Tang, C.; Wang, L.; Liu, Y. Fractal Analysis on Pore Structure and Modeling of Hydration of Magnesium Phosphate Cement Paste. *Fractal Fract.* **2022**, *6*, 337. [CrossRef]
61. Mustin, M. *Le Compost: Gestion de la Matière Organique*; Francois Dubusc: Paris, France, 1987.
62. Agostini, F.; Skoczylas, F.; Lafhaj, Z. About a possible valorisation in cementitious materials of polluted sediments after treatment. *Cem. Concr. Compos.* **2007**, *29*, 270–278. [CrossRef]
63. Ramaroson, J.; Dia, M.; Dirion, J.-L.; Nzihou, A.; Depelsenaire, G. Thermal Treatment of Dredged Sediment in a Rotary Kiln: Investigation of Structural Changes. *Ind. Eng. Chem. Res.* **2012**, *51*, 7146–7152. [CrossRef]
64. Dia, M.; Ramaroson, J.; Nzihou, A.; Zentar, R.; Abriak, N.E.; Depelsenaire, G.; Germeau, A. Effect of Chemical and Thermal Treatment on the Geotechnical Properties of Dredged Sediment. *Procedia Eng.* **2014**, *83*, 159–169. [CrossRef]
65. Dubois, V. Etude du Comportement Physico-Mécanique et Caractérisation Environnementale des Sédiments Marins—Valorisation en Technique Routière. Ph.D. Thesis, University of Artois, Arras, France, 2006.
66. Tran, N.T. Valorisation de Sédiments Marins et Fluviaux en Technique Routière. Ph.D. Thesis, University of Artois, Arras, France, 2009.
67. Care, S.; Baroghel-bouny, V.; De Larrard, F.; Charonat, Y. *Effet des Additions Minérales sur les Proprietes D'usage des Betons—Plan D'experience et Analyse Statistique*; LCPC: Paris, France, 2000.
68. Lothenbach, B.; Le Saout, G.; Gallucci, E.; Scrivener, K. Influence of limestone on the hydration of Portland cements. *Cem. Concr. Res.* **2008**, *38*, 848–860. [CrossRef]
69. Weerdt, K.D.; BK, S.; Rønning, U. *Ternary Blended Cements with Fly Ash and Limestone*; SINTEF: Trondheim, Norway, 2007.
70. Bertolini, L.; Carsana, M.; Gastaldi, M.; Lollini, F.; Redaelli, E. Binder characterisation of mortars used at different ages in the San Lorenzo church in Milan. *Mater. Charact.* **2013**, *80*, 9–20. [CrossRef]
71. Park, C.-K. Hydration and solidification of hazardous wastes containing heavy metals using modified cementitious materials. *Cem. Concr. Res.* **2000**, *30*, 429–435. [CrossRef]
72. Arliguie, G.; Grandet, J. Influence de la composition d'un ciment portland sur son hydration en presence de zinc. *Cem. Concr. Res.* **1990**, *20*, 517–524. [CrossRef]
73. Minocha, A.K.; Jain, N.; Verma, C.L. Effect of inorganic materials on the solidification of heavy metal sludge. *Cem. Concr. Res.* **2003**, *33*, 1695–1701. [CrossRef]
74. Malviya, R.; Chaudhary, R. Leaching behavior and immobilization of heavy metals in solidified/stabilized products. *J. Hazard. Mater.* **2006**, *137*, 207–217. [CrossRef] [PubMed]
75. Malviya, R.; Chaudhary, R. Factors affecting hazardous waste solidification/stabilization: A review. *J. Hazard. Mater.* **2006**, *137*, 267–276. [CrossRef]

76. Minocha, A.K.; Neeraj, J.; Verma, C.L. Effect of organic materials on the solidification of heavy sludge. *Constr. Build. Mater.* **2003**, *17*, 77–81. [CrossRef]
77. Benezet, J.C.; Benhassaine, A. The influence of particle size on the pozzolanic reactivity of quartz powder. *Powder Technol.* **1999**, *103*, 26–29. [CrossRef]
78. *NF P18-513*; Addition for Concrete—Metakaolin—Specifications and Conformity Criteria. AFNOR: Ile-de-France, France, 2012.
79. Cyr, M.; Lawrence, P.; Ringot, E. Efficiency of mineral admixtures in mortars: Quantification of the physical and chemical effects of fine admixtures in relation with compressive strength. *Cem. Concr. Res.* **2006**, *36*, 264–277. [CrossRef]
80. Walker, R.; Pavía, S. Physical properties and reactivity of pozzolans, and their influence on the properties of lime–pozzolan pastes. *Mater. Struct.* **2011**, *44*, 1139–1150. [CrossRef]
81. Fernandez, R.; Martirena, F.; Scrivener, K.L. The origin of the pozzolanic activity of calcined clay minerals: A comparison between kaolinite, illite and montmorillonite. *Cem. Concr. Res.* **2011**, *41*, 113–122. [CrossRef]
82. Limeira, J.; Etxeberria, M.; Agulló, L.; Molina, D. Mechanical and durability properties of concrete made with dredged marine sand. *Constr. Build. Mater.* **2011**, *25*, 4165–4174. [CrossRef]
83. Bouamrane, A.; Elouazzani, D.C.; Barna, L.T.; Mansouri, K. Valorisation des boues de papeterie comme matières premières secondaires dans les mortiers de ciment Portland. *J. Mater. Environ. Sci.* **2014**, *5*, 605–614.
84. Limeira, J.; Agullo, L.; Etxeberria, M. Dredged marine sand in concrete: An experimental section of a harbor pavement. *Constr. Build. Mater.* **2010**, *24*, 863–870. [CrossRef]
85. Hentschel, M.L.; Page, N.W. Selection of Descriptors for Particle Shape Characterization. *Part. Part. Syst. Charact.* **2003**, *20*, 25–38. [CrossRef]
86. Felekoglu, B. A new approach to the characterisation of particle shape and surface properties of powders employed in concrete industry. *Constr. Build. Mater.* **2009**, *23*, 1154–1162. [CrossRef]
87. Benkaddour, M.; Kazi Aoual, F.; Semcha, A. Durabilité des mortiers à base de pouzzolane naturelle et de pouzzolane artificielle. *Rev. Nat. Technol.* **2009**, *1*, 63–73.
88. *NF EN 196-3+A1*; Methods of Testing Cement—Part 3: Determination of Setting Times and Soundness. AFNOR: Ile-de-France, France, 2009.
89. Mostafa, N.Y.; Brown, P.W. Heat of hydration of high reactive pozzolans in blended cements: Isothermal conduction calorimetry. *Thermochim. Acta* **2005**, *435*, 162–167. [CrossRef]
90. Snelson, D.G.; Wild, S.; O'Farrell, M. Heat of hydration of Portland Cement–Metakaolin–Fly ash (PC–MK–PFA) blends. *Cem. Concr. Res.* **2008**, *38*, 832–840. [CrossRef]
91. Paria, S.; Yuet, P. Solidification-stabilization of organic and inorganic contaminants using portland cement: A literature review. *Environ. Rev.* **2006**, *14*, 217–255. [CrossRef]
92. Sora, I.N.; Pelosato, R.; Botta, D.; Dotelli, G. Chemistry and microstructure of cement pastes admixed with organic liquids. *J. Eur. Ceram. Soc.* **2002**, *22*, 1463–1473. [CrossRef]

materials

Article

Prediction of the Compressive Strength of Waste-Based Concretes Using Artificial Neural Network

Mouhamadou Amar [1,2,*], Mahfoud Benzerzour [1,2], Rachid Zentar [1,2] and Nor-Edine Abriak [1,2]

1 IMT Nord Europe, Institut Mines-Télécom, Centre for Materials and Processes, F-59000 Lille, France
2 Univ. Lille, Institut Mines-Télécom, Univ. Artois, Junia, ULR 4515—LGCgE—Laboratoire de Génie Civil et géoEnvironnement, F-59000 Lille, France
* Correspondence: mouhamadou.amar@imt-nord-europe.fr

Abstract: In the 21st century, numerous numerical calculation techniques have been discovered and used in several fields of science and technology. The purpose of this study was to use an artificial neural network (ANN) to forecast the compressive strength of waste-based concretes. The specimens studied include different kinds of mineral additions: metakaolin, silica fume, fly ash, limestone filler, marble waste, recycled aggregates, and ground granulated blast furnace slag. This method is based on the experimental results available for 1303 different mixtures gathered from 22 bibliographic sources for the ANN learning process. Based on a multilayer feedforward neural network model, the data were arranged and prepared to train and test the model. The model consists of 18 inputs following the type of cement, water content, water to binder ratio, replacement ratio, the quantity of superplasticizer, etc. The ANN model was built and applied with MATLAB software using the neural network module. According to the results forecast by the proposed neural network model, the ANN shows a strong capacity for predicting the compressive strength of concrete and is particularly precise with satisfactory accuracy (R^2 = 0.9888, MAPE = 2.87%).

Keywords: artificial neural network; concrete; mineral additions; prediction; formulation; compressive strength

Citation: Amar, M.; Benzerzour, M.; Zentar, R.; Abriak, N.-E. Prediction of the Compressive Strength of Waste-Based Concretes Using Artificial Neural Network. *Materials* **2022**, *15*, 7045. https://doi.org/10.3390/ma15207045

Academic Editor: Yuri Ribakov

Received: 12 September 2022
Accepted: 28 September 2022
Published: 11 October 2022

1. Introduction

According to the World Bank, the amount of global solid waste is currently 2.2 billion tons per year. This figure is likely to increase due to global demographic and economic growth. The overconsumption and inefficient use of materials also have a critical impact on the environment and climate. The annual cement needs in France are 16.9 million tons (in 2020), and in 2020 the need for aggregates was over 400 million tons, 96% of natural origin. Concrete is an ancient and widely used material because of its mechanical properties, which have long been appreciated. The Egyptians were already using it around 2600 BC in the pyramid of Abu Rawash. This material continued to evolve until the invention of reinforced concrete in 1867 by Joseph Monier (1823–1906). Concrete is a multiphase material consisting of a granular skeleton and a cementitious matrix. Due to chemical reactions of hydration followed by hardening, the assembly stiffens. This attributes some specific physicochemical properties to the material: elastic properties, high compressive strength, low tensile strength, permeability, etc. These properties are the result of a series of chemical reactions, triggered as soon as the anhydrous cement and water come into contact. These reactions thus give rise initially to portlandite ($Ca(OH)_2$), which acts as a trigger for setting, then to various hydrates (C-S-H, C-A-H, C-A-S-H, etc.) representing the "glue" of the matrix.

Concrete structures now represent more than 90% of modern structures. Concrete has thus become the most important building material on the planet, in terms of volume and turnover. Its success stems from, among other things, its extraordinary versatility and availability virtually anywhere on Earth, and its durability. It is at the dawn of the 21st century that mankind is confronted with an unprecedented paradigm: How to ensure the sustainability of nature and biodiversity for future generations and at the same time meet current growing economic needs: energy, materials, and resources of all kinds?

The construction sector is one of the most important in Europe. From an environmental standpoint, it represents 30% of carbon dioxide emissions (2009). Cement production is an important source of CO_2 (5 to 7% of worldwide emissions). The concrete industry worldwide consumes annually over 8–12 billion tons of natural aggregates [1]. This is why the use of binders based on additional byproduct materials, such as metakaolin, fly ash, sludge ash, blast-furnace slag, or silica fume, is gaining more and more interest and their efficiency has been proven [2–4]. Because of the environmental issues explained above and because they are a way to enhance the durability of structures exposed to harsh environments, their utility is fundamental [2].Several formulations and prediction models (Feret, Bolomey, Abram, Powers [4–7]) have proposed that the compressive strength R_c of concrete depends mainly on:

i. The grade of the cement, the age, and the method of curing;
ii. The water to binder ratio, as well as the compactness of the granular skeleton.

Currently, there is a lack of research on comprehensive methods for predicting the strength of byproduct-based concrete or mortar. Hence, authors usually use correlation or some empirical or non-adapted formulae.

Recently, numerous studies have forecast the compressive strength of cementitious materials using extrapolation methods, compressible packing models, regression analysis methods, genetic algorithms, fuzzy logic, and artificial neural networks (ANNs) [8,9]. However, among these approaches, the ANN seems a relevant and efficient method due to its ability to learn from input and output relationships in complex problems [10]. Moreover, the ANN is suitable for modeling different properties of concrete, for mapping its mechanical characteristics, including compressive and tensile strength, slump, filling capacity, and segregation, and for many types of concrete [8–10].

In recent years, several studies have reported that ANNs can be used to solve engineering problems. However, the required data may be complex or insufficient [11] for estimating the compressive strength of concretes [12,13]. These studies involved issues related to high performance [1,14–19], self-compacting concrete [20–23], and lightweight concretes [24–26], sulphate resistance in concrete [27,28], cyclic behavior of concretes [29], recycled aggregate [1,30,31] and waste material [27,32,33]. Machine learning and AI are steadily gaining interest and over 10,152 papers were published in 2021 in the material engineering field alone.

Faridehmihr et al. [34] explored the use of waste materials, including fly ash (FA), palm oil fly ash (POFA), waste ceramic powder (WCP), and granulated blast-furnace slag (GBFS) in alkali-activated materials. Properties such as the mechanical resistance can be properly predicted using an artificial neural network (ANN) combined with a metaheuristic Krill Herd algorithm (KHA) model. Faridehmihr et al. [34] showed that ANNs are efficient in investigating the cradle-to-gate life-cycle assessment (LCA) of ternary blended alkali-activated mortars. Mhaya [35] evaluated the performance of several modified rubberized concretes by exposing them to aggressive environments. The final mechanical properties were predicted using an ANN combined with particle swarm optimization (PSO). A similar work conducted by Golafshani [36] added to an ANN the use of a multi-objective multi-verse optimizer (MOMVO). Alabduljabbar et al. [37] adopted the same method as that of Sadowski et al. [38] and used an optimized ANN to estimate the mechanical properties in a wide experimental study on the sustainability of employing waste sawdust and supplementary cementitious material (SCM) to make high-performance cement-free lightweight concretes. Ray et al. [39] monitored the consequences of the incorporation of fine glass aggregate and condensed milk can fiber (Sn) on the compressive and splitting tensile strength at three curing ages using an ANN. The results showed very good accuracy.

It is clear that the prediction of concrete properties can be efficiently performed using machine learning technology [40,41].

In the early 2000s, several studies [42,43] showed the great potential to optimize mix proportioning and forecasting of concrete properties. Apostolopoulou et al. [44] investigated the use of ANNs to simulate the characteristics of lime-based mortars, such as compressive and flexural strength and consistency. The final results showed that the developed ANN models fit satisfactorily with the experimental data. Gupta et al. [45] used an ANN in a recent study since there was no mathematical model for the rapid prediction of mechanical properties of rubberized concrete. The trained network based on data compiled from recent research showed results that predicted compressive strength, modulus of elasticity (static and dynamic), and mass loss. Several other studies on this type of concrete led to similar conclusions, while others focused on predicting the properties of recycled aggregate-based concrete [35,46]. The test data are generally sets of compressive strength, splitting strength, porosity, the permeability coefficient of recycled aggregate, etc. Based on mean squared error (MSE), root mean square error (RMSE), and coefficient of regression (r^2), the results proved to have a very good fit, as stated by Dantas et al. [47]. It has been demonstrated that ANNs can predict the compressive and tensile strength of concretes containing construction and agricultural wastes [32,48], blast furnace slag [35], and alkali-activated mortars [34,37]. Some authors combined an ANN with other techniques, such as a genetic algorithm (GA) [35], statistics and holistic models [44], the cuckoo search method [49,50], ANFIS models [24,51], fuzzy logic models [52,53], and the Monte Carlo approach [54], to optimize the prediction results. Jiang et al. [53] and Farooq et al. [55] studied the prediction of mechanical properties of self-compacting concretes and high-performance concretes using an ANN on over 1030 datasets. The excellent findings obtained suggested that machine learning processes are quite robust and efficient, becoming indispensable for concrete property prediction. In addition, Asteris et al. [56] developed a methodology that predicts the effects of seismic loads on masonry structures. The authors were able to take into account the weakness, damage, fragility, and general properties of structures. Ray et al. [39], like Sadowski et al. [38], recently showed that ANN techniques are relevant in predicting properties of waste-based concretes and mineral admixtures such as metakaolin, silica fume, dust-based, filler-based, glass waste-based quartz mineral, and fibers. Bui et al. [57] used a whale optimization algorithm (WOA) coupled with a neural network (NN) with over 400 nodes to simulate the 28-day compressive strength of concrete. The results showed that the WOA-NN is reliable and has the highest correlation of 0.8976 when compared to different techniques of modeling. Other recent studies [58,59] conducted on the prediction of concrete's compressive strength used several methods (Support Vector Regression (SVR), Decision Tree Regression (DTR), Gradient Boosting Regression (GBR), and ANN) for comparative purposes. It was shown that SVR, DTR, and ANN were reliable methods.

2. Research Significance

More than 10 billion tons of concrete are currently used worldwide, and in the USA, for example, USD 9.4 billion would be needed to restore the country's 600,000 bridges. It is therefore important to emphasize that the emergence of innovative processes and techniques in the formulation and composition of concretes is very much needed. The use of supplementary cementitious material may be one of the best solutions for nature and resource preservation. One of the best-established approaches to reducing the impact of cement on the environment is the replacement of clinker with other materials. This method reduces energy consumption and increases production, without any additional industrial installation [60]. These substitutes are generally reactive byproducts from other industries: granulated blast furnace slag (GBFS), a byproduct of the iron industry; and fly ash (FA), generated by electricity production after the burning of coal. Moreover, natural materials such as calcined clays, pozzolans, and limestone fillers have proved suitable for concrete use. Several theoretical methods exist to predict concrete strength: those of Feret (1897),

Abrams (1920), Bolomey (1925) [3,6], etc. Depending on the case, certain preponderant parameters, such as the water/binder ratio (W/L), the substitution rate (p(%)), and the maturation time, may influence the final strength of the concrete. However, these existing formulae are not always adapted to the materials cited above. In this work, ANN modeling and results in predicting concrete properties were investigated. The aim was to develop and set up AI-based tools to predict the properties of concretes containing byproducts reused as supplementary cementitious materials. This research highlights the potential of using an ANN with satisfactory and reliable results in predicting the characteristics of environmentally friendly concretes [38,61]. The novelty of this article relies on the scale of the dataset used and its extensiveness. Contrarily to several studies, this study tested the concomitant set of multiple supplementary cementitious materials using machine learning.

3. Artificial Neural Networks

Artificial neural networks (ANNs), which are part of the machine learning process, involve mathematical techniques based on the conception of interconnected layers of nodes [62]. An artificial neural network (ANN) is an artificial intelligence system that focuses on the identification and solving of complex issues and phenomena. A parallel can be drawn with conventional digital computing techniques, yet neural networks have many additional assets. For instance, they use equivalent processing modes and distributed information storage, and also have high accuracy. Furthermore, these methods are very robust when operating following the training process and are flexible to new information and learning [63]. The ANN system is meant to recreate the biological characteristics of the nerve cell structure of the brain.

Usually, an ANN is made up of an input layer of neurons, which includes other layers within it. These neurons predict the process results [10]. The junction of the layers is based on link weights according to Rafiq et al. [64]. As a definition, it can be said that an ANN is a computing system composed of multiple simple units and highly interconnected processing elements. These elements analyze information through the dynamic state response to external inputs. An ANN is skilled in memorizing the characteristics or features of given data and can match or make connections from new data to old with different levels of success [62,65]. The hidden layers (HLs) play the role of connecter or information carrier. The structure then enables the nets to extract a non-linear correlation from the available dataset [24].

There are six main parts in an ANN around a considered neuron nj: inputs (pi), bias (bj), weights (wij), sum function (n)j, activation function (f), and outputs (aj), as displayed in Figure 1. Inputs can be defined as information considered to be decision variables coming from neurons or the external environment. Weights are values that convey the effect of inputs or process elements on each other. Random weight values can be triggered when the process starts. The sum function is an operation that reflects the whole effect of inputs and weights by taking into account a bias value on this process element [13,66] (Equation (1)).

$$(n)_j = \sum_{i=1}^{i=k} w_{ij} p_i + b_j \tag{1}$$

where:

i = [1;k] is the number of the ith input neuron

j = [1;m] is the number of the jth output neuron

k = number of units in the ith input vector.

b_j = value of bias (referred to as the activation threshold) associated with jth node.

The activation function or transfer function (usually the log-sigmoid function or the hyperbolic tangent [24]) is a function that processes the $(n)_j$ value and then determines the corresponding output value according to the formula in Equation (2) [16,67]. It also represents a way to simulate a phenomenon's reaction using input and output parameters [68].

$$(a)_j = f(n)_j = \frac{1}{1 + e^{-\alpha(n)_j}} \tag{2}$$

where $(a)_j$ is the output of the jth neuron and α is a constant used to control the slope of the semi-linear region [13], and usually $\alpha = 1$.

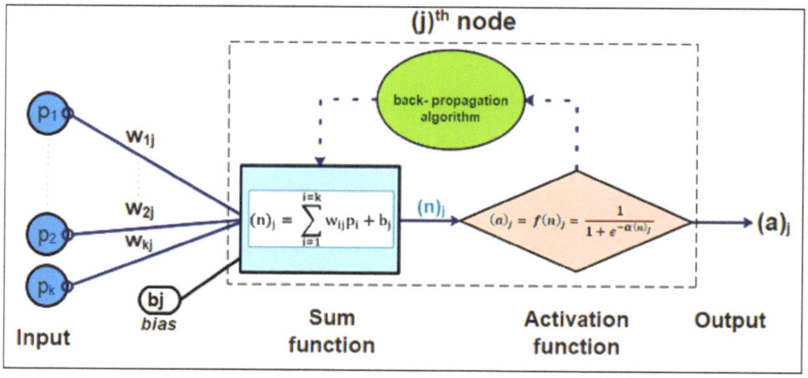

Figure 1. Architecture of artificial node and its interactions in the neural network.

3.1. Neuron Model (Logsig, Tansig, Purelin)

In an ANN, each input is weighted with an appropriate w. The sum of the weighted inputs and the bias forms the input to the transfer function $f(n)_j$. Multilayer networks often use the log-sigmoid transfer function logsig$(n)_j$. The function logsig$(n)_j$ generates outputs between 0 and 1 as the neuron's net input goes from negative to positive infinity. Alternatively, multilayer networks can use the tan-sigmoid transfer function tansig$(n)_j$ or purlin. Logsig$(n)_j$ appears to be more adapted to the current study as it was found to be more accurate for some predictions [69]. Several algorithms can be implemented in ANN modeling, such as Bayesian regularization, Scaled Conjugate Gradient, Levenberg–Marquardt, one-step secant, and some other combination rules [19]. The most popular ANN method is the feedforward multilayer perceptron (MLP) system. The general scheme of the adopted neural network system is given in Figure 2. Final weight values come at the end of the training process and their final value are defined based on how well the model was trained.

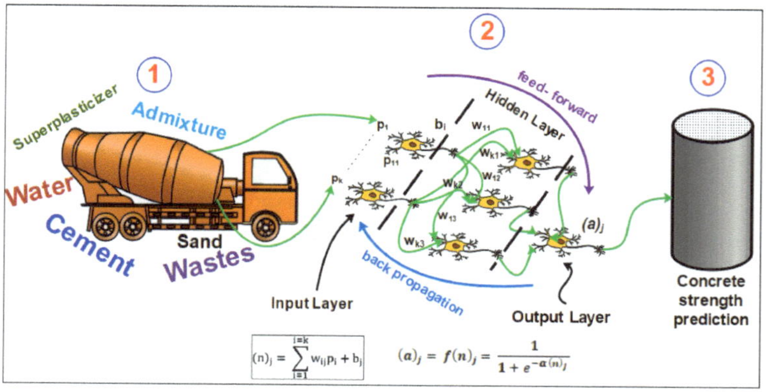

Figure 2. General scheme of the use of ANN for concrete design.

3.2. Training Methods

The neural network models applied in this study were developed using the Neural Network Toolbox in MATLAB software. The models were generated with 02 hidden layers and 10 neurons per hidden layer (Table 1). Of the total data, 70% was used for the training process. In our approach, 15% of the remaining data was used for testing and the other 15% for validation. The training process was operated using the Levenberg–Marquardt backpropagation algorithm (LMBPA), similar to Abu Yaman et al. [20] and Kumar et al. [70]. The LMBPA was chosen due to its simplicity of use. It was also shown that one of the most reliable ANN training algorithms is the backpropagation (BP) algorithm, which distributes the network error to arrive at the best fit or minimum error [71,72] and was, accordingly, used in this study.

Table 1. General characteristics of the ANN model.

Method Type	Regulation Technique	Learning Method	Training Method	Activation Function	Inputs	Hidden Layers (HL)	Neurons per HL	Outputs
Feedforward, back propagation network	Gradient descent	Supervised	Levenberg-Marquardt backpropagation algorithm (LMBPA)	Log-sigmoid (logsig)	18	2	10	1

3.3. Feedforward Network

A feedforward neural network was used in this study. This seems to be the most commonly used ANN architecture type. Feedforward networks have all their neurons classified into different layers. All neurons in each of the considered previous layers are connected to the neurons in the next layer. The multilayer architecture considered in this study, also called a multilayer perception [70], is given in Figure 3. There is no reliable method for deciding the number of neural units or layers required for a particular problem. This comes with experience and trials that are necessary to achieve the best network configuration [9].

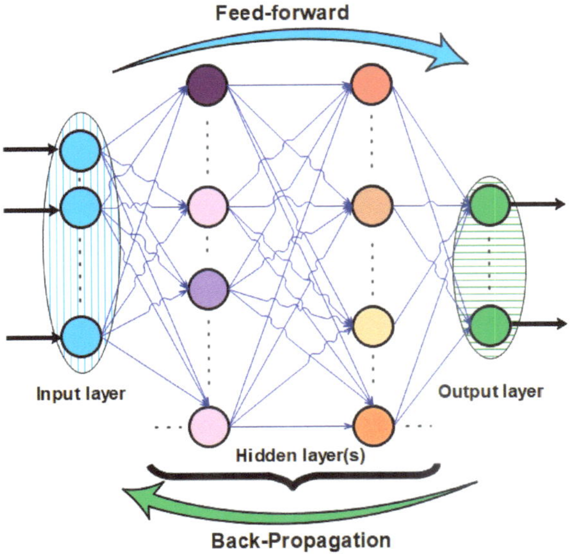

Figure 3. Architecture of applied neural network and learning process principle.

The structure using multiple layers of neurons creates nonlinear relationships between input and output vectors. The number of layers determines the complexity of the architecture and the forecast precision. When the training process is completed, a positive value of weight signifies that the corresponding feature is directly related to the output. On the other hand, a negative weight implies that the corresponding feature is inversely linked to the output. The more the weight related to a feature, the more the effect of the corresponding feature on the output.

4. The Learning and Testing Process

4.1. The Backpropagation Algorithm (BPA)

The term 'backpropagation' indicates a method in which a correction gradient is calculated for nonlinear multilayer networks [73]. This step is an essential part of the network learning process and is performed by the learning algorithm [64]. To assess the performance of the neural network model, an error measure such as root mean square error (RMS) can be used [24]. The determination of and reduction in the error value or cost function can be performed using the so-called generalized delta rule [11]. In fact, the error (which is the gap between forecast and actual values) is reduced using a backpropagation algorithm [9]. Then, during the BPA process, the neuron weights are subsequently adjusted (Figure 4). According to Oztas et al. [18], the BPA is one of the most famous and most widely used training algorithms [11]. In a multi-layer perceptron (MLP) this method corresponds to a gradient descent technique that minimizes the error or cost of the process.

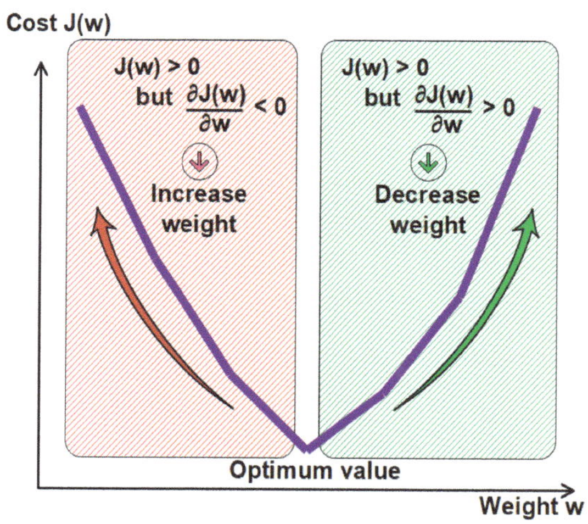

Figure 4. Relationship between cost and weight of the ANN.

LMBPA was used in this present study, as implemented in MATLAB and its neural network fitting module. The LMBPA is the fastest backpropagation algorithm for many engineering problems and is highly recommended as a first-choice supervised algorithm, according to Sobhani et al. [10]. However, it requires more memory than other algorithms such as the Momentum, Adagrad, and Rmsprop methods.

A cost function (error function) can be defined to quantify the difference between the actual value and desired (forecast) outputs (Equation (3)):

$$J\,(w_{ij}) = \frac{1}{2} * \left(\Sigma_i \Sigma_j (a_{PREDICT} - a_{TARGET})_j^{\,2}\right) \tag{3}$$

where:

$a_{PREDICT} = a_{(j,\, PREDICT)}$ is the forecast value,

$a_{TARGET} = a_{(j,\, TARGET)}$ is the experimental value.

Gradient descent is an optimization algorithm that approaches a local minimum of a function by taking steps proportional to the negative of the gradient of the function at the current point. The main objective of the algorithm is then to reduce the cost value and adjust the weight that must be updated very smoothly and slowly by iteration until convergence.

In the gradient descent technique, the adjusted weight can be expressed as (Equation (4)):

$$w_{ij}[n+1] = w_{ij}\,[n] + \tau * \frac{\nabla J\,(w_{ij})}{\nabla (w_{ij})}[n] \quad (4)$$
$$\text{new weight} = \text{old weight } - \text{derivative Rate } * \text{ learning rate}$$

where τ is known as the step-size parameter and affects the rate of convergence of the algorithm, and $\frac{\nabla J\,(w_{ij})}{\nabla (w_{ij})}$ is the derivative rate or gradient of the loss function $J\,(w_{ij})$.

The learning process consists of changing the weights in order to minimize this J(w) in a gradient descent technique. The training process is considered as successfully completed when the iterative process has converged [9].

4.2. Modeling Performance Criteria

The accuracy and error quantification of the proposed system was evaluated using performance parameters. The first parameter is the R^2 coefficient (coefficient of determination), which is the absolute fraction of variance of a variable. It is a measure of the proportion of the information in the data that is explained by the model [62]. The value of R^2 varies from 0 to 1. The closer R^2 is to 1, the closer the forecast value is to the experimental one, expressed as (Equation (5)):

$$R^2 = 1 - \left(\frac{\sum_{i=1}^{N} (a_{PREDICT} - a_{TARGET})^2}{\sum_{i=1}^{N} (a_{PREDICT})^2} \right) \quad (5)$$

The root mean square error (RMSE) is the square root of the mean square error and indicates the average distance of a data point (targeted) from the expected value (predicted) provided by the model. The lower the RMSE value, the better the model (Equation (6)):

$$RMSE = \sqrt{\left(\frac{1}{N}\right) * \left(\sum_{i=1}^{N} (a_{PREDICT} - a_{TARGET})^2\right)} \quad (6)$$

For a better understanding, RMSE can be normalized using the mean of the actual value. This can facilitate comparisons between datasets or models [20].

MAPE (Equation (7)) is the mean absolute percentage error and is a statistical value of prediction accuracy. It indicates a better model fit through a percentage value. However, MAPE places a heavier penalty on negative errors than positive errors due to the division by the factor $a_{PREDICT}$.

$$MAPE = \left(\frac{100}{N}\right) * \left(\sum_{i=1}^{N} \left(\frac{|a_{PREDICT} - a_{TARGET}|}{|a_{PREDICT}|}\right)\right) \quad (7)$$

MAE is the mean absolute error formula and is given by Equation (8):

$$MAE = \left(\frac{1}{N}\right) * \left(\sum_{i=1}^{N} (|a_{PREDICT} - a_{TARGET}|)\right) \quad (8)$$

In all the formulae above:

- N is the number of experiments,
- $a_{PREDICT} = a_{(j,\,PREDICT)}$ is the predicted value for the jth neuron
- $a_{TARGET} = a_{(j,\,TARGET)}$ is the experimental value for the jth neuron.

5. Bibliographic Dataset and Data Preparation

The comprehensiveness, structure, and volume of the data used for training are vital to building an effective network. This is what must lead to better learning, testing, and validating for the network and accurate prediction of all aspects of the relationship between inputs and outputs [20].

Experimental datasets from different sources were used. The notation used is given in Table 2. This is an inhomogeneous collection from the experimental data of some previous research work. The present database was built from the literature and includes a total of 1303 concrete formulations from 22 different studies. We used a large dataset to minimize the lack of data that causes informational uncertainty and to minimize model accuracy problems.

Table 2. Data notation and abbreviations.

C	W	W/B	Specimen Compression Type	S	CA	SP	
Cement	Water	Water/Binder	1 = cubic 2 = cylindrical	Sand	Coarse aggregates	Superplasticizer	
MK	LF	SF	GGBFS	FA	MW	RA	Rc (MPa)
Metakaolin	Limestone filler	Silica fume	Ground granulated blast furnace slag	Fly ash	Marble waste	Recycled aggregates	Compressive strength

Data were assembled from the bibliography and 18 selected inputs were considered: the content of water, cement, and fine and coarse aggregate; admixture; age; and water/binder ratio; superplasticizer; the slump, etc. One output was considered, which is compressive strength.

This study takes into account a very wide spectrum of materials and quantities, as shown in Table 3. In Table 4, we also present an excerpt of concrete mixes from the dataset used. Part of the extensive list of formulations used for the ANN training - testing is given in the Supplementary Materials.

Table 3. General characteristics of concrete formulations used in the dataset.

Parameter	C	E	MK	LF	SF	GGBFS	FA	MW	RA	W/B	Age (days)	Rc (MPa)
Minimum (Kg/m^3)	70.0	95.0	0.0	0.0	0.0	0.0	0.0	0.0	0.0	0.08	1.0	1.0
Maximum (Kg/m^3)	833.3	319.0	104.1	317.0	208.3	360.0	544.0	500.0	1772.0	0.72	365.0	123.0

Table 4. Excerpt of dataset used in AI training and validation processes.

Mix N°	Author	C (Kg/m^3)	W (Kg/m^3)	W/B	Specimen Compression Type	S (Kg/m^3)	Coarse Aggregates (Kg/m^3)	SP (%)	Slump/Flow (mm)	MK (Kg/m^3)	LF (Kg/m^3)	SF (Kg/m^3)	GGBFS (Kg/m^3)	FA (Kg/m^3)	MW (Kg/m^3)	RA (Kg/m^3)	Age (days)	Rc (MPa)
	[74]																	
1		280.0	202.0	0.72	2	777.0	988.0	0.0	160.0	0.0	0.0	0.0	0.0	0.0	0.0	0.0	1	5.3
2		224.0	185.0	0.66	2	788.0	1003.0	0.2	170.0	0.0	0.0	0.0	0.0	56.0	0.0	0.0	1	5.0
3		168.0	157.0	0.56	2	802.0	1041.0	0.8	180.0	0.0	0.0	0.0	0.0	112.0	0.0	0.0	1	3.9
4		112.0	124.0	0.44	2	801.0	1106.0	1.4	210.0	0.0	0.0	0.0	0.0	168.0	0.0	0.0	1	2.6
5		112.0	150.0	0.27	2	418.0	1101.0	0.7	220.0	0.0	0.0	0.0	0.0	448.0	0.0	0.0	1	2.3
6		340.0	203.0	0.60	2	737.0	977.0	0.1	220.0	0.0	0.0	0.0	0.0	0.0	0.0	0.0	1	7.6
7		272.0	188.0	0.55	2	743.0	985.0	0.2	210.0	0.0	0.0	0.0	0.0	68.0	0.0	0.0	1	7.6
	[10]																	
8		350.0	95.2	0.27	1	575.9	1273.0	0.0	—	0.0	0.0	0.0	0.0	0.0	0.0	0.0	28	61.1
9		350.0	98.5	0.28	1	558.2	1325.4	0.0	—	0.0	0.0	0.0	0.0	0.0	0.0	0.0	28	54.0
10		339.5	97.7	0.28	1	655.3	1273.0	0.0	—	0.0	10.5	0.0	0.0	0.0	0.0	0.0	28	65.7
11		339.5	97.6	0.28	1	535.0	1247.0	0.0	—	0.0	10.5	0.0	0.0	0.0	0.0	0.0	28	62.2
12		336.0	97.6	0.28	1	535.0	1247.0	0.0	—	0.0	14.0	0.0	0.0	0.0	0.0	0.0	28	54.5
13		332.5	97.7	0.28	1	655.3	1273.0	0.0	—	0.0	17.5	0.0	0.0	0.0	0.0	0.0	28	63.1
14		329.0	97.6	0.28	1	535.0	1247.0	0.0	—	0.0	21.0	0.0	0.0	0.0	0.0	0.0	28	52.2
	[75]																	
15		350.2	157.60	0.45	1	810.4	1200.6	0.0	5.0	0.0	0.0	0.0	0.0	0.0	0.0	0.0	1	19.07
16		332.2	157.30	0.45	1	809.2	1198.9	0.6	10.0	17.5	0.0	0.0	0.0	0.0	0.0	0.0	1	21.50
17		314.2	157.10	0.45	1	808.0	1197.0	1.2	15.0	34.9	0.0	0.0	0.0	0.0	0.0	0.0	1	22.43
18		296.3	156.90	0.45	1	806.8	1195.3	1.8	25.0	52.3	0.0	0.0	0.0	0.0	0.0	0.0	1	20.23
19		278.5	156.70	0.45	1	805.6	1193.6	2.4	75.0	69.6	0.0	0.0	0.0	0.0	0.0	0.0	1	19.33
20		260.7	156.40	0.45	1	804.5	1191.8	3.0	75.0	86.9	0.0	0.0	0.0	0.0	0.0	0.0	1	15.73
21		243.0	156.20	0.45	1	803.3	1190.0	3.6	90.0	104.1	0.0	0.0	0.0	0.0	0.0	0.0	1	14.53
22		350.2	157.60	0.45	1	810.4	1200.6	0.0	5.0	0.0	0.0	0.0	0.0	0.0	0.0	0.0	7	50.23
23		332.2	157.30	0.45	1	809.2	1198.9	0.6	10.0	17.5	0.0	0.0	0.0	0.0	0.0	0.0	7	53.80

Table 4. *Cont.*

Mix N°	Author	C (Kg/m³)	W (Kg/m³)	W/B	Specimen Compression Type	S (Kg/m³)	Coarse Aggregates (Kg/m³)	SP (%)	Slump/Flow (mm)	MK (Kg/m³)	LF (Kg/m³)	SF (Kg/m³)	GGBFS (Kg/m³)	FA (Kg/m³)	MW (Kg/m³)	RA (Kg/m³)	Age (days)	Rc (MPa)
	[76]																	
24		300.0	165.0	0.41	1	1095.0	722.0	1.0	30.0	0.0	0.0	0.0	0.0	0.0	100.0	0.0	7	25.8
25		300.0	165.0	0.41	1	1095.0	722.0	2.0	57.0	0.0	0.0	0.0	0.0	0.0	100.0	0.0	7	30.7
26		300.0	165.0	0.41	1	1095.0	722.0	3.0	58.0	0.0	0.0	0.0	0.0	0.0	100.0	0.0	7	22.2
27		300.0	180.0	0.45	1	1071.0	706.0	1.0	43.0	0.0	0.0	0.0	0.0	0.0	100.0	0.0	7	25.8
28		300.0	180.0	0.45	1	1071.0	706.0	2.0	60.0	0.0	0.0	0.0	0.0	0.0	100.0	0.0	7	28.9
29		300.0	189.0	0.47	1	1055.0	696.0	2.0	66.0	0.0	0.0	0.0	0.0	0.0	100.0	0.0	7	27.6
30		300.0	201.0	0.50	1	1039.0	685.0	2.0	68.0	0.0	0.0	0.0	0.0	0.0	100.0	0.0	7	26.2
	[77]																	
38		553.5	161.6	0.10	2	734.2	0.0	0.0	0.0	0.0	0.0	0.0	0.0	0.0	0.0	1145.9	7	51.6
39		524.5	205.5	0.13	2	695.7	0.0	0.0	55.0	0.0	0.0	0.0	0.0	0.0	0.0	1085.8	7	40.4
40		498.3	245.2	0.16	2	661.0	0.0	0.0	179.0	0.0	0.0	0.0	0.0	0.0	0.0	1031.6	7	28.9
41		474.7	280.9	0.19	2	629.7	0.0	0.0	531.0	0.0	0.0	0.0	0.0	0.0	0.0	982.7	7	24.6
42		553.5	152.0	0.08	2	734.2	0.0	0.0	0.0	0.0	0.0	0.0	0.0	0.0	0.0	1351.5	7	62.1
43		524.5	196.4	0.11	2	695.7	0.0	0.0	32.0	0.0	0.0	0.0	0.0	0.0	0.0	1280.6	7	46.4
44		498.3	229.4	0.13	2	661.0	0.0	0.0	180.0	0.0	0.0	0.0	0.0	0.0	0.0	1216.7	7	33.6
45		474.7	272.7	0.17	2	629.7	0.0	0.0	563.0	0.0	0.0	0.0	0.0	0.0	0.0	1159.1	7	27.3

6. Results and Discussion

The structure of the ANN applied in this study is shown in Figures 5 and 6. The network consists of 18 inputs, two hidden layers, and one output, and was used for 1310 data values.

The results in Figure 7 show the model performance results measured through error minimization techniques. During the learning process, the error drops as the network is continuously trained. The patterns in Figure 7 are respectively training, validation, and testing relative to model error.

- Pattern 1 (blue, Training) describes the training error obtained from 70% of the samples and improves the model's fit by adjusting the network according to its error.
- Pattern 2 (green, Validation) fits the network generalization ability that instructed the network on when to stop the training process. Pattern 2 represents the ability of the model to predict new data [32] (predictive performance). The training process is halted when validation error stops decreasing, which inherently avoids over-fitting.
- Pattern 3 (red, Testing) does not affect training and is an independent measure of network performance. This error measured on the test data indicates how well the model is generalized to the data during and after training.

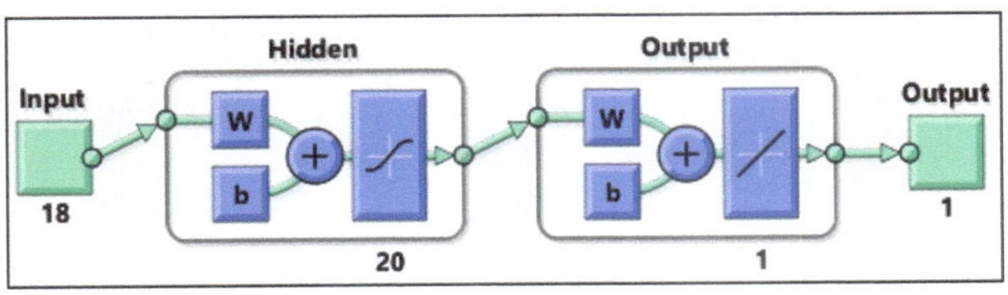

Figure 5. Artificial neural network architecture from MATLAB software (R2022a).

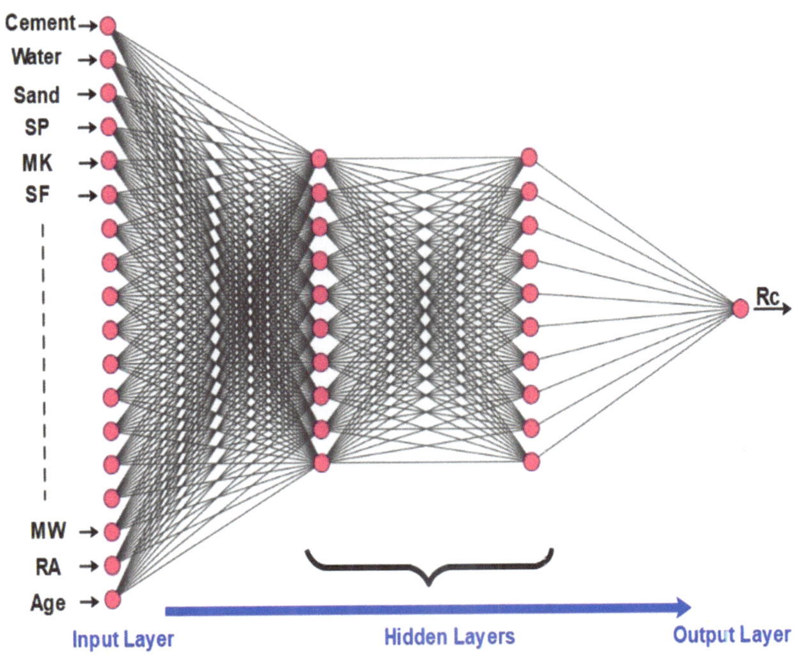

Figure 6. Architecture of the artificial neural network in this study.

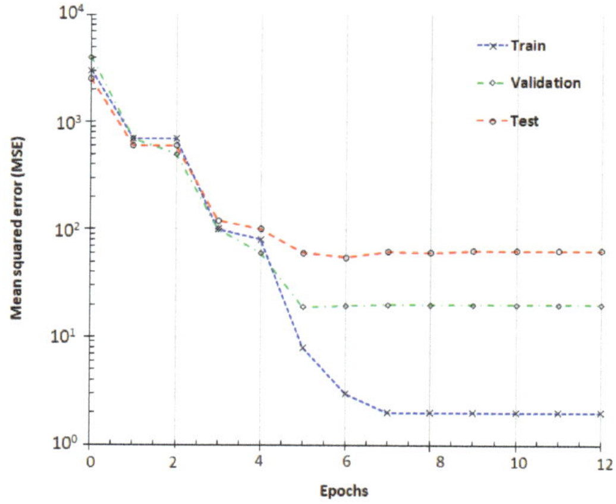

Figure 7. Model performance plot based on mean squared error (MSE) analysis.

In Figures 7 and 8, the results clearly demonstrate that the gradient begins to stabilize when the epoch equals 6. The values of the coefficient of determination R^2 are 0.9982, 0.9763, and 0.9566 for training, validation, and testing respectively. The histogram shown in Figure 9 shows that the error value is lowering from training to testing. This corresponds to an improvement of the model throughout the processing and precision. The high values of R^2 in Figure 10 mean that the model seems to have sufficient accuracy and is also well trained.

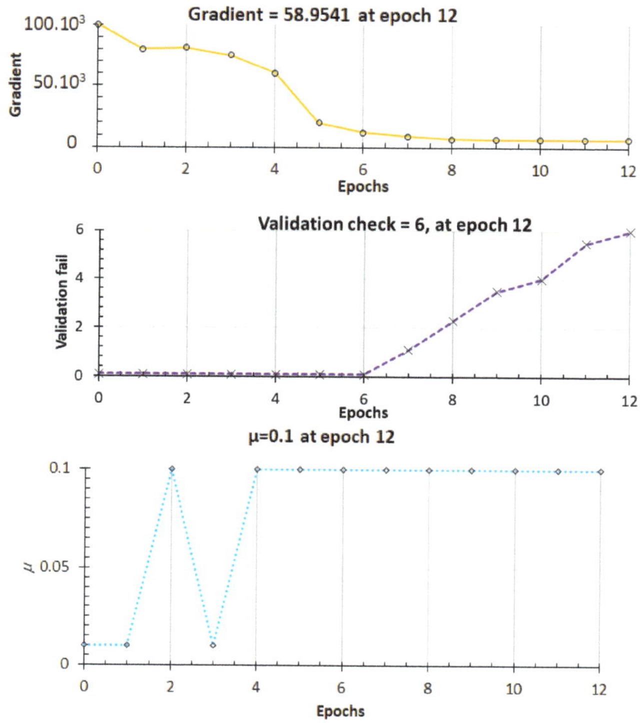

Figure 8. Plot of model performance evolution regarding the number of epochs.

Figure 9. Training state results for performance checking.

In Figure 10, the network outputs with respect to targets for training, validation, and test sets are plotted against the target values. For a perfect fit, the data should fall along a 45° line, where the network outputs are equal to the targets. For this problem, the fit is very good for all datasets, with R^2 values of 0.95 or above in each case.

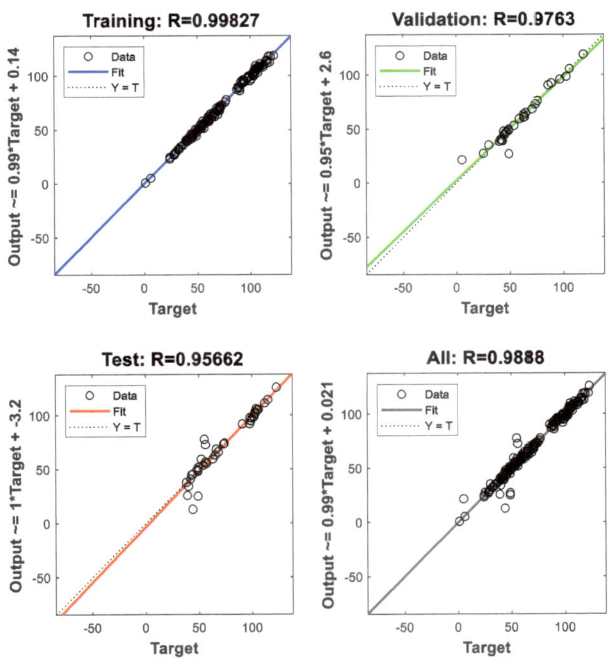

Figure 10. Regression analysis results.

The performance indicators for compressive strength accuracy are given in Table 5. The value of RMSE, which is 2.91 MPa, shows that the gap between predicted and experimental values is small. MAPE shows that the predicted compressive strength deviated on average by 2.87% from the experimental data. This indicates that the differences between forecast and actual results were negligible. All these points indicate that the ANN strength predictive model was able to reproduce the experimental compressive strength results with high accuracy. These results are comparable to those obtained in similar studies [29,51]. In the same order, the determination coefficients reached by [39,78] are between 0.9443 and 0.9836.

Table 5. Values of model performance accuracy parameters.

Parameters	Performance Indicators				
	RMSE (MPa)	R^2	MSE (MPa)	MAE (MPa)	MAPE (%)
Values	2.91	0.9888	8.4689	1.7463	2.87

7. Conclusions

This study aimed to use an artificial neural network to predict the compressive strength of waste-based concretes. The methodology, architecture, and learning methods were explained, based on feedforward and backpropagation techniques. A bibliographic dataset compiled from the literature was then used, including a total of 1303 concrete formulations from 22 different studies. The important conclusions that can be drawn from this work are:

- The ANN model can predict compressive strength with high accuracy by learning the deep features of the water–cement ratio, the cement and admixture content, the age of the concrete, etc.
- The results have demonstrated that multilayer feedforward artificial neural networks are practicable methods to forecast compressive strength in concretes.
- Errors of the model calculated from R^2, MSE, MAPE and MAE show small gaps between experimental and forecast values.

The above results suggest that the use of ANN is suitable for concrete compressive strength prediction. A coming study that we are undertaking will test the use of Decision Tree Regression (DTR) in the prediction of concrete properties. This machine learning method has been stated to be a very efficient approach.

Supplementary Materials: The following supporting information can be downloaded at: https://www.mdpi.com/article/10.3390/ma15207045/s1. The supplementary materials file contains part of the formulations used in the ANN training are collected from [20,67,79–83].

Author Contributions: Conceptualization, MA.; methodology, MA.; software, M.A.; validation, M.B. and N.-E.A.; investigation, M.A.; resources, M.A.; data curation, M.B and R.Z.; writing—original draft preparation, M.A.; writing—review and editing, M.A., M.B., and R.Z.; supervision, N.-E.A. All authors have read and agreed to the published version of the manuscript.

Funding: This research received no external funding.

Institutional Review Board Statement: This study did not require ethical approval.

Informed Consent Statement: Not applicable.

Data Availability Statement: The data used in this study are available from the corresponding author on submission of a reasonable request.

Acknowledgments: This study is financially supported by the ECOSED industrial chair. The authors of this paper express their sincere appreciation for their help and support.

Conflicts of Interest: The authors declare that there is no conflict of interest regarding the publication of this paper.

References

1. Tu, T.Y.; Chen, Y.Y.; Hwang, C.L. Properties of HPC with recycled aggregates. *Cem. Concr. Res.* **2006**, *36*, 943–950. [CrossRef]
2. Gastaldini, A.; Hengen, M.; Gastaldini, M.; Amaral, F.D.; Antolini, M.; Coletto, T. The use of water treatment plant sludge ash as a mineral addition. *Constr. Build. Mater.* **2015**, *94*, 513–520. [CrossRef]

3. Bolomey, J. Granulation et prévision de la résistance probable des bétons. *Travaux* **1935**, *19*, 228–232. [CrossRef]
4. Powers, T.; Brownyard, T. Studies of the Physical Properties of Hardened Portland Cement Paste. *J. Proc.* **1946**, *43*, 101–132. Available online: https://www.concrete.org/publications/internationalconcreteabstractsportal (accessed on 19 August 2016).
5. Feret, R. Sur la Compacité des Mortiers Hydrauliques. 1892. Available online: http://www.bcin.ca/Interface/openbcin.cgi?submit=submit&Chinkey=123645.
6. Bolomey, J. Détermination de la résistance à la compression des mortiers et bétons. *Bull. Tech. la Suisse Rom.* **1925**, *51*, 126–132. [CrossRef]
7. Nagaraj, T.; Banu, Z. Generalization of Abrams' law. *Cem. Concr. Res.* **1996**, *26*, 933–942. [CrossRef]
8. Naderpour, H.; Mirrashid, M. An innovative approach for compressive strength estimation of mortars having calcium inosilicate minerals. *J. Build. Eng.* **2018**, *19*, 205–215. [CrossRef]
9. Sarıdemir, M. Prediction of compressive strength of concretes containing metakaolin and silica fume by artificial neural networks. *Adv. Eng. Softw.* **2009**, *40*, 350–355. [CrossRef]
10. Sobhani, J.; Najimi, M.; Pourkhorshidi, A.R.; Parhizkar, T. Prediction of the compressive strength of no-slump concrete: A comparative study of regression, neural network and ANFIS models. *Constr. Build. Mater.* **2010**, *24*, 709–718. [CrossRef]
11. Ince, R. Prediction of fracture parameters of concrete by Artificial Neural Networks. *Eng Fract. Mech.* **2004**, *71*, 2143–2159. [CrossRef]
12. Onal, O.; Ozturk, A.U. Artificial neural network application on microstructure–compressive strength relationship of cement mortar. *Adv. Eng. Softw.* **2010**, *41*, 165–169. [CrossRef]
13. Azimi-Pour, M.; Eskandari-Naddaf, H. ANN and GEP prediction for simultaneous effect of nano and micro silica on the compressive and flexural strength of cement mortar. *Constr. Build. Mater.* **2018**, *189*, 978–992. [CrossRef]
14. Khan, M.I. Predicting properties of High Performance Concrete containing composite cementitious materials using Artificial Neural Networks. *Autom. Constr.* **2011**, *22*, 516–524. [CrossRef]
15. Prasad, B.R.; Eskandari, H.; Reddy, B.V. Prediction of compressive strength of SCC and HPC with high volume fly ash using ANN. *Constr. Build. Mater.* **2009**, *23*, 117–128. [CrossRef]
16. Bui, D.K.; Nguyen, T.; Chou, J.S.; Nguyen-Xuan, H.; Ngo, T.D. A modified firefly algorithm-artificial neural network expert system for predicting compressive and tensile strength of high-performance concrete. *Construct. Build. Mater.* **2018**, *180*, 320–333. [CrossRef]
17. Khosravani, M.R.; Nasiri, S.; Anders, D.; Weinberg, K. Prediction of dynamic properties of ultra-high performance concrete by an artificial intelligence approach. *Adv. Eng. Softw.* **2018**, *127*, 51–58. [CrossRef]
18. Öztaş, A.; Pala, M.; Özbay, E.; Kanca, E.; Çağlar, N.; Bhatti, M.A. Predicting the compressive strength and slump of high strength concrete using neural network. *Constr. Build. Mater.* **2006**, *20*, 769–775. [CrossRef]
19. Chithra, S.; Kumar, S.S.; Chinnaraju, K.; Ashmita, F.A. A comparative study on the compressive strength prediction models for High Performance Concrete containing nano silica and copper slag using regression analysis and Artificial Neural Networks. *Constr. Build. Mater.* **2016**, *114*, 528–535. [CrossRef]
20. Abu Yaman, M.; Elaty, M.A.; Taman, M. Predicting the ingredients of self compacting concrete using artificial neural network. *Alex. Eng. J.* **2017**, *56*, 523–532. [CrossRef]
21. Uysal, M.; Tanyildizi, H. Predicting the core compressive strength of self-compacting concrete (SCC) mixtures with mineral additives using artificial neural network. *Constr. Build. Mater.* **2011**, *25*, 4105–4111. [CrossRef]
22. Vakhshouri, B.; Nejadi, S. ANFIS application to predict the compressive strength of lightweight self-compacting concrete. In Proceedings of the 2016 Future Technologies Conference (FTC), San Francisco, CA, USA, 6–7 December 2016; pp. 28–35. [CrossRef]
23. Mashhadban, H.; Kutanaei, S.S.; Sayarinejad, M.A. Prediction and modeling of mechanical properties in fiber reinforced self-compacting concrete using particle swarm optimization algorithm and artificial neural network. *Constr. Build. Mater.* **2016**, *119*, 277–287. [CrossRef]
24. Sadrmomtazi, A.; Sobhani, J.; Mirgozar, M. Modeling compressive strength of EPS lightweight concrete using regression, neural network and ANFIS. *Constr. Build. Mater.* **2013**, *42*, 205–216. [CrossRef]
25. Bingöl, A.F.; Tortum, A.; Gül, R. Neural networks analysis of compressive strength of lightweight concrete after high temperatures. *Mater. Des.* **2013**, *52*, 258–264. [CrossRef]
26. Altun, F.; Kişi, Ö.; Aydin, K. Predicting the compressive strength of steel fiber added lightweight concrete using neural network. *Comput. Mater. Sci.* **2008**, *42*, 259–265. [CrossRef]
27. Hendi, A.; Behravan, A.; Mostofinejad, D.; Moshtaghi, S.M.; Rezayi, K. Implementing ANN to minimize sewage systems concrete corrosion with glass beads substitution. *Constr. Build. Mater.* **2017**, *138*, 441–454. [CrossRef]
28. Hodhod, O.; Salama, G.A. Analysis of sulfate resistance in concrete based on artificial neural networks and USBR4908-modeling. *Ain Shams Eng. J.* **2013**, *4*, 651–660. [CrossRef]
29. Abambres, M.; Lantsoght, E.O. ANN-Based Fatigue Strength of Concrete under Compression. *Materials* **2019**, *12*, 3787. [CrossRef]
30. Duan, P.; Shui, Z.; Chen, W.; Shen, C. Effects of metakaolin, silica fume and slag on pore structure, interfacial transition zone and compressive strength of concrete. *Constr. Build. Mater.* **2013**, *44*, 1–6. [CrossRef]
31. Deng, F.; He, Y.; Zhou, S.; Yu, Y.; Cheng, H.; Wu, X. Compressive strength prediction of recycled concrete based on deep learning. *Constr. Build. Mater.* **2018**, *175*, 562–569. [CrossRef]

32. Getahun, M.A.; Shitote, S.M.; Gariy, Z.C.A. Artificial neural network based modelling approach for strength prediction of concrete incorporating agricultural and construction wastes. *Constr. Build. Mater.* **2018**, *190*, 517–525. [CrossRef]
33. Dehghan, S.; Sattari, G.; Chelgani, S.C.; Aliabadi, M. Prediction of uniaxial compressive strength and modulus of elasticity for Travertine samples using regression and artificial neural networks. *Min. Sci. Technol. (China)* **2010**, *20*, 41–46. [CrossRef]
34. Faridmehr, I.; Bedon, C.; Huseien, G.; Nikoo, M.; Baghban, M. Assessment of Mechanical Properties and Structural Morphology of Alkali-Activated Mortars with Industrial Waste Materials. *Sustainability* **2021**, *13*, 2062. [CrossRef]
35. Mhaya, A.M.; Huseien, G.F.; Faridmehr, I.; Abidin, A.R.Z.; Alyousef, R.; Ismail, M. Evaluating mechanical properties and impact resistance of modified concrete containing ground Blast Furnace slag and discarded rubber tire crumbs. *Constr. Build. Mater.* **2021**, *295*, 123603. [CrossRef]
36. Golafshani, E.M.; Behnood, A. Predicting the mechanical properties of sustainable concrete containing waste foundry sand using multi-objective ANN approach. *Constr. Build. Mater.* **2021**, *291*, 123314. [CrossRef]
37. Alabduljabbar, H.; Huseien, G.; Sam, A.; Alyouef, R.; Algaifi, H.; Alaskar, A. Engineering Properties of Waste Sawdust-Based Lightweight Alkali-Activated Concrete: Experimental Assessment and Numerical Prediction. *Materials* **2020**, *13*, 5490. [CrossRef]
38. Sadowski, Ł.; Piechówka-Mielnik, M.; Widziszowski, T.; Gardynik, A.; Mackiewicz, S. Hybrid ultrasonic-neural prediction of the compressive strength of environmentally friendly concrete screeds with high volume of waste quartz mineral dust. *J. Clean. Prod.* **2019**, *212*, 727–740. [CrossRef]
39. Ray, S.; Haque, M.; Ahmed, T.; Nahin, T.T. Comparison of artificial neural network (ANN) and response surface methodology (RSM) in predicting the compressive and splitting tensile strength of concrete prepared with glass waste and tin (Sn) can fiber. *J. King Saud Univ. Eng. Sci.* **2021**, *4*, 144–169. [CrossRef]
40. Yeşilmen, S.; Tatar, B. Efficiency of convolutional neural networks (CNN) based image classification for monitoring construction related activities: A case study on aggregate mining for concrete production. *Case Stud. Constr. Mater.* **2022**, *17*, e01372. [CrossRef]
41. Almasaeid, H.H.; Suleiman, A.; Alawneh, R. Assessment of high-temperature damaged concrete using non-destructive tests and artificial neural network modelling. *Case Stud. Constr. Mater.* **2022**, *16*, e01080. [CrossRef]
42. Yang, D.S.; Park, S.K.; Lee, J.H. A prediction on mix proportion factor and strength of concrete using neural network. *KSCE J. Civ. Eng.* **2003**, *7*, 525–536. [CrossRef]
43. Kim, J.I.; Kim, D.K. Application of neural networks for estimation of concrete strength. *KSCE J. Civ. Eng.* **2002**, *6*, 429–438. [CrossRef]
44. Apostolopoulou, M.; Asteris, P.G.; Armaghani, D.J.; Douvika, M.G.; Lourenço, P.B.; Cavaleri, L.; Bakolas, A.; Moropoulou, A. Mapping and holistic design of natural hydraulic lime mortars. *Cem. Concr. Res.* **2020**, *136*, 106167. [CrossRef]
45. Gupta, T.; Patel, K.; Siddique, S.; Sharma, R.K.; Chaudhary, S. Prediction of mechanical properties of rubberised concrete exposed to elevated temperature using ANN. *Measurement* **2019**, *147*, 106870. [CrossRef]
46. Topçu, I.B.; Sarıdemir, M. Prediction of properties of waste AAC aggregate concrete using artificial neural network. *Comput. Mater. Sci.* **2007**, *41*, 117–125. [CrossRef]
47. Dantas, A.T.A.; Leite, M.B.; de Jesus Nagahama, K. Prediction of compressive strength of concrete containing construction and demolition waste using artificial neural networks. *Constr. Build. Mater.* **2013**, *38*, 717–722. [CrossRef]
48. Shaik, S.B.; Karthikeyan, J.; Jayabalan, P. Influence of using agro-waste as a partial replacement in cement on the compressive strength of concrete – A statistical approach. *Constr. Build. Mater.* **2020**, *250*, 118746. [CrossRef]
49. Faridmehr, I.; Nehdi, M.; Nikoo, M.; Huseien, G.; Ozbakkaloglu, T. Life-Cycle Assessment of Alkali-Activated Materials Incorporating Industrial Byproducts. *Materials* **2021**, *14*, 2401. [CrossRef]
50. Catherina Vasanthalin, P.; Chella Kavitha, N. Prediction of compressive strength of recycled aggregate concrete using artificial neural network and cuckoo search method. *Mater. Today: Proc.* **2021**, *46*, 8480–8488. [CrossRef]
51. Armaghani, D.J.; Asteris, P.G. A comparative study of ANN and ANFIS models for the prediction of cement-based mortar materials compressive strength. *Neural Comput. Appl.* **2020**, *33*, 4501–4532. [CrossRef]
52. Özcan, F.; Atis, C.; Karahan, O.; Uncuoğlu, E.; Tanyildizi, H. Comparison of artificial neural network and fuzzy logic models for prediction of long-term compressive strength of silica fume concrete. *Adv. Eng. Softw.* **2009**, *40*, 856–863. [CrossRef]
53. Jiang, H.; Liu, G.; Alyami, H.; Alharbi, A.; Jameel, M.; Khadimallah, M.A. Intelligence decision mechanism for prediction of compressive strength of self-compaction green concrete via neural network. *J. Clean. Prod.* **2022**, *340*, 130580. [CrossRef]
54. Irandegani, M.A.; Zhang, D.; Shadabfar, M. Probabilistic assessment of axial load-carrying capacity of FRCM-strengthened concrete columns using artificial neural network and Monte Carlo simulation. *Case Stud. Constr. Mater.* **2022**, *17*. [CrossRef]
55. Farooq, F.; Ahmed, W.; Akbar, A.; Aslam, F.; Alyousef, R. Predictive modeling for sustainable high-performance concrete from industrial wastes: A comparison and optimization of models using ensemble learners. *J. Clean. Prod.* **2021**, *292*, 126032. [CrossRef]
56. Asteris, P.G.; Moropoulou, A.; Skentou, A.D.; Apostolopoulou, M.; Mohebkhah, A.; Cavaleri, L.; Rodrigues, H.; Varum, H. Stochastic Vulnerability Assessment of Masonry Structures: Concepts, Modeling and Restoration Aspects. *Appl. Sci.* **2019**, *9*, 243. [CrossRef]
57. Bui, D.T.; Abdullahi, M.M.; Ghareh, S.; Moayedi, H.; Nguyen, H. Fine-tuning of neural computing using whale optimization algorithm for predicting compressive strength of concrete. *Eng. Comput.* **2019**, *37*, 701–712. [CrossRef]
58. Garg, A.; Aggarwal, P.; Aggarwal, Y.; Belarbi, M.O.; Chalak, H.D.; Tounsi, A.; Gulia, R. Machine learning models for predicting the compressive strength of concrete containing nano silica. *Comput. Concr.* **2022**, *30*, 33–42. [CrossRef]

59. Zhuang, X.; Zhou, S. The Prediction of Self-Healing Capacity of Bacteria-Based Concrete Using Machine Learning Approaches. *Comput. Mater. Contin.* **2019**, *59*, 57–77. [CrossRef]
60. Scrivener, K.L.; John, V.M.; Gartner, E.M. *Eco-Efficient Cements: Potential Economically Viable Solutions for a Low-CO2 Cement-Based Materials Industry*; United Nations Environment Program: Nairobi, Kenya, 2016.
61. Asteris, P.G.; Koopialipoor, M.; Armaghani, D.J.; Kotsonis, E.A.; Lourenço, P.B. Prediction of cement-based mortars compressive strength using machine learning techniques. *Neural Comput. Appl.* **2021**, *33*, 13089–13121. [CrossRef]
62. DeRousseau, M.; Kasprzyk, J.; Srubar, W. Computational design optimization of concrete mixtures: A review. *Cem. Concr. Res.* **2018**, *109*, 42–53. [CrossRef]
63. Agrawal, V.; Sharma, A. Prediction-of-Slump-in-Concrete-using-Artificial-Neural-Networks. *World Acad. Sci. Eng. Technol.* **2010**, *4*, 279–286.
64. Rafiq, M.; Bugmann, G.; Easterbrook, D. Neural network design for engineering applications. *Comput. Struct.* **2001**, *79*, 1541–1552. [CrossRef]
65. Flood, I.; Kartam, N. Neural Networks in Civil Engineering. I: Principles and Understanding. *J. Comput. Civ. Eng.* **1994**, *8*, 131–148. [CrossRef]
66. KewalRamani, M.A.; Gupta, R. Concrete compressive strength prediction using ultrasonic pulse velocity through artificial neural networks. *Autom. Constr.* **2006**, *15*, 374–379. [CrossRef]
67. Duan, Z.; Kou, S.; Poon, C. Prediction of compressive strength of recycled aggregate concrete using artificial neural networks. *Constr. Build. Mater.* **2013**, *40*, 1200–1206. [CrossRef]
68. Patil, S.R. Regionalization of an Event Based Nash Cascade Model for Flood Predictions in Ungauged Basins. Ph.D. Thesis, University of Stuttgart, Stuttgart, Germany, July 2008.
69. Dorofki, M.; Elshafie, A.H.; Jaafar, O.; Karim, O.A.; Mastura, S. Comparison of Artificial Neural Network Transfer Functions Abilities to Simulate Extreme Runoff Data. *Int. Proc. Chem. Biol. Environ. Eng.* **2012**, *33*, 39–44.
70. Kumar, G.S.; Rajasekhar, K. Performance analysis of Levenberg—Marquardt and Steepest Descent algorithms based ANN to predict compressive strength of SIFCON using manufactured sand. *Eng. Sci. Technol. Int. J.* **2017**, *20*, 1396–1405. [CrossRef]
71. Bishop, C.M. *Pattern Recognition and Machine Learning*; Springer: Berlin/Heidelberg, Germany, 2006.
72. Kubat, M. Neural networks: A comprehensive foundation by Simon Haykin, Macmillan, 1994, ISBN 0-02-352781-7. *Knowl. Eng. Rev.* **1998**, *13*, 409–412. [CrossRef]
73. Chen, Q.; Norcio, A.F.; Wang, J. Neural Network Based Stereotyping for User Profiles. *Neural Comput. Appl.* **2000**, *9*, 259–265. [CrossRef]
74. Huang, C.-H.; Lin, S.-K.; Chang, C.-S.; Chen, H.-J. Mix proportions and mechanical properties of concrete containing very high-volume of Class F fly ash. *Constr. Build. Mater.* **2013**, *46*, 71–78. [CrossRef]
75. Wild, S.; Khatib, J.; Jones, A. Relative strength, pozzolanic activity and cement hydration in superplasticised metakaolin concrete. *Cem. Concr. Res.* **1996**, *26*, 1537–1544. [CrossRef]
76. Alyamaç, K.E.; Ince, R. A preliminary concrete mix design for SCC with marble powders. *Constr. Build. Mater.* **2009**, *23*, 1201–1210. [CrossRef]
77. Li, L.G.; Lin, C.J.; Chen, G.M.; Kwan, A.K.H.; Jiang, T. Effects of packing on compressive behaviour of recycled aggregate concrete. *Constr. Build. Mater.* **2017**, *157*, 757–777. [CrossRef]
78. Naderpour, H.; Rafiean, A.H.; Fakharian, P. Compressive strength prediction of environmentally friendly concrete using artificial neural networks. *J. Build. Eng.* **2018**, *16*, 213–219. [CrossRef]
79. Jalal, M.; Mansouri, E. Effects of Fly Ash and Cement Content on Rheological, Mechanical, and Transport Properties of High-Performance Self-Compacting Concrete. *Sci. Eng. Compos. Mater.* **2012**, *19*, 393–405. [CrossRef]
80. Irfan Che-Ani, A.; Siddiqi, Z.; M Zain, M.F.; Abd, S.M.; Sopian, K.; Jamil, M. Mathematical Regression Model for the Prediction of Concrete Strength. *academia.edu* **2007**, *5*, 3096–3402.
81. Jiang, L.H.; Malhotra, V.M. Reduction in Water Demand of Non-Air-Entrained Concrete Incorporating Large Volumes of Fly Ash. *Cem. Concr. Res.* **2000**, *30*, 1785–1789. [CrossRef]
82. McCarthy, M.J.; Dhir, R.K. Development of High Volume Fly Ash Cements for Use in Concrete Construction. *Fuel* **2005**, *84*, 1423–1432. [CrossRef]
83. Bouzoubaâ, N.; Lachemi, M. Self-Compacting Concrete Incorporating High Volumes of Class F Fly Ash: Preliminary Results. *Cem. Concr. Res.* **2001**, *31*, 413–420. [CrossRef]

materials

MDPI

Review

Recycling of Mining Waste in the Production of Masonry Units

Nicoleta Cobîrzan [1], Radu Muntean [2], Gyorgy Thalmaier [3] and Raluca-Andreea Felseghi [1,4,*]

1 Faculty of Civil Engineering, Technical University of Cluj-Napoca, Constantin Daicoviciu Street, No. 15, 400020 Cluj-Napoca, Romania; Nicoleta.Cobarzan@ccm.utcluj.ro
2 Faculty of Civil Engineering, Transilvania University of Brașov, Bd. Eroilor, No. 29, 500036 Brașov, Romania; radu.m@unitbv.ro
3 Faculty of Materials and Environmental Engineering, Technical University of Cluj-Napoca, Bd. Muncii, No. 103–105, 400641 Cluj-Napoca, Romania; Gyorgy.Thalmaier@sim.utcluj.ro
4 Faculty of Electrical Engineering and Computer Science, "Ștefan cel Mare" University of Suceava, Universității Street, No. 13, 720229 Suceava, Romania
* Correspondence: andreea.felseghi@ccm.utcluj.ro

Abstract: Masonry units made of clay or Autoclaved Aerated Concrete (AAC) are widely used in constructions from Romania and other countries. Masonry units with superior mechanical and thermal characteristics can improve the energy efficiency of buildings, especially when they are used as the main solutions for building envelope construction. Their production in recent years has increased vertiginously to meet the increased demand. Manufactured with diversified geometries, different mechanical and/or thermal characteristics have a high volume in the mass of the building and a major influence in their carbon footprint. Starting from the current context regarding the target imposed by the long-term strategy of built environment decarbonization, the aim of the paper is to analyze the potential of reusing mining waste in the production of masonry units. Mining waste represents the highest share of waste generated at national level and may represent a valuable resource for the construction industry, facilitating the creation of new jobs and support for economic development. This review presents the interest in integrating mining wastes in masonry unit production and the technical characteristics of the masonry units in which they have been used as raw materials in different percentages. Critical assessment framework using SWOT analysis highlights the key sustainability aspects (technical, environmental, social, economic) providing a comprehensive and systematic analysis of the advantages and disadvantages regarding the integration of mining waste as secondary raw materials into masonry units production.

Keywords: circular economy; construction materials; greener environment; masonry; mining waste; resources conservation; value-added properties

Citation: Cobîrzan, N.; Muntean, R.; Thalmaier, G.; Felseghi, R.-A. Recycling of Mining Waste in the Production of Masonry Units. *Materials* **2022**, *15*, 594. https://doi.org/10.3390/ma15020594

Academic Editor: Yuri Ribakov

Received: 11 November 2021
Accepted: 9 January 2022
Published: 13 January 2022

Publisher's Note: MDPI stays neutral with regard to jurisdictional claims in published maps and institutional affiliations.

1. Introduction

Construction, building operation, and the construction industry in general contribute more than 39% of total global CO_2 emissions [1], playing a key role in achieving the long-term goals set out in the Paris Agreement (2015) [2] to *"keep the temperature rise to 1.5 °C compared to the industrialization period"* and to ensure the transition to climate neutrality by 2050 [3].

To meet these goals, the EU's development priorities set out in the "European Green Deal" [3] call for increasing the rate of building renovation making them more energy efficient and development of the industrial sector based on sustainable principles to support the transition from linear to circular economy [4,5]. The circular economy is a new paradigm that aims to keep materials in the economy for a very long time, thus limiting the generation of waste and exploitation of primary natural resources [6,7].

The circular economy requires recycling of materials considered waste and leads to major changes in products and systems design, re-adaptation of technological flow (supply and distribution chains), or company-specific business model [3,8]. Extension of digital passport for all products to sustain waste circularity [9] and population education about the advantages obtained by using materials made of recycled waste are also imperative in this regard.

Recycling of waste to preserve the natural resources and to reduce disposal [10] will reduce the environmental impact by contributing to the growth of circular economy above the value of 9% currently recorded [11], which without climate neutrality cannot be ensured. Promoting waste recycling to reduce excessive consumption of natural resources in the U.E. and/or dependence on imports emerges also from the Raw Materials Initiative (RMI) launched in 2008 to identify the level of needs and mineral resources for each EU country [12].

Romania is a country rich in non-metallic mineral resources (sand, clay minerals, bentonite, limestone, graphite, perlite, diatomite) and metalliferous (copper ore, lead, zinc, aluminum, gold, manganese chromium, nickel, tungsten) [13], whose exploitation has generated large amounts of waste over time (Figure 1). Mining of copper, manganese, zinc, gold, uranium has greatly diminished since the 1990s as a result of industrial restructuring and the closure of most of the operational mines, declared inefficient [14,15]. The strategic objectives provided by the "Mining Industry Strategy developed for the period 2017–2035" aim to increase the use of domestic mining resources to reduce dependence on imports of mineral raw materials, which are vital for the development and sustainable growth of the Romanian economy [15], but also for development of mining areas that have now become disadvantaged areas being characterized by a low living standard and lack of jobs. It is necessary to use innovative technologies of "mineral resources" starting from cradle to exploitation, processing, or recovery of waste.

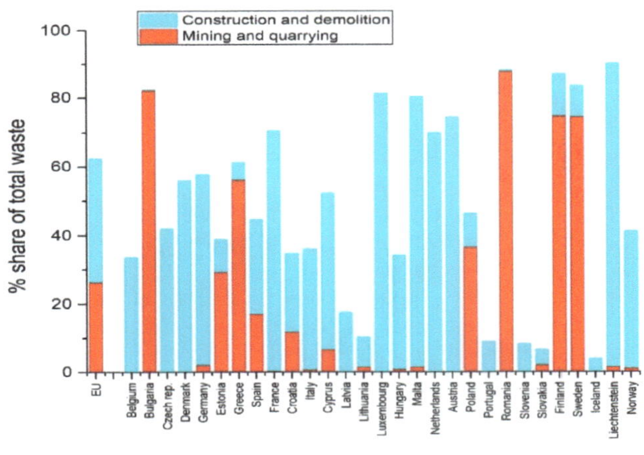

Figure 1. Waste generation by European countries, 2018 [16].

The mining industry, during its existence has significantly affected the environment, a supplementary reason for which the recovery of mining waste in various areas are stringent issues raised nowadays. An important direction in this regard is to find new techniques and technologies for the recovery of waste, exploiting this valuable resource in terms of economic growth [15].

According to statistical data valid for 2018 and published by Eurostat [16], the quantity of waste generated in the European Union by all economic activities and households was over 2300 million tons. As a percentage, the waste from construction and demolition and those from mining and quarrying activities represent the largest share, 35.9% and 26.6%, respectively (Figure 1).

Much waste from mining and construction activities are classified as major mineral wastes, being different from other types of wastes, and represent about three quarters (74%

or 5.2 tons per capita) of total waste generated in 2018 at the EU level. The higher levels of major mineral waste recorded in EU countries are recorded in areas with important mining and quarrying activities, such as Romania, Finland, Sweden, or Bulgaria.

According to Figure 1, Romania generates approximately 88% of mining waste from exploitation and processing activities of mineral resources (mining, quarrying activities) and only 0.3% from construction activities (from all data recorded), out of a total of 203 million tons of waste.

The distribution of waste generated by various economic and household activities at the national level is presented in Figure 2.

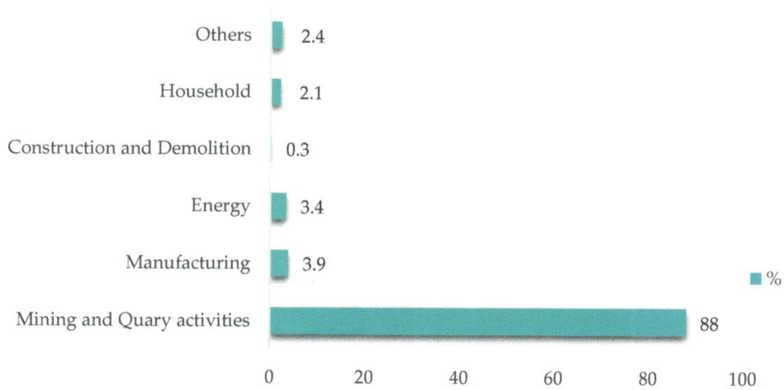

Figure 2. Percent of waste by types of activities in Romania, 2018 [16].

In concrete figures, in 2018, a quantity of 178.6 million tons of mining waste was generated, increasing compared to the previous years, for which, only 8.9 million tons were recovered in various forms [17].

The use of mining waste in construction may play a significant role for the conservation of natural resources, but also for reducing the impact on the environment or human health due to inadequate storage.

Extractive industrial waste was defined and classified initially by Directive 75/442/EEC [18] and later by Directive 2014/955/EU [19] in four major groups: tailings; waste rock; sewage sludge; topsoil.

Tailings are solid wastes or sludges that remain after mineral treatment by separation processes (e.g., grinding, crushing, sizing, flotation, and other physico-chemical techniques) to extract valuable minerals from a lesser rock, while waste rocks are economically unprofitable rocks that are excavated to gain access to ore. Sewage sludge results during exploitation by sedimentation in flooded quarries or tailings ponds and may contain large amounts of heavy metals, non-metals (arsenic), and neutral salts (sulfate, calcium). The vegetal soil is the upper layer of the land surface (after removing the roots and capitalizing of wood) and the lower layer of soil up to the rock layer; although identified as waste, these soil materials can be stored in stacks and reused for vegetation resettlement. Hazardous mining waste can contain harmful substances and toxic metals [20], which can cause environmental disasters and endanger human health or habitat destruction. In contrast, in the case of non-hazardous and inert mining waste, the quantities of heavy metals are reduced and may represent a valuable resource for the production of new materials [20] after pre-treatment.

By analyzing the current literature, where information is dissipated in various sources, there is a critical necessity to investigate, compile, and elect the main information reachable to assist further development of mining waste circularity. Therefore, this review is intended to emphasize comprehensive information on mining waste used in the construction materials sector, specifically for masonry units' production, and identifies the key sustainability aspects from technical, economic, social, and environmental points of view.

The major contributions of this review are:

- comprehensive review on integrating mining waste used for masonry unit's production (clay masonry units, autoclaved aerated concrete units, and geopolymers masonry units);
- using the "SWOT" model to conduct a critical evaluation framework;
- overall, this review article may represent a preliminary study for academics, scientific researchers, and industry to further identify the feasibility parameters and limitations for exploiting mining waste as secondary raw materials in the production of masonry units.

Further, the main highlights of this work can be summarized as follows:

- reasons for recycling mining waste;
- overview of the masonry units on the transition to the circular economy context in the brick industry to produce eco-responsible construction materials;
- discussions regarding the influence of the mining waste chemical and mineralogical composition on the manufacturing technological process of new construction materials;
- discussed the effect of incorporating mining waste into masonry unit's production (clay masonry units, autoclaved aerated concrete units, and polymers masonry units);
- over 100 scientific papers were reviewed to present an overview of masonry developments over the time and to the discuss the role of mining waste as a raw material in the production of masonry units.

The structure of this paper is described as follows. First, a general overview on masonry units in transition to the circular economy is addressed, making assessments of clay masonry, autoclaved aerated concrete, and geopolymers masonry units, but also on chemical composition and mineralogical composition of mining waste. Afterward, the various approaches in the literature related to mining waste for clay masonry units' production, mining waste for autoclaved aerated concrete production, and mining waste for geopolymer masonry units' production were presented and discussed. Following these discussions, a critical evaluation framework was developed that uses SWOT analysis to identify the advantages and pitfalls of the integration of mining waste as secondary raw materials into masonry units' body. Finally, conclusions were drawn that summarize the main appraisals on mining waste as secondary raw materials used for the production of masonry units based on the current discussion. This review provides a good basis and guidelines for encouraging the wide adoption of mining waste circularity in masonry unit production for sustainable construction.

2. Masonry Units: An Overview

Masonry units (clay brick and AAC) are used for the construction of structural or non-structural walls based on their mechanical properties. The group of masonry unit, the wall density, volume of cavities (%), seismic area, and mechanical characteristics are defining factors in selecting the height regime of masonry buildings. These criteria are regulated by the Romanian codes P100-1/2013 [21], CR6-2013 [22], and must be analyzed in the preliminary stage of the building design.

In the case of framed structures, the walls being infilled can be constructed with masonry units with lower compressive strength, if adequate constructive measures are provided to prevent their failure during a strong earthquake.

The restrictions given by design codes limit the use of masonry units with lower compressive strength than 5 MPa to the construction of masonry buildings placed in seismic areas, which are specific to Romanian territory.

The content of soluble salts and the highwater absorption of units can affect the durability of masonry work and their lifespan, making necessary the rehabilitation work that has the effect the increasing embodied energy of building.

The transition to climate neutrality involves a drastic reduction of CO_2 emissions in construction sector, especially in buildings, so that the uses of materials with improved thermal characteristics or/and low embodied energy for construction of envelope members are required and have become a necessity.

The development of building materials with high mechanical strength and/or thermal properties and/or low embodied energy has now become a major challenge for masonry unit manufacturing. Replacing conventional with secondary raw materials resulted from industrial and mining activities can be an effective solution in ensuring climate neutrality targets.

2.1. Clay Masonry Unit

Fired bricks are materials with wide applications in the construction sector, influencing energy consumption in buildings, greenhouse gasses emissions, but also the costs related to heating and cooling the buildings during operation. Fired at temperatures up to 900 °C required for mineralogical transformation of raw materials and for brick sintering, they have a major impact on the environment.

It is imperative in the transition towards a circular economy in the brick industry to produce materials with high thermal characteristics and/or low embodied energy, long lifespan, and possibilities for reusing or recycling at the end of lifecycle so as to keep the product in economy as long as possible. Clay bricks have a long lifecycle up to 150 years [23,24] and can be reused after building demolition for new masonry works if their physical and mechanical characteristics are adequate. Recycling the secondary raw materials in the brick and other industries may also be a viable strategy for "closing the loop".

Physical, mechanical, and thermal characteristics of clay units are in accordance with chemical composition or mineralogical content of raw materials (primary or secondary), materials granulometry, firing temperature, porosity, and forming technology (pressure, extrusion, geo-polymerization) [25]. Based on targeted characteristics, the composition of the clay matrix can be optimized so as to meet the required criteria in terms of sustainability and the circular economy.

In recent decades, the brick industry has evolved continuously, producing a wide range of masonry units (solid and perforated bricks, blocks with vertical or horizontal cavities) [23,24], whose thermal characteristics have been improved to meet the thermal requirements imposed by norms. As can be seen in the scientific literature, the thermal properties of bricks was optimized by increasing the volume of cavities, by changing the cavity geometry or the profile of exterior surface [26–29], and by filling the cavities with different thermal insulation materials [30–34]. Another solution studied and adopted for ceramic product optimization in terms of embodied energy or thermal performance was the incorporation in the clay matrix, organic or inorganic waste [35,36], resulted from different activities.

Organic waste (agricultural, food, industry waste, etc.) incorporated in different percentages in clay body as pore forming and/or fluxing have been investigated by different authors [36–39] highlighting economic, mechanical, or thermal advantages, respectively, and the caloric value [40] to contribute to circular economy.

Inorganic waste with role of degreaser and/or fluxing being used in the clay matrix to reduce plasticity [41,42] and shrinkage to drying or firing or to increase porosity have a lower impact on the environment, but require higher water consumption. Construction and demolition waste [43,44], cutting and processing of natural stone [45–47], mining activities [20,41,48–63], or industrial waste are an alternative solution with high potential in the manufacturing of clay units.

Many of these measures are aimed at increasing thermal performance and/or reducing the embodied energy of the products. The limitations given by the geometric features and characteristics, especially when are used in seismic areas or by the mechanical strength have been intensively analyzed in the literature. Some of the identified solutions can be implemented at the producer level, but not all of them are efficient and economic. Finding feasible solutions with wide applicability in the field is becoming more stringent to reduce the environmental impact and to preserve the primary resources.

2.2. AAC Masonry Units

AAC units are produced by autoclaving and have a high porosity of up to 50–60% from mass. They are usually made of cement, lime, silicious materials, gypsum, and aluminum powder [64] as pore forming agent [65].

Depending on the mechanical and geometrical characteristics, AAC units can be used to construct the structural wall in case of masonry buildings with lower height regime or to construct non-structural wall (for infill and partitioning). They have good thermal characteristics and low density, which can contribute in the reduction of the building mass and the related seismic force. ACC bricks are considered eco-friendly materials [66] due to the lower embodied energy compared to that of fired masonry units.

The use of waste in the production of AAC units has been analyzed by various researchers to reduce CO_2 emissions and to conserve natural resources. They have been used as substitutes for silicious materials, cement, or aluminum [67] in various percentages, so that their mechanical properties to remain within the acceptable limits provided by standards or design codes.

The potential of using waste in autoclaved aerated concrete has been studied by [68], coal bottom ash by [69], silica fume and fly ash by [70], industrial and agricultural waste by [71], perlite waste by [72], rice husk ash by [73], graphite tailings by [74], copper tailings by [75,76], hematite tailings by [77], iron tailings by [78–81], silicon tailings by [82], coal gangue by [83], etc. Studies have shown the feasibility of using secondary raw materials as a substitute for virgin materials.

The mechanical characteristics of AAC units are influenced by the materials' chemical and mineralogical content, but also by the pore distribution and the porosity of the final product.

The formation of crystalline phases (Tobermorite—$Ca_5Si_6O_{16}(OH)_2 \cdot 4H_2O$) is favored in the autoclaving process. The increase of soluble SiO_2 favors the reaction with $Ca(OH)_2$, thus contributing to the appearance of Calcium-Silicate-Hidrate (C-S-H) gel. Dual-alkali silicates reacts with SiO_2 to form tobermorite with a role in increasing the material mechanical parameters and formation of C-S-H gel [66,79].

Thermal characteristics are influenced not only by the type or pore distribution, but also by the chemical content of raw materials. Low specific weight materials can significantly contribute to the reduction of thermal conductivity [66]. Also, the raw materials that favor the appearance of phase composition (crystalline or amorphous) have a favorable effect in increasing the thermal performance of materials.

2.3. Geopolymers Masonry Units

An alternative solution to fired clay bricks is represented by geopolymer masonry units. Geopolymers, invented and developed by Joseph Davidovits since the 1970s [84], are now also used for the production of ecological concrete or masonry units [85].

Geopolymers are obtained following a chemical reaction between aluminosilicates oxides and hydroxide and/or alkaline silicate solutions (NaOH; KOH) [35,86], processed at a temperature below 100 °C to reduce water evaporation [87]. The advantages of using geopolymers consist in saving energy and CO_2 emissions [86].

The mechanical characteristics of the product are mainly influenced by the used raw materials, the calcination temperature, the curing time, or the used additives [88,89].

Geopolymerized bricks made of different waste have been intensively studied in recent years, such as industrial aggregates [90], fly ash, construction and demolition waste [89], tailings [91–96], etc.

3. Mining Waste Used for Masonry Unit's Production

Knowledge about chemical content and mineralogical composition are essentially in the selection stage of secondary raw materials. They must be similar and compatible in regards to chemical and mineralogical composition with conventional ones if used as substitutes or additive in the production of new buildings materials [97] with conventional technology. Re-adaptation of technological flow to new supply chains involves technological changes and investment cost. The chemical and mineralogical content of mining waste investigated by the authors [41,48,50–52,54,56,62,75–78,82,83] is shown in Table 1, Figures 3 and 4.

Table 1. Mineralogical composition of selected mining waste.

Authors	Mine Waste	Mineralogical Content	Ref.
Ettoumi et al., 2021	Phosphate sludge	calcite, dolomite, bassanite, heulandite, vermiculite, quartz, hematite, fluorapatite	[52]
Yang et al., 2014	Iron tailings	quartz, calcite, hematite, clinochlore, pyrite, amphibole	[54]
Wei et al., 2021	Gold tailings	quartz, sanidine, mica, pyrite, montmorillonite	[62]
Fang et al., 2011	Copper tailings	andradite, quartz	[76]
Zhao et al., 2012	Hematite tailings	hematite, quartz, chlorite, calcite	[77]
Ma et al., 2016	Iron tailings	gypsum quartz, albite, muscovite, calcite, terranovaite	[78]
Wang et al., 2016	Coal gangue	quartz, siderite, illite, kaolinite, montmoriolite, anorthite, muscovite	[83]

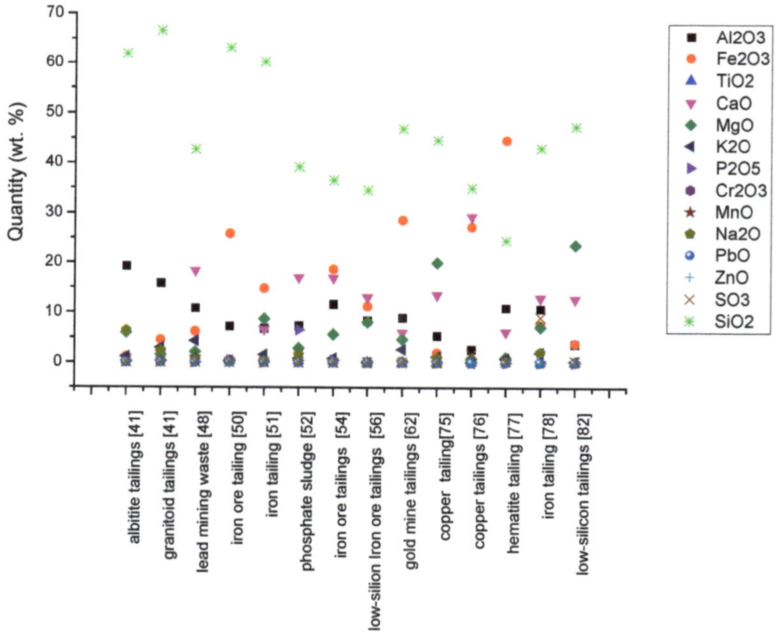

Figure 3. Chemical composition of waste.

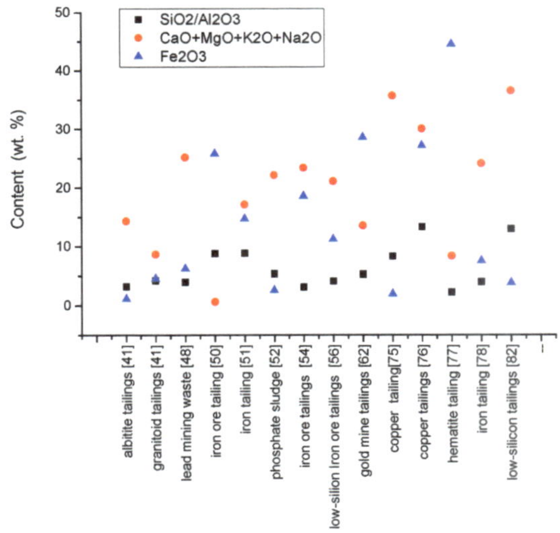

Figure 4. Ratio of SiO_2/Al_2O_3, Fe_2O_3 and fluxing materials.

3.1. Chemical Composition of Mining Waste

In order to produce geopolymeric materials, the SiO_2 and Al_2O_3 content of the raw materials used must be over 70% [98].

Clay used in manufactured of bricks usually contains SiO_2 (50–60%), Al_2O_3 (10–20%), CaO, Fe_2O_3, and other oxides in low quantities [99]. The presence of quartz in the clay mass might cause the appearance of cracks during cooling, while alumina ensures resistance and quality to the final product after the mullitization process. The presence of Fe_2O_3 in the brick mass gives the desired color (from red to dark brown), but a high content, usually over than 9% [100], can lead to the appearance of the "black core", when the oxygen rate is insufficient during firing. The formation of mullite (above 950 °C) causes the color of the product to become yellow or white [101] due to the Fe_2O_3 reaction and consumption during the mullitization process.

Also, the high content of soluble salt can favor the appearance of efflorescence, with a negative impact on durability of masonry units and works. Mining waste has different chemical composition, depending on their nature (Table 1 and Figure 3). Iron tailings have a high content of SiO_2 and Fe_2O_3 and a lower content of Al_2O_3 and CaO [50,51]. In contrast, the granitoid and albitite tailings [41] are rich in SiO_2 and Al_2O_3 and low in Fe_2O_3 and CaO. The SiO_2/Al_2O_3 ratio represents the amount of free silica contained in the material [50]. As can be observed from Figure 3, for mining waste, the SiO_2/Al_2O_3 ratio is between 2 and 13 (Figure 4) well above the kaolinite [100]. The Fe_2O_3 content is in the range of 1 ÷ 44% and the content of CaO + MgO + K_2O + Na_2O varies from 0.6 ÷ 36%.

In the case of the clays, the CaO content classifies the materials in the category of calcareous materials. A content over 6% of CaO indicates a calcareous material [101], while a content of K_2O, Fe_2O_3, CaO, MgO, and TiO_2 over 9% indicates low refractive materials.

The compositional quality of secondary raw materials may be different or lower than primary ones [102]. They can contain various pollutant soluble salts that arise the necessity to use different treatments or to use additional additives to improve the quality [44] prior to the production of the new materials.

Raw materials must be classified according to the standards and be followed by declaration of performance [97].

3.2. Mineralogical Composition of Mining Waste

The mineralogical composition of mining waste (Table 1) from raw materials influences the energy consumption during the manufacturing, but also the porosity and durability of products. Raw materials rich in calcite require higher firing temperature above 1100 °C [103] to complete the phase transformation of minerals. At a temperature of about 800 °C, calcite is transformed into calcium oxide, increasing the brick porosity [103,104] and decreasing the thermal conductivities and density. Great attention must be paid to the size of clay granules responsible for the transformation of calcium oxides into calcium hydroxide, a consequence of which is the volume increase and mechanical strength decrease.

Another important parameter in selecting the clay brick units is the color. The presence of iron oxide in clay matrix is essential in obtaining the reddish color. Other parameters that influence the brick color are temperature and firing time [105].

Studies carried out in recent years highlighted the potential of using mining waste in the production of all types of masonry units (Tables 2–4). Depending on the chemical and mineralogical characteristics, secondary raw materials replaced partially or totally the conventional raw materials (clay, sand, cement, etc.).

Table 2. Types of mining waste used in bricks.

Authors	Material	Composition	Technology/Firing/ Curing Temperature	Origin of Waste	Ref.
Marrocchino et al., 2021	bricks	plastic clay B (60–100%) [a] metamorphic eluvium R (0–40%) [a] Granitoid eluvium (0–20%) [a] Granitoid tailings (0–20%) [a] Albitite tailings (0–20%) [a]	extrusion/1000 °C	Italy	[41]
Suarez-Marcias et al., 2020	bricks	lead mine tailing (0–100%) clay (100–0%)	pressing 50 ± 1 MPa/950 ± 5 °C	Spain	[48]
Loutou et al., 2019	bricks	red clay (100%)	molded <6 MPa/900 °C, 1000 °C, and 1100 °C	Morocco	[49]
Mendes et al., 2019	bricks	iron ore tailing (0–40%) [b] grey clay (30–70%) [b] yellow clay (30–70%) [b]	pressing/extrusion 40–70 MPa/850 °C, 950 °C, and 1050 °C	Brazil	[50]
Li et al., 2019	porous bricks	100% iron tailing [b]	foam-gel casting/ 1070–1120 °C	China	[51]
Ettoumi et al., 2021	brick	100% phosphate sludge	pressing 6 MPa/900 °C, 1000 °C, and 1100 °C	Tunisia	[52]
Bayoussef et al., 2021	bricks	red clay (70–100%) fly ash (0–30%)	pressing/1100 °C	Morocco	[53]
Yang et al., 2014	bricks	Low silicon iron ore tailing (80–100%) [a] fly ash (0–20%) [a]	pressing 20 MPa/900 °C, 950 °C, and 1000 °C	China	[54]
Vilela et al., 2020	soil-cement bricks	soil partial substituted with iron ore tailings 0–40% Portland cement (10%) ratio soil: cement (9:1)	pressing/curing 20 ± 2 °C (28 days)	Brazil	[55]
Weishi et al., 2018	brick	low-silicon iron ore tailings (75%) curing agent (fly ash, lime, gypsum) (25%)	molding pressure 50 MPa/curing 30–60 °C	China	[56]
da Silva et al., 2014	red ceramic	stearic acid emulsion iron tailings (0–5%)[b] clay (95–100%)[b]	pressed 20 MPa/950 °C	Brazil	[57]
Luo et al., 2020	sintered brick	iron ore tailings (54%)[b] shale (10%) [b] coal gangue powder (30%) [b] sewage sludge (0–12%) [b]	pressing 20 MPa/950–1100 °C	China	[58]

Table 2. *Cont.*

Authors	Material	Composition	Technology/Firing/ Curing Temperature	Origin of Waste	Ref.
Wang et al., 2019	brick	Iron tailings (40–70%) [a] Fly ash (20–50%) [a] Kaolin (10%) [a]	Pressing 20 MPa/1000, 1050, 1100, 1150, or 1200 °C)	China	[59]
Chen et al., 2011	brick	hematite tailings (77–100%), fly ash (0–8%) [a] clay (0–15%)	pressing 20–25 MPa/850, 900, 950, 980, 1000, 1030, and 1050 °C	China	[60]
Yonggang et al., 2011	bricks	fine gold tailings (60–100%) [b] medium gold tailings (10–30%) [b] clays (10–40%) [b]	pressed 5–20 MPa/900–1050 °C	NA	[61]
Wei et al., 2021	sintered bricks	gold mine tailing (60–100%) +clay (0–40%)	pressing/900–1050 °C	China	[62]

(wt %) [a] (mass %) [b].

Table 3. Mining waste used for AAC product.

Authors	Materials	Composition	Origin of Waste	Ref.
Huang et al., 2012	AAC	copper tailings (30%) [a] blast furnace slag (35%) [a] quartz sand (20%) [a] cement clinker (10%) [a] gypsum (5%) [a]	China	[75]
Fang et al., 2011	Autoclaved sand-lime brick	copper tailing (0–88%) [b] sand powder (0–15%) [b] river sand (0–88%) [b] lime (6.7–13.3%) [b]	China	[76]
Zhao, Y. 2012	Autoclaved bricks	*Optimum mixture:* hematite tailings:lime:sand ratio (70:15:15) [b]	China	[77]
Ma et al. 2016	AAC blocks	iron tailings (0–68%) [b] cement (8%) [b] quicklime (19–27%) [b] silicon sand (0–68%) [b] gypsum (3%) [b] Al powder (0.14%) [b]	China	[78]
Liang et al., 2019	AAC	iron tailing (30–55%) [b] silica sand (5–30%) [b] lime (20–30%) [b] ordinary Portland cement (5–15%) [b] flue gas desulfurization gypsum (5%) [b]	China	[79]
Cai et al., 2016	AAC blocks	iron tailings (0–68%) [b] cement (8%) [b] quicklime (21%) [b] silicon sand (0–68%) [b] gypsum (3%) [b] al powder (0.14%) [b]	China	[80]
Zhao et al., 2009	autoclaved brick	low-silicon tailings (83%) [b] alkali-activated slag/fly ash	China	[82]
Wang et al., 2016	AAC	coal gangue (1–40%) [b] iron ore tailing (20–59%) [b] lime (25%) [b] cement (10%) [b] gypsum (5%) [b] Al powder (0.06%) [b]	China	[83]

(wt %) [a] (mass %) [b].

Table 4. Types of mining waste used for geopolymer.

Authors	Materials	Composition	Technology/ Curing Temperature	Origin of Waste	Ref.
Ahmari, S. and Zhang, L., 2012	geopolymer bricks	copper mine tailings NaOH solution (10–15 M)	forming pressure (0–35 MPa)/60 to 120 °C	Arizona	[95]
Ahmari S. and Zhang, L., 2013	geopolymer bricks	copper mine tailings, sodium hydroxide NaOH (15 M)	forming pressure (0–35 MPa)/60 to 120 °C	Arizona	[92]
Beulah et al., 2021	geopolymer bricks	iron ore tailings (50–90%); GGBS (10–50%) and red mud (50–90%) GGBS (10–50%) NaOH (8 M)	NA	India	[93]
Zhang et al. 2021	geopolymer	gold mine tailing NaOH solutions (4–12 M)	molding/75 °C	Peru	[94]

3.3. Mining Waste for Clay Masonry Units

The authors [41,48–50,53,57,59–62] studied the mining waste in the mixture of clay (Table 2), highlighting their microstructural and technical characteristics.

The authors Marrocchino et al. [41] used waste from mining (granitoid, albitite tailings) and from construction activities as secondary raw materials for brick manufacturing. The addition of 20% waste acts as degreaser and consequently contributes to the reduction of plasticity index (PI) to 22% in the case of granitoid tailings, respectively, to 20% in case of albitite tailings, reaching the optimum extrusion interval. Drying and firing shrinkage was also reduced, which can have a beneficial effect in reducing costs and the risk of cracking.

Iron tailings were analyzed by the authors [50,51,54–59] as a partial or total substitute of clay and sand in the production of bricks and ceramic materials. The chemical content of mining waste is close to the chemical content of ceramic materials and consequently can be an alternative solution to conventional raw materials [81]. The bricks were formed by uniaxial pressing or extrusion at 20–70 MPa and fired at temperatures between 850–1200 °C.

Studies conducted by Mendes et al. [50], showed that iron ore tailings (IOT) can be used in the clay matrix with a degreaser role, improving the extrusion parameters and thus contributing to the reduction of the cracks resulted during drying and firing. The optimal percentage of IOT in the brick mass was 29.1%, determined from experimental results on ceramic materials and statistical data processing. The average compressive strength of small-scaled perforated bricks made from the optimized mixture was 4.3 MPa with a water absorption of 20.9%. In terms of environmental performance, IOT based bricks can be included in the category of inert materials.

Li et al. [51] used the iron tailings (100%) to produce porous bricks, with improved thermal performance compared to dense materials. The low value of thermal conductivity of 0.032 W/(m·K), measured in the case of bricks with a porosity of 89%, highlights the potential of using the bricks in energy efficient buildings.

Yang et al. [54] investigated the low-silicon IOT bricks with fly ash in different percentage, showing the possibility of using the two recycled materials in the brick-making industry. The mechanical strengths of the samples decreased with the increase of the fly ash content, being above the minimum value required by standards (10 MPa).

Vilela et al. [55] analyzed the possibility of partially substituting the soil from the cement soil bricks with up to 40% iron ore mining waste. The density of the samples at 28 days varied between 1670 and 1750 (kg·m^{-3}), with a figure 5% higher in the case of mixture with 40% addition of mining waste compared with reference. An insignificant decrease in compressive strength from 2.79 MPa (0% mining waste) to 2.66 MPa (40% mining waste) was recorded in the case of samples with an addition of 40%. The thermal conductivity increases from 1.42 W/m·K in the case of samples without mining waste to 1.59 W/m·K, due to the filling of the pore with mining waste.

Weishi et al. [56] studied low-silicon IOT bricks with addition of curing agent, obtaining at 28 days a compressive strength of up to 32 MPa, depending on initial curing temperature and the content of the stearic acid emulsion.

da Silva et al. [57] used 5% of iron ore tailings in the ceramic mass, obtaining increases in the value of flexural strength from 6.5 to 6.7 MPa, compared with the samples without additive whose flexural strength was 5.1 MPa. The porosity increases from 25.5% to 42.5% and water absorption decreased from 31% to 25.5% compared with clay samples. After firing, the color of the samples with the addition of mining waste was reddish, close to that of traditional ceramics, unlike the samples without mining waste, whose color was lighter (reddish-orange). Waste requires low or no grinding, thus contributing positively to production costs and environmental impact.

Luo et al. [58] investigated the sintered bricks made of mining waste. The optimal ratio of raw materials in the brick mass was selected based on the technical criteria as 54:30:10:6 (iron ore tailings:coal gangue:shale:sludge). It was found that material density decreases from 1740 to 1638 kg/m^3 with the variation of the sludge ratio and the water absorption increased from 14.2% to 17.5%, being below 18%, which is the maximum accepted value. The compressive strength decreased from 18.2 to 14.2 MPa. Heavy metals (Cu, Zn, Cr, Pb) in the leachates were immobilized in the clay matrix during sintering, resulting in a leaching content in sintered brick lower than in the leaching of raw materials. The leaching efficiency was high, thus fulfilling the requirements provided in national standards.

Wang et al. [59] studied the bricks with optimal mass ratio 60:30:10 (IOT:flyash:kaolin). Based on XRD analysis, they studied the effect of heavy metals in bricks, highlighting the formation of new crystalline structure (spinel, silicate, aluminosilicate), having the effect of "fixing heavy metal" in the system and decreasing the amount of leaching (Cu, Zn, Cd, Pb). The leaching content decreases with increasing the sintering temperature. The result indicated that Zinc was immobilized at 850 °C, while Pd and Cd required increasing the temperature over 1000 °C.

The gold mine tailings rich in SiO_2 and Fe_2O_3 were investigated by Wei et al. [62] to produce clay bricks. The results showed that mechanical properties increased with the increases of clay content and technological conditions (temperature and firing). At a temperature of 1000 °C, the mechanical strength of the sample with the addition of 40% clay was about 21 MPa and the water absorption decrease below the limit value accepted by the standard.

Gold tailing have also been studied by Yanggang et al. [61] in fired bricks composition. The amount of waste range between 60–90% of the clay mass, depending on the material fineness. Values over 10 MPa were obtain in the case of bricks with 60–90% fine tailings (less than 0.6 mm) and 10–25% medium tailings.

Suárez-Macías et al. [48] analyzed the possibility of introducing the tailings from a lead mine in the clay matrix. An increase of open porosity was observed with the increase of the tailing percent or the absorption of water and the decrease of compressive strength. Water absorption increased from 10.61 (0% tailing) to 29.52% (100% tailing) and the density decreased from 2010 to 1510 kg/m^3. Compressive strength decreased from 118 to 7.2 MPa. The color of the brick varied from red to brown depending on the chemical composition and sintering temperature. The leachate from the metallic elements (Fe, Pb, Mn, and Ti) was substantially reduced with the incorporation into the clay mass, but Al, Mg, Br, Cu were not completely retained, which may require higher firing temperatures.

Loutou et al. [49], studied the possibility of valorization mining waste (residual rocks) resulted from the exploitation of phosphate ores in the production of fired clay bricks.

Phosphate sludge has been investigated by Ettoumi et al. [52] as a potential substitute of clays in the production of ceramic bricks, in proportion of up to 100%. Flexural (bending) strength was up to 13.4 MPa with a water absorption of about 12.5 for the samples fired at a temperature of 1100 °C. Increasing the firing temperature to 1100 °C was necessary to reduce the leachate content of (Na, K), respectively, to 1100 °C to reduce the content of Cr (from leachate) below the limits provided by the standard.

Bayoussef et al. [53] analyzed the micro and macrostructural characteristics of brick made of red clays resulted from phosphate mining activities and fly ash. The compressive strength of the bricks increased from 77 to 98 MPa in the case of samples with 10% admixture of fly ash, recording decreases of up to 73 and 57 MPa for samples with an addition of 20% and 30% fly ash, respectively. This is due to the growth of micropores and the appearance of microfissures in the materials mass. The density of the samples increased from 1840 to 1970 kg/m^3 in the case of the sample with 30% addition of fly ash.

Zhu et al. [63] investigated the impermeable bricks made of tailing and gangue as mining waste from a feldspar mine in China, obtaining compressive strength of about 40 MPa in the case of specimens with addition of 60 wt% of gangue and 40 wt% of tailings. The values of compressive strength were reduced up to 5 MPa (90 wt%) with the increase of gangue content. Following the micro and macrostructural investigations performed on the permeable bricks, the optimal technological parameters were established, which lead to results above the values provided by standards. The optimum percent of residual mining waste was 80–90% (tailings 20% and 60–70% gangue) with addition of 10–20% waste ceramics to optimize the technological parameters (permeability and compressive strength). The increase of the sintering temperature led to the increase of the liquid phase generated by tailings and to the densification of the bricks with the effect of reducing their permeability.

3.4. Mining Waste for AAC Units

The potential of using mining waste in the production of AAC (75–80, 82, 83) has been analyzed by scientific researchers (Table 3). The complexity of the investigations performed revealed a series of advantages, but also disadvantages of using mining waste in the production of AAC. Chemical, mineralogical, and/or particle size distribution of raw materials were analyzed in the preliminary stage, then, based on preliminary results, the optimal percentage of mining waste, which can substitute the conventional raw materials established.

The macroscopical (mechanical strength, physical characteristics) and microscopical characterization (chemical content, mineralogical composition, microstructural, and thermal analysis) on the final products showed that the iron, copper, and hematite tailings can replace conventional raw materials in different percentage from products volume/weight having the mechanical characteristics in the acceptable limits required by standards.

The authors [78,80] found that the amount of C-S-H and the compressive strength of products decreases with the increase of the IOT content and their particle size in the AAC blocks' composition. Mechanical properties of AAC investigated by Ma et al. [78] vary between 1.65 and 3.15 MPa, depending on the IOT content and silicon and tailings fineness. The increase of quartz finesse has a positive effect in the formation of hydration products (C-S-H and tobermorite), but leads to increases in energy demand for manufacturing. A recommended ratio for investigated AAC was 8:21–27:62–68:40:60:3:0.14% (cement:quiklime:siliceous:iron tailings:gypsum:Al powder), in which the compressive strength was about 2.5 MPa.

Mining waste such "coal gangue" and iron tailings have also been studied as secondary raw materials in the production of AAC. Studies have shown that the optimal percentage of mining waste in AAC mass is 40% iron tailings and 20% coal gangue [83], obtaining a compressive strength of 3.7 MPa.

Liang et al. [79] used IOT in a percentage of 30–55% in the production of AAC, with values of density and compressive strength influenced by percentage of mining waste from 604 to 590 kg/m^3, respectively from 4.17 to 3.2 MPa. A compressive strength of about 4.17 MPa was recorded in the case of samples with 40% IOT from mass.

Zhao et al. [77] used hematite tailings to produce autoclaved bricks. The optimal ratio of hematite tailings:lime:sand is 70:15:15 [%] of the mixture mass, obtaining the values of compressive strength of 21.2 MPa.

The authors Huang et al. [75] analyzed the AAC with 30% copper tailings as a lime substitute, obtaining compressive strength of 4 MPa for a density of 610 kg/m^3. The replacement of lime with skarn-type copper tailing (SCT) contributes to savings of CO_2 emissions that would result from calcination.

Fang et al. [76] have analyzed the autoclaved sand lime bricks with copper tailing admixture in percentages from 40% to 88% from product mass. The results show that compressive strength of samples was reduced from 24.3 to 4.6 MPa in the case of specimens with 88% of tailings compared with reference. For specimens with 50% copper tailings, the values of compressive strength were increased from 10.2 up to 20.3 MPa due to increase of lime/sand ratio.

Zhao et al. [82] have studied bricks made of low silicon tailings in 85% and cementing materials 15% from their mass. The results shows that mechanical characteristics of specimens were about 16 MPa for compressive strength and 3.8 MPa for the bending.

3.5. Mining Waste for Geopolymer Masonry Units

As in the previous cases, different mining wastes were utilized in manufacturing geopolymer masonry units, as presented in Table 4. Copper mine tailings have been analyzed by authors Ahmari and Zhang [92,95] as an alternative to the production at low temperature of eco-friendly bricks through geopolymerization technology. It has been shown that the compressive strength of bricks was dependent on technological conditions (forming, pressing, curing temperature), but also on the NaOH concentration. The heavy metals have been immobilized in the network of geopolymers.

The authors Beulah et al. [93] studied the geopolymer bricks made of IOT and red bricks with admixture of ground granulated blast furnace slag (GGBS). The optimum mix of bricks consisted of 70% of red clay/IOT and 30% GGBS. The values of compressive strength were 28.23 MPa (70% IOT + 30% GGBS) and 16.67 MPa (70% red clay + 30% GGBS), respectively, at 28 days. The density was 1080 kg/m^3 and only 888 kg/m^3 in case of samples with IOT, respectively, for red clay.

The authors Zhang et al. [94] in their research investigated the influence of NaOH molarity in geopolymerization degree and their failure process under monotonic and cyclic loading.

4. Critical Assessment Discussions

As can be observed in Tables 2–4, mining waste has been investigated by many researchers, showing the potential in using as secondary raw materials for the production of masonry units by substituting totally or partially the natural raw materials (clays, cement, aggregates). Investigations have been performed on waste collected from different countries, especially in areas with mining deposits and/or are scarce in natural resources.

The feasibility of the integration of mining waste as a raw material in manufacturing of masonry units is gained from the SWOT analysis based on the four key aspects (technical, environmental, social, economic) to develop the critical assessment framework (Table 5).

Table 5. Strengths–Weaknesses–Opportunities–Threats (SWOT) analysis.

Strengths	Weaknesses
S1. Technical strengths: ✓ base of raw materials with relatively long depletion periods, consisting of qualitatively important resources with diverse chemical compositions; ✓ chemical and mineralogical characteristics close to those of raw materials, which allows the substitution of conventional raw materials in diversified percentages depending on the technical characteristics targeted; ✓ the values of the mechanical resistances on laboratory scale are good, being dependent on the content of the mining waste from the sample mass; ✓ the physical-mechanical characteristics of samples with the addition of mining waste often exceed the limit values imposed by the design codes; ✓ samples with low content of mining waste or those burned at higher temperatures have a low content of heavy metals, thus classifying the finished products in the category of inert materials; ✓ mining tailings materials do not require grinding and screening, which can lead to increased productivity and reduced production costs.	**W1. Technical weaknesses:** ✓ the coarse mining waste requires mechanical treatment (grinding, drying, collecting) in the preliminary stage, which can increase the energy consumption and CO_2 emissions; ✓ high content of toxic substances and hazardous elements, which limit their uses in a high percent as conventional material substitution; ✓ lower characteristics compared with conventional raw materials; ✓ requires development of the existing technology, which can increase the investment cost; ✓ studies must be extended from laboratory to large-scale products to validate their technical properties; ✓ the current technical endowment does not ensure economic efficiency; ✓ inadequate access to transport infrastructure and utilities, cantonment of productive capacities in isolated, mono-industrial areas; ✓ lack of specialized post-extraction processing industries.
S2. Environmental strengths: ✓ can substitute total or partially the virgin raw material contributing to the preservation of natural resources; ✓ increasing the efficiency in the manufacturing process of the final products by consuming less natural and material resources; ✓ reduce the water, air, and land pollution, but also the disposal cost.	**W2. Introduction risks:** ✓ high content of toxic substances or hazardous elements may endanger the lives of employees in brick factories; ✓ difficult operating conditions.
S3. Social strengths: ✓ new job opportunity in the recycling supply chains; ✓ low cost of the product will increase material affordability; ✓ there is a higher and professional education infrastructure capable of preparing qualified personnel for the activity of reusing mining waste in construction.	**W3. Lack of support from the government:** ✓ lack of national regulation for secondary materials; ✓ lack of financial support for technological investments; ✓ non-existence at the national level of up-to-date and transparent statistical databases, referring to: deposits in operation, concessioned or potentially concessionable, volumes and quantities of reserves exploited annually and, implicitly, of resulting waste, economic agents carrying out activities of extraction of useful mineral substances, and waste storage; ✓ weak interest of local authorities, due to the absence of legislative provisions through which part of the fees or royalties paid by operating license holders should be directed to environmental funds, programs to restore affected areas, reuse of waste and its reintegration in the value chain.
S4. Economic strengths: ✓ may increase the business competitivity due to product attractivity and low cost, especially in areas with scarce natural resources; ✓ emergence of potential suppliers, demanders, and end-users;	**W4. Environmental weaknesses:** ✓ treatment process requires energy consumption and CO_2 emissions, especially for coarse waste; ✓ may contain high or medium percent of toxic substances and hazardous elements.
S5. Diversity in resources harnessing: ✓ decreases the dependence on the exploitation of natural resources (clay, limestone, sand); ✓ harnessing waste as it is possible to produce new products from waste resources; ✓ harnessing diversity for global business performance; ✓ they can be an alternative to conventional materials with a higher price [106].	**W5. Economic weaknesses:** ✓ the treatment process especially of coarser waste may increase the production cost; ✓ require financial supports to innovate the supply chain.

Table 5. *Cont.*

Opportunities	Threats
O1. Technical opportunities: ✓ the existence of a natural competitive advantage in markets in neighboring countries that do not have mineral waste reserves and the expansion of the market with the accession of new member states to the EU; ✓ the perspective of implementing new technologies for reuse of mining waste. **O2. Environmental opportunities:** ✓ recycling of mining waste to support the transition from a linear to circular economy; ✓ reuse of waste resulting in the production stage of masonry units or their recycling in the construction and demolition stages in the production of new construction materials is a viable solution to ensure the circularity of the materials; ✓ the waste resulting in the construction and demolition stages can also be reused in the construction of masonry units in the case of new buildings if they meet the quality criteria, but also in other complementary areas [107]. **O3. Social opportunities:** ✓ the existence of a potential for capitalization of some activities related to the exploitation of mining waste; ✓ collaboration opportunities between academic and research institutes or/and economic operators to contribute and to transfer the know-how to the last one [108]; ✓ creation of new jobs, especially in disadvantaged areas due to the decrease/closure of mining operations. **O4. Economic opportunities:** ✓ increased cooperation with mining industry to contribute to waste circularity and to ensure industrial symbiosis; ✓ the "green" transition is a major opportunity by creating markets for clean technologies and products, as well as creating new value chains in the construction material sectors; ✓ implementation of digital, intelligent technologies that can ensure the transition to the circular economy can effectively contribute to the sustainable development of enterprises to increase competitiveness, create jobs, and reduce the impact on the environment.	**T1. Technical threats:** ✓ content of useful substance; ✓ the existence of other types of waste possible to be used as secondary materials in construction that can lead to the development of products with technical characteristics similar or even superior to those obtained on the basis of mining waste; ✓ insufficient study of the products at the producer level, on a real scale; ✓ the need to develop pilot factories to validate the results obtained at the laboratory level; ✓ the aggregates resulting from the recycled materials have an increased porosity, so that the maximum allowed percentage is limited, so as not to affect the technical qualities of the final product. **T2. Environmental threats:** ✓ adopting mandatory reduction policies. **T3. Social threats:** ✓ increasing the health risks for the labor force involved; ✓ lack of specialized labor force; ✓ immaturity of the legislative framework; ✓ job creation in the areas of mining involved in waste recycling will lead to income security and increased living standards. **T4. Economic threats:** ✓ increasing the costs of transporting mining waste from the storage area to the factory; ✓ the need to invest in adapting the technological flow to the current requirements; ✓ the existence of other waste (organic or inorganic) generated as a result of local industrial activities can reduce the attractiveness of mining waste; ✓ low costs for the exploitation of conventional natural resources (clay, limestone, sand); ✓ sufficient fiscal instruments to support investment programs in the recycling of mining waste are not developed.

5. Conclusions

The main issues defined by the Ellen MacArthur Foundation [4] and targeted in the circular economy are to preserve the quality and economic value of products for as long as possible [109]. Masonry units can contribute substantially to the energy efficiency of buildings regardless of their destination, climatic or seismic zoning. The reduction of embodied energy, of exploitation or of the recurrent or post-use energy is necessary in order to ensure the transition from the linear to the circular economy, but also in order to achieve the long-term objectives [3,110].

The studies carried out by different authors presented and analyzed in the present synthesis highlight the feasibility of using mining waste in the production of clay and geopolymer bricks or ACC units due to their characteristics and good mechanical properties.

The use of residual waste in the production of masonry units requires the development of standards [111] based on which to establish for secondary raw materials the maximum permissible concentrations of heavy metals, soluble salt content, chemical composition, particle size, and water absorption. The optimization of the maximum percentage of waste in the mass of the material must be correlated with chemical and mineralogical characteristics of the raw materials used. The results performed at the laboratory level on small samples must be validation in the pilot factories where the products will be tested on a real scale, before their production on a large-scale. Significant variations in mechanical properties may occur for bricks/block with different geometrical features or with cavities.

The construction materials industry should focus on re-adaptation of the technological flow so that it can respond to new trends. Some advanced and innovative technological solutions are required to contribute to the reduction of pollution and CO_2 emissions.

To increase the attractiveness of the use of secondary raw materials in the production of new ones, they must be competitive with natural resources [107,112] both economically and qualitatively. The costs associated with the recycling process (sorting, crushing, grinding, transport) must be low enough so that the cost of materials with addition of mining waste to not exceed the cost of conventional materials.

Creating government policies, supported by financial facilities to stimulate the producers of masonry units to invest in modernizing the existing technological flow, but also creating an industrial symbiosis so that the waste from mining activities is redirected in real time to construction industry or other complementary sectors. Integrating the transition process to eliminate the costs of waste storage and to reduce the impact on the environment.

In Romania, there are large deposits of mining waste (Figures 1 and 2) with high potential for recovery in the construction industry. An industrial symbiosis between the mining and construction industry are imperative to reduce the environmental impact and to optimize the material properties and technological flows.

6. Recommendations for Further Research

Quantifying the main findings of this review investigation, the recommendation for future research directions is to address topics in dedicated works, are as follows:

- necessity for additional examinations, research, and comparative analysis having as objective the concrete effect determination of the integration of mining waste in different proportions in the masonry units, in terms of resistance to aggressive environment, non-destructive testing, microstructural studies, etc.;
- Life cycle assessment and life-cycle cost analysis of production processes that integrate mining waste into masonry units vs. classical production technologies of masonry units;
- assessment of average energy consumption and transport cost needed for using different types of mining wastes for masonry units;
- large-scale validation on samples of real dimensions and in technological production conditions similar to those existing at the manufacturer, in order to ensure the research results transfer to the economic operator;
- elaboration of materials containing technical data to inform end users about the advantages of using waste-based masonry units;
- conducting surveys (social analysis) at the level of economic and administrative agents, communities in mining areas on the feasibility, opportunity and sustainability of recycling and environmentally responsible use of mining waste to create new value chains in the construction materials industry.

Thus, more focused research is absolutely necessary to be developed in this area, and mining waste as secondary raw materials seems to be a promising contribution towards the sustainability of the construction materials industry.

Author Contributions: Conceptualization, N.C., R.M., G.T. and R.-A.F.; Methodology, N.C. and R.-A.F.; Validation, N.C., R.M., G.T. and R.-A.F.; Formal analysis, N.C. and G.T.; Investigation, N.C. and R.M.; Resources, N.C., R.M., G.T. and R.-A.F.; Data curation, R.M. and G.T.; Writing—original draft preparation, N.C., R.M., G.T. and R.-A.F.; Writing—review and editing, N.C., R.M., G.T. and R.-A.F.;

Visualization, R.M. and G.T.; Supervision, N.C. and G.T.; Project administration, N.C.; Funding acquisition, N.C., R.M., G.T. and R.-A.F. All authors have read and agreed to the published version of the manuscript.

Funding: This research received no external funding.

Institutional Review Board Statement: Not applicable.

Informed Consent Statement: Not applicable.

Data Availability Statement: Not applicable.

Acknowledgments: This work was supported by a grant of the Romanian Ministry of Research and Innovation, CCCDI-UEFISCDI, project number PN-III-P1-1.2-PCCDI-2017-0776/No.36 PCCDI/15.03.2018, within PNCDI III.

Conflicts of Interest: The authors declare no conflict of interest.

References

1. IEA. Global Status Report for Buildings and Construction. 2019. Available online: https://www.iea.org/reports/global-status-report-for-buildings-and-construction-2019 (accessed on 10 July 2021).
2. Paris Agreement. UN Framework Convention on Climate Change–EUR-Lex. Available online: https://eur-lex.europa.eu/content/paris-agreement/paris-agreement.html (accessed on 10 July 2021).
3. Communication from the Commission to the European Parliament; The European Council; The Council; The European Economic and Social Committee and the Committee of the Regions. The European Green Deal COM/2019/640 Final. Available online: https://eur-lex.europa.eu/legal-content/EN/TXT/HTML/?uri=CELEX:52019DC0640&from=EN (accessed on 10 July 2021).
4. Ellen MacArthur Foundation towards the Circular Economy: Economic and Business Rationale for an Accelerated Transition. 2012. Available online: http://circularfoundation.org/sites/default/files/ (accessed on 10 July 2021).
5. Communication from the Commission to the European Parliament; The Council; The European Economic And Social Committee and the Committee of the Regions a New Circular Economy. Action Plan for a Cleaner and More Competitive Europe COM/2020/98 Final. Available online: https://eur-lex.europa.eu/legal-content/EN/TXT/HTML/uri=CELEX:52020DC0098&from=EN (accessed on 10 July 2021).
6. Norouzi, M.; Châfer, M.; Cabeza, L.F.; Jiménez, L.; Boer, D. Circular economy in the building and construction sector: A scientific evolution analysis. *J. Build. Eng.* **2021**, *44*, 102704. [CrossRef]
7. Velenturf, A.P.M.; Archer, S.A.; Gomes, H.I.; Christgen, B.; Lag-Brotons, A.J.; Purnell, P. Circular economy and the matter of integrated resources. *Sci. Total Environ.* **2019**, *689*, 963–969. [CrossRef] [PubMed]
8. Smol, M.; Kulczycka, J.; Henclik, A.; Gorazda, K.; Wzorek, Z. The possible use of sewage sludge ash (SSA) in the construction industry as a way towards a circular economy. *J. Clean. Prod.* **2015**, *95*, 45–54. [CrossRef]
9. Atta, I.; Bakhoum, E.S.; Marzouk, M.M. Digitizing material passport for sustainable construction projects using BIM. *J. Build. Eng.* **2021**, *43*, 103233. [CrossRef]
10. Communication from the Commission to the European Parliament; The Council; The European Economic and Social Committee and the Committee of the Regions. Towards a Circular Economy: A Zero Waste Program for Europe/COM/2014/398 Final. Available online: https://www.oecd.org/env/outreach/EC-Circular-econonomy.pdf (accessed on 6 July 2021).
11. Circle Economy the Circularity Gap Report: Closing the Circularity Gap in a 9% World. 2019, p. 56. Available online: https://www.circulareconomyclub.com/listings/the-circularity-gap-report-our-world-is-only-9-circular/ (accessed on 6 July 2021).
12. European Commission. The Raw Materials Initiative—Meeting Our Critical Needs for Growth and Jobs in Europe. Available online: https://eur-lex.europa.eu/legal-content/EN/TXT/?uri=CELEX:52008DC0699 (accessed on 16 July 2021).
13. Available online: http://www.namr.ro (accessed on 16 July 2021).
14. Marinescu, M.; Kriz, A.; Tiess, G. The necessity to elaborate minerals policies exemplified by Romania. *Resour. Policy* **2013**, *38*, 416–426. [CrossRef]
15. The Ministry of Economy. Strategia Minieră a României 2017–2035 (Romania's Mining Strategy 2017–2035). 2017. Available online: http://www.economie.gov.ro/images/resurse-minerale/STRATEGIE%20MINIERA%20draft%20final%2024%20OCT%202016.pdf (accessed on 16 July 2021).
16. Eurostat Waste Generation by Economic Activity and Households. Available online: https://ec.europa.eu/eurostat/statistics-explained/index.php?title=Waste_statistics (accessed on 18 July 2021).
17. National Agency for Environmental Protection. Raport Privind Starea Mediului in Romania 2019 (Annual Report on the State of the Environment in Romania for 2019). 2019. Available online: http://www.anpm.ro/ro/raport-de-mediu or http://www-old.anpm.ro/upload/181393_RSM%202019_final.pdf (accessed on 20 July 2021).
18. Council Directive 75/442/EEC of 15 July 1975 on Waste Management. 1975. Available online: https://eur-lex.europa.eu/eli/dir/1975/442 (accessed on 20 July 2021).
19. European Commission Commission Decision of 18 December 2014 amending Decision 2000/532/EC on the list of waste pursuant to Directive 2008/98/EC of the European Parliament and of the Council. *Off. J. Eur. Union* **2014**, *7*, 44–86.

20. Veiga Simão, F.; Chambart, H.; Vandemeulebroeke, L.; Cappuyns, V. Incorporation of sulphidic mining waste material in ceramic roof tiles and blocks. *J. Geochem. Explor.* **2021**, *225*, 106741. [CrossRef]
21. *Code P100-1/2013*; Seismic Design Code—Part I—Provisions for the Design of Buildings. Indicative P100-1 (In Romanian) Elaborated by UTCB, Endorsed by MDRAP. Part I, n. 558 of 03. 09. 2013. Part I, n. 558 of 03. 09. 2013; Official Journal of Romania: Bucharest, Romania, 2013.
22. Cod de Proiectare Pentru Structuri Din Zidărie, Indicativ CR6–2013 (Design Code for Masonry Structures, Indicative CR6–2013). 2013, pp. 2–285. Available online: https://www.mdlpa.ro/userfiles/consultari_publice/24_04_13/act_16.pdf (accessed on 24 July 2021).
23. BDA Brick Development Association. Available online: https://brick.org.uk/ (accessed on 24 July 2021).
24. Ceramie-Unie Ceramic Roadmap, Paving the Way to 2050. Available online: http://cerameunie.eu/topics/cerame-unie-sectors/cerame-unie/ceramic-industry-roadmap-paving-the-way-to-2050/ (accessed on 24 July 2021).
25. Murmu, A.L.; Patel, A. Towards sustainable bricks production: An overview. *Constr. Build. Mater.* **2018**, *165*, 112–125. [CrossRef]
26. Morales, M.P.; Juárez, M.C.; López-Ochoa, L.M.; Doménech, J. Study of the geometry of a voided clay brick using rectangular perforations to optimize its thermal properties. *Appl. Therm. Eng.* **2011**, *31*, 2063–2065. [CrossRef]
27. Morales, M.P.; Juárez, M.C.; Muñoz, P.; Gómez, J.A. Study of the geometry of a voided clay brick using non-rectangular perforations to optimise its thermal properties. *Energy Build.* **2011**, *43*, 2494–2498. [CrossRef]
28. Sutcu, M.; del Coz Díaz, J.J.; Álvarez Rabanal, F.P.; Gencel, O.; Akkurt, S. Thermal performance optimization of hollow clay bricks made up of paper waste. *Energy Build.* **2014**, *75*, 96–108. [CrossRef]
29. Morales, M.P.; Juárez, M.C.; López-Ochoa, L.M.; Muñoz, P. Influence of tongue and groove system on the thermal properties of large-format voided clay bricks for single-leaf walls. *Constr. Build. Mater.* **2012**, *30*, 169–173. [CrossRef]
30. Pavlík, Z.; Jerman, M.; Fořt, J.; Černý, R. Monitoring Thermal Performance of Hollow Bricks with Different Cavity Fillers in Difference Climate Conditions. *Int. J. Thermophys.* **2015**, *36*, 557–568. [CrossRef]
31. Kočí, J.; Maděra, J.; Jerman, M.; Černý, R. Computational assessment of thermal performance of contemporary ceramic blocks with complex internal geometry in building envelopes. *Energy Build.* **2015**, *99*, 61–66. [CrossRef]
32. Li, J.; Meng, X.; Gao, Y.; Mao, W.; Luo, T.; Zhang, L. Effect of the insulation materials filling on the thermal performance of sintered hollow bricks. *Case Stud. Therm. Eng.* **2018**, *11*, 62–70. [CrossRef]
33. Zukowski, M.; Haese, G. Experimental and numerical investigation of a hollow brick filled with perlite insulation. *Energy Build.* **2010**, *42*, 1402–1408. [CrossRef]
34. Lai, M.K.; Maged, A.S.B. Study of Sawdust and Expanded Polystyrene as Cavity Filler Material on the Effect of Thermal Conductivity in Perforated Clay Brick. *Mater. Sci. Forum* **2017**, *890*, 411–414. [CrossRef]
35. Salleh, S.Z.; Awang Kechik, A.; Yusoff, A.H.; Taib, M.A.A.; Mohamad Nor, M.; Mohamad, M.; Tan, T.G.; Ali, A.; Masri, M.N.; Mohamed, J.J.; et al. Recycling food, agricultural, and industrial wastes as pore-forming agents for sustainable porous ceramic production: A review. *J. Clean. Prod.* **2021**, *306*, 127264. [CrossRef]
36. Heniegal, A.M.; Ramadan, M.A.; Naguib, A.; Agwa, I.S. Study on properties of clay brick incorporating sludge of water treatment plant and agriculture waste. *Case Stud. Constr. Mater.* **2020**, *13*, e00397. [CrossRef]
37. Kazmi, S.M.S.; Munir, M.J.; Patnaikuni, I.; Wu, Y.F.; Fawad, U. Thermal performance enhancement of eco-friendly bricks incorporating agro-wastes. *Energy Build.* **2018**, *158*, 1117–1129. [CrossRef]
38. Muñoz, P.; Mendívil, M.A.; Letelier, V.; Morales, M.P. Thermal and mechanical properties of fired clay bricks made by using grapevine shoots as pore forming agent. Influence of particle size and percentage of replacement. *Constr. Build. Mater.* **2019**, *224*, 639–658. [CrossRef]
39. Anjum, F.; Naz, M.Y.; Ghaffar, A.; Shukrullah, S.; AbdEl-Salam, N.M.; Ibrahim, K.A. Study of thermal and mechanical traits of organic waste incorporated fired clay porous material. *Phys. B Condens. Matter* **2020**, *599*, 412479. [CrossRef]
40. Demir, I.; Baspinar, M.S.; Orhan, M. Utilization of kraft pulp production residues in clay brick production. *Build. Environ.* **2005**, *40*, 1533–1537. [CrossRef]
41. Marrocchino, E.; Zanelli, C.; Guarini, G.; Dondi, M. Recycling mining and construction wastes as temper in clay bricks. *Appl. Clay Sci.* **2021**, *209*, 106152. [CrossRef]
42. Rehman, M.U.; Ahmad, M.; Rashid, K. Influence of fluxing oxides from waste on the production and physico-mechanical properties of fired clay brick: A review. *J. Build. Eng.* **2020**, *27*, 100965. [CrossRef]
43. dos Reis, G.S.; Cazacliu, B.G.; Cothenet, A.; Poullain, P.; Wilhelm, M.; Sampaio, C.H.; Lima, E.C.; Ambros, W.; Torrenti, J.M. Fabrication, microstructure, and properties of fired clay bricks using construction and demolition waste sludge as the main additive. *J. Clean. Prod.* **2020**, *258*, 120733. [CrossRef]
44. Cobîrzan, N.; Balog, A.-A.; Thalmaier, G.; Nasui, M.; Munteanu, C.; Babota, F. Microscopical and Macroscopical Analysis of Recovered Bricks for Assessing Their Reusability in Masonry Buildings. *Procedia Manuf.* **2020**, *46*, 144–149. [CrossRef]
45. Saboya, F.; Xavier, G.C.; Alexandre, J. The use of the powder marble by-product to enhance the properties of brick ceramic. *Constr. Build. Mater.* **2007**, *21*, 1950–1960. [CrossRef]
46. Cobo-Ceacero, C.J.; Cotes-Palomino, M.T.; Martínez-García, C.; Moreno-Maroto, J.M.; Uceda-Rodríguez, M. Use of marble sludge waste in the manufacture of eco-friendly materials: Applying the principles of the Circular Economy. *Environ. Sci. Pollut. Res.* **2019**, *26*, 35399–35410. [CrossRef]

47. Menezes, R.R.; Ferreira, H.S.; Neves, G.A.; de L. Lira, H.; Ferreira, H.C. Use of granite sawing wastes in the production of ceramic bricks and tiles. *J. Eur. Ceram. Soc.* **2005**, *25*, 1149–1158. [CrossRef]
48. Suárez-Macías, J.; Terrones-Saeta, J.M.; Iglesias-Godino, F.J.; Corpas-Iglesias, F.A. Retention of Contaminants Elements from Tailings from Lead Mine Washing Plants in Ceramics for Bricks. *Minerals* **2020**, *10*, 576. [CrossRef]
49. Loutou, M.; Taha, Y.; Benzaazoua, M.; Daafi, Y.; Hakkou, R. Valorization of clay by-product from Moroccan Phosphate Mines for the Production of Fired Bricks. *J. Clean. Prod.* **2019**, *229*, 169–179. [CrossRef]
50. Mendes, B.C.; Pedroti, L.G.; Fontes, M.P.F.; Ribeiro, J.C.L.; Vieira, C.M.F.; Pacheco, A.A.; de Azevedo, A.R.G. Technical and environmental assessment of the incorporation of iron ore tailings in construction clay bricks. *Constr. Build. Mater.* **2019**, *227*, 116669. [CrossRef]
51. Li, R.; Zhou, Y.; Li, C.; Li, S.; Huang, Z. Recycling of industrial waste iron tailings in porous bricks with low thermal conductivity. *Constr. Build. Mater.* **2019**, *213*, 43–50. [CrossRef]
52. Ettoumi, M.; Jouini, M.; Neculita, C.M.; Bouhlel, S.; Coudert, L.; Taha, Y.; Benzaazoua, M. Characterization of phosphate processing sludge from Tunisian mining basin and its potential valorization in fired bricks making. *J. Clean. Prod.* **2021**, *284*, 124750. [CrossRef]
53. Bayoussef, A.; Oubani, M.; Loutou, M.; Taha, Y.; Benzaazoua, M.; Manoun, B.; Hakkou, R. Manufacturing of high-performance ceramics using clays by-product from phosphate mines. *Mater. Today Proc.* **2021**, *37*, 3994–4000. [CrossRef]
54. Yang, C.; Cui, C.; Qin, J.; Cui, X. Characteristics of the fired bricks with low-silicon iron tailings. *Constr. Build. Mater.* **2014**, *70*, 36–42. [CrossRef]
55. Vilela, A.P.; Eugênio, T.M.C.; de Oliveira, F.F.; Mendes, J.F.; Ribeiro, A.G.C.; de, S. Brandão Vaz, L.E.V.; Mendes, R.F. Technological properties of soil-cement bricks produced with iron ore mining waste. *Constr. Build. Mater.* **2020**, *262*, 120883. [CrossRef]
56. Weishi, L.; Guoyuan, L.; Ya, X.; Qifei, H. The properties and formation mechanisms of eco-friendly brick building materials fabricated from low-silicon iron ore tailings. *J. Clean. Prod.* **2018**, *204*, 685–692. [CrossRef]
57. da Silva, F.L.; Araújo, F.G.S.; Teixeira, M.P.; Gomes, R.C.; von Krüger, F.L. Study of the recovery and recycling of tailings from the concentration of iron ore for the production of ceramic. *Ceram. Int.* **2014**, *40*, 16085–16089. [CrossRef]
58. Luo, L.; Li, K.; Weng, F.; Liu, C.; Yang, S. Preparation, characteristics and mechanisms of the composite sintered bricks produced from shale, sewage sludge, coal gangue powder and iron ore tailings. *Constr. Build. Mater.* **2020**, *232*, 117250. [CrossRef]
59. Wang, G.; Ning, X.; Lu, X.; Lai, X.; Cai, H.; Liu, Y.; Zhang, T. Effect of sintering temperature on mineral composition and heavy metals mobility in tailings bricks. *Waste Manag.* **2019**, *93*, 112–121. [CrossRef]
60. Chen, Y.; Zhang, Y.; Chen, T.; Zhao, Y.; Bao, S. Preparation of eco-friendly construction bricks from hematite tailings. *Constr. Build. Mater.* **2011**, *25*, 2107–2111. [CrossRef]
61. Yonggang, Y.; Shenhong, Z.; Qiuyi, L.; Benju, Y.; Yu, C. Research on Making Fired Bricks with Gold Tailings. In Proceedings of the 2011 International Conference on Computer Distributed Control and Intelligent Environmental Monitoring, Changsha, China, 19–20 February 2011; pp. 1687–1690.
62. Wei, Z.; Zhao, J.; Wang, W.; Yang, Y.; Zhuang, S.; Lu, T.; Hou, Z. Utilizing gold mine tailings to produce sintered bricks. *Constr. Build. Mater.* **2021**, *282*, 122655. [CrossRef]
63. Zhu, M.; Wang, H.; Liu, L.; Ji, R.; Wang, X. Preparation and characterization of permeable bricks from gangue and tailings. *Constr. Build. Mater.* **2017**, *148*, 484–491. [CrossRef]
64. Kalpana, M.; Mohith, S. Study on autoclaved aerated concrete: Review. *Mater. Today Proc.* **2020**, *22*, 894–896. [CrossRef]
65. Li, X.G.; Liu, Z.L.; Lv, Y.; Cai, L.X.; Jiang, D.B.; Jiang, W.G.; Jian, S. Utilization of municipal solid waste incineration bottom ash in autoclaved aerated concrete. *Constr. Build. Mater.* **2018**, *178*, 175–182. [CrossRef]
66. Qu, X.; Zhao, X. Previous and present investigations on the components, microstructure and main properties of autoclaved aerated concrete—A review. *Constr. Build. Mater.* **2017**, *135*, 505–516. [CrossRef]
67. Song, Y.; Li, B.; Yang, E.H.; Liu, Y.; Ding, T. Feasibility study on utilization of municipal solid waste incineration bottom ash as aerating agent for the production of autoclaved aerated concrete. *Cem. Concr. Compos.* **2015**, *56*, 51–58. [CrossRef]
68. Drochytka, R.; Zach, J.; Korjenic, A.; Hroudová, J. Improving the energy efficiency in buildings while reducing the waste using autoclaved aerated concrete made from power industry waste. *Energy Build.* **2013**, *58*, 319–323. [CrossRef]
69. Kurama, H.; Topçu, I.B.; Karakurt, C. Properties of the autoclaved aerated concrete produced from coal bottom ash. *J. Mater. Process. Technol.* **2009**, *209*, 767–773. [CrossRef]
70. Serhat Baspinar, M.; Demir, I.; Kahraman, E.; Gorhan, G. Utilization potential of fly ash together with silica fume in autoclaved aerated concrete production. *KSCE J. Civ. Eng.* **2014**, *18*, 47–52. [CrossRef]
71. El-Didamony, H.; Amer, A.A.; Mohammed, M.S.; El-Hakim, M.A. Fabrication and properties of autoclaved aerated concrete containing agriculture and industrial solid wastes. *J. Build. Eng.* **2019**, *22*, 528–538. [CrossRef]
72. Różycka, A.; Pichór, W. Effect of perlite waste addition on the properties of autoclaved aerated concrete. *Constr. Build. Mater.* **2016**, *120*, 65–71. [CrossRef]
73. Kunchariyakun, K.; Asavapisit, S.; Sombatsompop, K. Properties of autoclaved aerated concrete incorporating rice husk ash as partial replacement for fine aggregate. *Cem. Concr. Compos.* **2015**, *55*, 11–16. [CrossRef]
74. Peng, Y.; Liu, Y.; Zhan, B.; Xu, G. Preparation of autoclaved aerated concrete by using graphite tailings as an alternative silica source. *Constr. Build. Mater.* **2021**, *267*, 121792. [CrossRef]

75. Huang, X.Y.; Ni, W.; Cui, W.H.; Wang, Z.J.; Zhu, L.P. Preparation of autoclaved aerated concrete using copper tailings and blast furnace slag. *Constr. Build. Mater.* **2012**, *27*, 1–5. [CrossRef]
76. Fang, Y.; Gu, Y.; Kang, Q.; Wen, Q.; Dai, P. Utilization of copper tailing for autoclaved sand–lime brick. *Constr. Build. Mater.* **2011**, *25*, 867–872. [CrossRef]
77. Zhao, Y.; Zhang, Y.; Chen, T.; Chen, Y.; Bao, S. Preparation of high strength autoclaved bricks from hematite tailings. *Constr. Build. Mater.* **2012**, *28*, 450–455. [CrossRef]
78. Ma, B.; Cai, L.-X.; Li, X.; Jian, S. Utilization of iron tailings as substitute in autoclaved aerated concrete: Physico-mechanical and microstructure of hydration products. *J. Clean. Prod.* **2016**, *127*, 162–171. [CrossRef]
79. Liang, X.; Wang, C.; Zhan, J.; Cui, X.; Ren, Z. Study on preparation of eco-friendly autoclaved aerated concrete from low silicon and high iron ore tailings. *J. New Mater. Electrochem. Syst.* **2019**, *22*, 224–230. [CrossRef]
80. Cai, L.; Ma, B.; Li, X.; Lv, Y.; Liu, Z.; Jian, S. Mechanical and hydration characteristics of autoclaved aerated concrete (AAC) containing iron-tailings: Effect of content and fineness. *Constr. Build. Mater.* **2016**, *128*, 361–372. [CrossRef]
81. Zhang, S.; Xue, X.; Liu, X.; Duan, P.; Yang, H.; Jiang, T.; Wang, D.; Liu, R. Current situation and comprehensive utilization of iron ore tailing resources. *J. Min. Sci.* **2006**, *42*, 403–408. [CrossRef]
82. Zhao, F.; Zhao, J.; Liu, H. Autoclaved brick from low-silicon tailings. *Constr. Build. Mater.* **2009**, *23*, 538–541. [CrossRef]
83. Wang, C.L.; Ni, W.; Zhang, S.Q.; Wang, S.; Gai, G.S.; Wang, W.K. Preparation and properties of autoclaved aerated concrete using coal gangue and iron ore tailings. *Constr. Build. Mater.* **2016**, *104*, 109–115. [CrossRef]
84. Davidovits, J. Geopolymers. *J. Therm. Anal.* **1991**, *37*, 1633–1656. [CrossRef]
85. Youssef, N.; Lafhaj, Z.; Chapiseau, C. Economic Analysis of Geopolymer Brick Manufacturing: A French Case Study. *Sustainability* **2020**, *12*, 7403. [CrossRef]
86. Zhao, J.; Tong, L.; Li, B.; Chen, T.; Wang, C.; Yang, G.; Zheng, Y. Eco-friendly geopolymer materials: A review of performance improvement, potential application and sustainability assessment. *J. Clean. Prod.* **2021**, *307*, 127085. [CrossRef]
87. Mohsen, Q.; Mostafa, N.Y. Investigating the possibility of utilizing low kaolinitic clays in production of geopolymer bricks. *Ceram.-Silik.* **2010**, *54*, 160–168.
88. Amin, S.K.; El-Sherbiny, S.A.; El-Magd, A.A.M.A.; Belal, A.; Abadir, M.F. Fabrication of geopolymer bricks using ceramic dust waste. *Constr. Build. Mater.* **2017**, *157*, 610–620. [CrossRef]
89. Komnitsas, K.; Zaharaki, D.; Vlachou, A.; Bartzas, G.; Galetakis, M. Effect of synthesis parameters on the quality of construction and demolition wastes (CDW) geopolymers. *Adv. Powder Technol.* **2015**, *26*, 368–376. [CrossRef]
90. Madani, H.; Ramezanianpour, A.A.; Shahbazinia, M.; Ahmadi, E. Geopolymer bricks made from less active waste materials. *Constr. Build. Mater.* **2020**, *247*, 118441. [CrossRef]
91. Kang, X.; Gan, Y.; Chen, R.; Zhang, C. Sustainable eco-friendly bricks from slate tailings through geopolymerization: Synthesis and characterization analysis. *Constr. Build. Mater.* **2021**, *278*, 122337. [CrossRef]
92. Ahmari, S.; Zhang, L. Durability and leaching behavior of mine tailings-based geopolymer bricks. *Constr. Build. Mater.* **2013**, *44*, 743–750. [CrossRef]
93. Beulah, M.; Sudhir, M.R.; Mohan, M.K.; Gayathri, G.; Jain, D. Mine Waste-Based Next Generation Bricks: A Case Study of Iron Ore Tailings, Red Mudand GGBS Utilization in Bricks. *Adv. Mater. Sci. Eng.* **2021**, *2021*, 9499613. [CrossRef]
94. Zhang, N.; Hedayat, A.; Bolaños Sosa, H.G.; González Cárdenas, J.J.; Salas Álvarez, G.E.; Ascuña Rivera, V.B. Damage evaluation and deformation behavior of mine tailing-based Geopolymer under uniaxial cyclic compression. *Ceram. Int.* **2021**, *47*, 10773–10785. [CrossRef]
95. Ahmari, S.; Zhang, L. Production of eco-friendly bricks from copper mine tailings through geopolymerization. *Constr. Build. Mater.* **2012**, *29*, 323–331. [CrossRef]
96. Zhang, N.; Tang, B.; Liu, X. Cementitious activity of iron ore tailing and its utilization in cementitious materials, bricks and concrete. *Constr. Build. Mater.* **2021**, *288*, 123022. [CrossRef]
97. Cicek, B.; Karadagli, E.; Duman, F. Valorisation of boron mining wastes in the production of wall and floor tiles. *Constr. Build. Mater.* **2018**, *179*, 232–244. [CrossRef]
98. Yun-Ming, L.; Cheng Yong, H.; Mustafa, A.B.M.; Hussin, K. Structure and Properties of Clay-Based Geopolymer Cements: A Review. *Prog. Mater. Sci.* **2016**, *83*, 595–629.
99. Muñoz Velasco, P.; Morales Ortíz, M.P.; Mendívil Giró, M.A.; Muñoz Velasco, L. Fired clay bricks manufactured by adding wastes as sustainable construction material–A review. *Constr. Build. Mater.* **2014**, *63*, 97–107. [CrossRef]
100. Muñoz, P.; Letelier, V.; Zamora, D.; Morales, M.P. Feasibility of using paper pulp residues into fired clay bricks. *J. Clean. Prod.* **2020**, *262*, 121464. [CrossRef]
101. Görhan, G.; Şimşek, O. Porous clay bricks manufactured with rice husks. *Constr. Build. Mater.* **2013**, *40*, 390–396. [CrossRef]
102. Steinmann, Z.J.N.; Huijbregts, M.A.J.; Reijnders, L. How to define the quality of materials in a circular economy? *Resour. Conserv. Recycl.* **2019**, *141*, 362–363. [CrossRef]
103. Elert, K.; Cultrone, G.; Navarro, C.R.; Pardo, E.S. Durability of bricks used in the conservation of historic buildings—influence of composition and microstructure. *J. Cult. Herit.* **2003**, *4*, 91–99. [CrossRef]
104. García-Ten, J.; Orts, M.J.; Saburit, A.; Silva, G. Thermal conductivity of traditional ceramics. *Ceram. Int.* **2010**, *36*, 2017–2024. [CrossRef]

105. De Bonis, A.; Cultrone, G.; Grifa, C.; Langella, A.; Leone, A.P.; Mercurio, M.; Morra, V. Different shades of red: The complexity of mineralogical and physico-chemical factors influencing the colour of ceramics. *Ceram. Int.* **2017**, *43*, 8065–8074. [CrossRef]
106. Megyesi, E.; Brumaru, M.; Naghiu, G.S.; Felseghi, R.A. Choosing the optimal type of external wall constructions for application in the field of passive houses. In Proceedings of the 14th International Multidisciplinary Scientific GeoConference SGEM, Albena, Bulgaria, 19–25 July 2014; pp. 65–72, ISBN 978-619-7105-21-6.
107. Horckmans, L.; Nielsen, P.; Dierckx, P.; Ducastel, A. Recycling of refractory bricks used in basic steelmaking: A review. *Resour. Conserv. Recycl.* **2019**, *140*, 297–304. [CrossRef]
108. Cretu, M.; Ceclan, A.; Czumbil, L.; Şteţ, D.; Bârgăuan, B.; Micu, D.D. Key performance indicators (KPIs) for the evaluation of the demand response in the Technical University of Cluj-Napoca buildings. In Proceedings of the 8th International Conference on Modern Power Systems (MPS), Cluj Napoca, Romania, 21–23 May 2019; pp. 1–4.
109. European Commission. Closing the Loop–An EU Action Plan for the Circular Economy COM/2015/0614 Final. Available online: http://eur-lex.europa.eu/legal-content/EN/TXT/?uri=CELEX:52015DC0614 (accessed on 15 August 2021).
110. Communication from the Commission to the European Parliament; The European Council; The Council; The European Economic and Social Committee; The Committee of the Regions and the European Investment Bank. A Clean Planet for All. A European Strategic Long-Term Vision for a Prosperous, Modern, Competitive and Climate Neutral Economy COM/2018/773. Available online: https://eur-lex.europa.eu/legal-content/EN/TXT/PDF/?uri=CELEX:52018DC0773 (accessed on 10 November 2021).
111. Zhang, L. Production of bricks from waste materials–A review. *Constr. Build. Mater.* **2013**, *47*, 643–655. [CrossRef]
112. Huysman, S.; Debaveye, S.; Schaubroeck, T.; De Meester, S.; Ardente, F.; Mathieux, F.; Dewulf, J. The recyclability benefit rate of closed-loop and open-loop systems: A case study on plastic recycling in Flanders. *Resour. Conserv. Recycl.* **2015**, *101*, 53–60. [CrossRef]

materials

MDPI

Article

Shear Strength Characteristics of Recycled Concrete Aggregate and Recycled Tire Waste Mixtures from Monotonic Triaxial Tests

Katarzyna Gabryś [1,*], Algirdas Radzevičius [2], Alojzy Szymański [3] and Raimondas Šadzevičius [2]

[1] Water Centre WULS, Warsaw University of Life Sciences—WULS, 6 Ciszewskiego Street, 02-787 Warsaw, Poland

[2] Department of Water Engineering, Faculty of Engineering, Vytautas Magnus University, Universiteto Street 10, Akademija, LT-53361 Kaunas, Lithuania; algirdas.radzevicius@vdu.lt (A.R.); raimondas.sadzevicius@vdu.lt (R.Š.)

[3] Department of Geotechnics, Warsaw University of Life Sciences—WULS, 159 Nowoursynowska Street, 02-787 Warsaw, Poland; alojzy_szymanski@sggw.edu.pl

* Correspondence: katarzyna_gabrys@sggw.edu.pl

Abstract: Recycled concrete aggregate (RCA) is a promising substitute for natural aggregates and the reuse of this material can benefit construction projects both economically and environmentally. RCA has received great attention in recent years in the form of aggregate as well as a geotechnical material of sand size. Next to RCA, another recycled material, which reduces the waste volume and is a part of the present challenges in civil engineering, is tire waste. Despite the good engineering properties of recycled tire waste (RTW), its use is still limited, even after almost 30 years since they were first introduced. To broaden the applicability of reused concrete and rubber, a further understanding of their properties and engineering behavior is required. For this reason, the main subject of this paper is composite materials that consist of anthropogenic soil recycled concrete aggregate (RCA) and crushed pieces of recycled tire waste (RTW). In this study, a series of isotropic consolidated drained triaxial tests were undertaken to characterize the shear strength of eight mixtures of variable grain-size distribution, rubber inclusion (RC), and fine fraction (FF) content. The results show that the introduction of rubber waste leads to changes in the strength parameters of the tested mixtures. Improvements in RCA shear strength were observed, the largest for the mixture M7 with 10% of recycled tire waste. Similarly, the effect of fine fraction content on the angle of internal friction and cohesion was found. Dilation characteristics were observed in all analyzed composites. Based on the results of all tests performed, including physical, geometric, chemical, and mechanical properties of the created composites, it can be stated that the samples would meet local road authority requirements for sub-base applications.

Keywords: recycling; environment; waste; anthropogenic soil; laboratory tests; mechanical properties

Citation: Gabryś, K.; Radzevičius, A.; Szymański, A.; Šadzevičius, R. Shear Strength Characteristics of Recycled Concrete Aggregate and Recycled Tire Waste Mixtures from Monotonic Triaxial Tests. *Materials* **2021**, *14*, 7400. https://doi.org/10.3390/ma14237400

Academic Editor: Yuri Ribakov

Received: 11 November 2021
Accepted: 30 November 2021
Published: 2 December 2021

1. Introduction

The modern civil engineering industry requires the continuous development of new technologies and materials. Because of the already existing deficiencies of aggregates and natural soil in various regions of Poland, as well as unfavorable forecasts for the market, the possibility of replacing natural aggregate with recycled materials becomes a good alternative.

Recycled concrete is a material that can successfully merge economic and environmental interests. The current prevalence of recycled concrete and the scale of demolition together with modernization work in the construction industry indicate that this material will long be a potential material for earth structures [1]. There are researchers that emphasize the fairly large aggregates recycling acquisition costs, the lack of homogeneity, and the need to introduce additional processes that significantly affect the economics of commodity production [2]. On the other hand, the use of recycled materials is undoubtedly a good way to go, because of the aforementioned environmental protection or the high energy expenditure of extracting natural aggregates, and the high price of transporting them.

A huge problem in the world, as well as in Poland, is the storage and management of rubber waste [3]. Currently, there are two groups of rubber waste, in which the first group includes vehicle tires (about 70%), whereas the second group is industrial waste and conveyor belts (around 30%) [4]. Worn tires are reconditioned, recycled, or burned in cement plants. At present, it is estimated that there are over 29 million tons of used tires in the world, 23% of which are recycled.

Shredded used tires are now being used in landfill engineering as subgrade reinforcement for constructing roads over soft soil, as well as aggregate in leach beds for septic systems, and as a substitute for leachate collection stone in landfills [5]. Crumbed or shredded used tires are applied as an energy-producing material, an admixture in bituminous concrete, and low-grade rubber products, e.g., truck-bed liners, doormats, and cushioning forms [6]. Whole rubber tires can be also used as reinforcement in the construction of retaining walls and slopes [7]. Another possible practical application of shredded tires alone or mixed with soil is as a lightweight material for embankment fill. In general, the use of shredded rubber waste depends on the degree of shredding. In the European Standard EN-14243, materials obtained from the recycling of tires are classified as shown in Table 1.

Table 1. Classification of materials obtained from waste tires according to EN-14243 [8].

Type of Shredded Rubber Waste	Particle Size in (mm)
Tire cuts (halves and smaller pieces)	>300
Shreds	50–300
Chips	10–50
Granulate	1–10
Powder	<1
Fine powder	<0.5
Scrap (a byproduct of tire retreading)	0–40

It is worth emphasizing that the economic aspects are at the moment a barrier to the wide introduction of the utilization of rubber waste from worn vehicle tires to the Polish market. However, in connection with the introduction of a ban on burning waste, the situation may radically change shortly.

In Table 2, a comparison between the basic parameters of recycled tire waste in the form of rubber aggregate and mineral soil packages is presented. To reduce the size of landfills and thus reduce the high fire risk, tires began to be packaged in packages reducing the volume of stored materials by more than four times. An attractive construction form of recycled material in the form of rubber packages (gabions) was obtained to employ in many various constructions [9].

Table 2. Comparison of characteristics of recycled rubber materials with mineral aggregate [9].

Properties	Rubber Waste (Rubber Aggregate)	Rubber Packages	Mineral Aggregate
Bulk density (kN/m^3)	5–7	7	18–20
Friction angle (°)	19–38	35–37	35–40
Cohesion (kPa)	0–11.5	0	0
Deformation modulus (MPa)	0.8–1.3	0.8–1.0	40–100
Water permeability (cm/s)	2–10	2–4	10–100
Thermal insulation (W/(m·K))	0.07	0.07	0.4–0.7
Vibration isolation	high	high	low

This study examined only the shear strength characteristics of recycled concrete aggregate (RCA) with recycled tire waste (RTW). Shear strength is a fundamental mechanical property that determines, e.g., the stability of embankment structures. The higher the value of the friction angle, the higher the soil shear resistance, the lower the earth pressure, the higher the stability. Moreover, recent geotechnical investigations on soil-rubber mixtures focus generally on the effects of rubber on the mechanical behavior of natural clean sands, silts, low-plasticity clays, gravels, and anthropogenic fly ashes. Several studies have been conducted as well to evaluate the effect of crumb rubber on the properties of cohesive, frequently expansive, soils. Unfortunately, very few tests have been done to examine the properties of recycled concrete aggregate and rubber waste composites.

Experiments on sand mixtures with rubber chips and rubber shreds by the international community date from the 1990s, when Edil and Bossher [10] and Wu et al. [11] examined the strength of those materials in triaxial apparatus. Mechanical properties of crushed tires were also later evaluated by Yang et al. [12] with isotropic and confined compression, using direct shear and triaxial tests. Various attempts to obtain optimum tire content with the maximum shear strength of the mixture resulted in nearly 35% by weight [13].

The inclusion of shredded tires to clay soils showed a significant decline of friction angle by approximately 22%, according to consolidated-drained triaxial tests [14]. In the case of rubber addition to noncohesive soils, like sand, a decrease of friction angles up to even 30% was observed in the direct shear apparatus tests [15]. The change in friction angle was accompanied by changes in cohesion. Bałachowski and Gotteland [16] detected an increase in the apparent cohesion of the sand–rubber mixtures, by approximately 50% of rubber waste content. Similar observations had Kowalska and Chmielewski [15] but at 30% of rubber content. When these values were exceeded, cohesion, after an initial increase, began to show a decreasing tendency. Thus, to achieve the highest strength of the mixture, the proper proportion between the values of strength parameters and the content of the rubber additive is always searched for.

When it comes to mixing RCA with RTW, in 2015, our group [17] began to study the physical and mechanical properties of geocomposites composed of recycled concrete aggregate and crushed car tires at different concentrations, namely 0%, 0.5%, 1% by weight. However, these studies were performed under cyclic loading conditions (cyclic CBR), and the results indicated an increase in the elasticity modulus (E and M_r) and a decrease in plastic deformations with the rubber content in the analyzed material. The subsequent studies concerned further mixtures of RCA with rubber waste [18], but this time RCA was improved with recycled rubber powder in the amount of 0% and 15%. The results showed that addition of tires significantly decreased the value of shear wave velocity—and consequently the values of the shear modulus—of the modified RCA.

To the authors' knowledge, very little attention has been paid so far to the strength and deformation characteristics of modified RCA by shredded rubber tires. Therefore, the primary purpose of the presented paper is to evaluate the friction angle (φ) and cohesion (c) of the recycled concrete aggregate and granulated rubber geocomposite with various rubber contents/sizes and under different normal stresses. This study focuses on the experimental investigation of strength characteristics of eight different mixtures. Isotropic consolidated drained (CID) tests were performed on samples with the different mixing ratios of RCA and shredded rubber tires.

2. Materials and Methods

2.1. Description of Materials

In this study, demolished concrete from building demolition sites and industry-produced scrap tires from Warsaw, Poland were used. Concrete aggregates were an element of concrete floors, especially concrete curbs from the demolition of roads. This material is about 20 years old. The strength class of concrete from which the aggregates were formed was estimated between C16/20 and C30/35. The aggregates were 99% broken cement concrete by weight and 1% glass and brick (Σ(Rb, Rg, X) $\leq$ 1% m/m), following the Polish Committee for Standardization [19]. They did not contain asphalt or any tar elements. The basic properties of the original concrete were determined for the authors' earlier study and can be found in their other publications [20]. Concrete aggregates required preliminary research on saturation to prevent the movement of water necessary for hydration [20]. The material did not show any binding properties. Recycled concrete aggregates (RCA) needed for the specimens are shown in Figure 1.

Figure 1. Grain fractions of concrete aggregate, from the smallest from the left: <0.063 mm; 0.063–0.125 mm; 0.125–0.25 mm; 0.25–0.50 mm; 0.5–1.0 mm; 1.0–2.0 mm.

After obtaining sufficient quantities of the selected fractions (Figure 1), four different mixtures of RCA were created with appropriate compositions and the following percentages of the fine fraction (FF), i.e., <0.063 mm:

- M1_RCA_0FF,
- M2_RCA_10FF,
- M3_RCA_20FF,
- M4_RCA_30FF.

The particle size distribution curves of all four mixtures are shown in Figure 2.

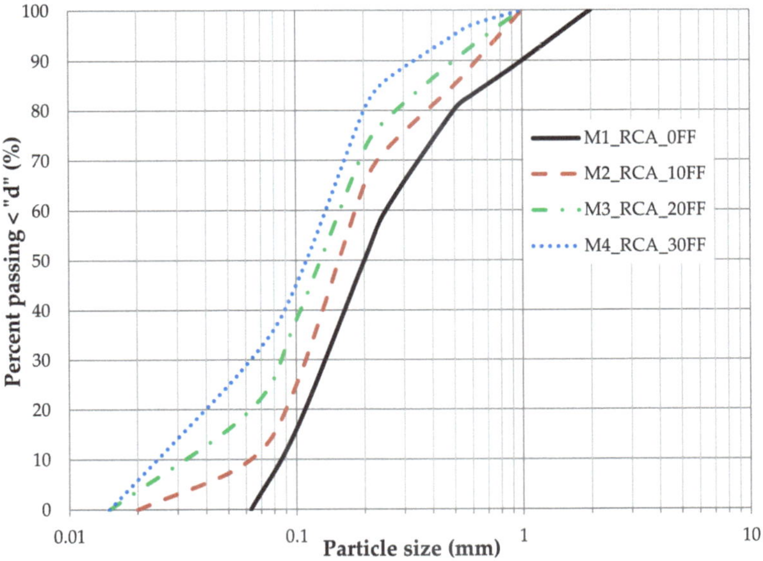

Figure 2. The particle size distribution curves of M1, M2, M3, and M4 used for this study.

The soil was classified as sand (Sa) for mixtures M1_RCA_0FF and M2_RCA_10FF and sand with silt (siSa) for mixtures M3_RCA_20FF and M4_RCA_30FF, following PN-EN ISO 14688-2:2006 [21]. The main properties of these mixtures of RCA are given in Table 3. The specific gravity for all used materials was found to be 2.62.

Table 3. Specifications of RCA mixtures.

Description	Mixtures No			
	1	2	3	4
Effective size, d_{10} (mm)	0.086	0.063	0.033	0.025
d_{30} (mm)	0.14	0.12	0.085	0.064
Mean size, d_{50} (mm)	0.20	0.16	0.14	0.12
d_{60} (mm)	0.25	0.18	0.17	0.14
Coefficient of uniformity, C_u	2.91	2.86	5.15	5.60
Coefficient of curvature, C_c	0.91	1.27	1.29	1.17
Minimum dry density, $\rho_{d\ min}$ (g/cm^3)	1.24	1.22	1.19	1.17
Maximum dry density, $\rho_{d\ max}$ (g/cm^3)	1.60	1.61	1.62	1.64

When evaluating the tested four RCA mixtures for coefficients of the grain size distribution (C_u, C_c, d_{10}), the two first (M1 and M2) can be classified as poorly graded sands. Mixtures M3 and M4 are classified as well-graded sands with silt. All tested mixtures are of uniform grain size.

In addition, selected physical and geometric tests were performed on the properties of the concrete aggregates used to create all the mixtures analyzed in this study. The results of these tests are presented in Table 4.

Table 4. Summary of test results for selected properties of concrete aggregate.

Description	Unit	Value
Sand index, SE	-	82
Shredding resistance, LA	%	39, LA_{40}
Abrasion resistance, M_{DE}	%	28, $M_{DE}35$
Frost resistance, F	%	9.64, F10
CBR value	%	60
Flatness index, FI	%	42, FI_{50}
Shape index, SI	%	23, SI_{40}
Content of grains with crushed or broken surface	%	78
Grain density	Mg/m^3	2.51
Water absorption	%	6.55
Bulk density loose	Mg/m^3	1.41
Methylene blue	g/kg aggregate	1.66

The laboratory physical properties evaluation also included compaction tests to characterize the optimum moisture content (OMC) and the maximum dry density ($\rho_{d\ max}$) of RCA by the Proctor method. The details regarding this method have been presented in [22]. The following results were obtained: the average value of OMC is 7.4%, the uncertainty of measurement is 6.3%, and the average value of $\rho_{d\ max}$ is equal to 1.69. The uncertainty of measurement is 2.5%.

The chemical compositions of RCA blends are listed in Table 5. In addition, the analysis of concentrations of chlorides, sulfates, and heavy metals in all tested materials was carried out. The methodology of these studies has been described in [23]. In Table 6 the results from conducted leaching tests of RCA blends are presented.

Table 5. Chemical properties of RCA mixtures.

Spectrum	O	Na	Mg	Al	Si	P	S	K	Ca	Ti	Fe
Mass percent (%)											
Mean value	38.22	0.31	1.09	3.43	15.28	0.02	0.69	1.27	37.03	0.19	2.48
Standard deviation	1.62	0.13	0.20	0.34	3.38	0.02	0.15	0.16	4.41	0.03	0.29
Mean standard deviation	0.57	0.05	0.07	0.12	1.20	0.01	0.05	0.06	1.56	0.01	0.10

Table 6. Leachate concentration from RCA mixtures.

Element	Co (mg/L)	Ni (mg/L)	Cu (mg/L)	Zn (mg/L)	Cd (mg/L)	Pb (mg/L)	Cr (mg/L)	Sulphates (mg/L)	Chlorides (mg/L)	Specific Conductivity (µS/cm)	pH
Value											
M1_RCA_0FF	0.018	<0.015	0.061	0.546	0.036	0.035	<0.03	43	5.0	144.2	9.55
M2_RCA_10FF	0.018	<0.015	0.067	0.541	<0.008	0.015	<0.03	55	15.0	456	9.70
M3_RCA_20FF	0.013	<0.015	0.045	0.520	<0.008	<0.015	<0.03	155	7.0	444	10.2
M4_RCA_30FF	0.015	<0.015	0.075	0.852	<0.008	<0.015	<0.03	250	8.5	666	9.79
Acceptance criteria *	1	0.5	0.5	2	0.05	0.5	0.5	500	1000		

* Official Gazette of the Republic of Poland, Regulation of the Minister of the Environment of 18 November 2014 on the conditions to be met for the introduction of sewage into waters and to land and on substances particularly harmful to the aquatic environment.

Considering the physical, geometric, and chemical properties of the recycled aggregate used to create the M1–M4 mixes, the potential application of the proposed mixtures could be as a supportive road subbase.

Ground rubber waste from worn vehicle tires applied in the presented research was shredded and processed in the local tire manufacturing plant. Two sizes of tire waste were used: powder with the diameter size respectively 0.5–1.0 mm (Figure 3a) and granulate with dimensions 1.0–2.0 mm (Figure 3b). Both rubber additives did not contain any amounts of textile parts or steel belts. Both applied rubber wastes are not harmful to health. According to the manufacturer's information, these products have stable composition and do not pose a threat during storage and transport. They have been positively assessed in terms of health by the national institute of hygiene.

(a) (b)

Figure 3. Rubber waste: (**a**) powder 0.5–1.0 mm; (**b**) granulate 1.0–2.0 mm.

The basic components of used waste tires are (% mass/tire): natural rubber 15%, SBR 20%, BR 10%, IIR/XIIR 5%, silica 15%, carbon black 15%, sulfur 2%, resin 2%, mineral and vegetable oils 10%, others (zinc oxide, stearic acid) 6%. The presence of heavy metals in tires is related to the technological process. Here, these are at very negligible amounts.

2.2. RCA–RTW Mixtures

To prepare the appropriate compositions of ingredients, the exact amount of RCA and RTW was estimated. Four mixtures were developed with varying percentages of recycled rubber content. All samples were prepared by careful hand mixing of rubber with recycled concrete aggregates, so the mass of granulate and/or powder was appropriately 20%, 15%, 10%, and 5% of the total mass. The authors limited this study to mixtures with maximum rubber content equal to 20% by weight, because it has been reported in the literature that for rubber content more than 30–35% by weight, the mixtures exhibit in general rubberlike behavior. This is mainly due to the predominant development of rubber-to-rubber interfaces and thus the overall static and dynamic responses of the mixtures are then mainly controlled by the rubber part [24].

The basis for forming new blends discussed here were the previously mentioned four RCA compositions (M1, M2, M3, and M4). The new blends, shown in Figure 4, were created in such a way that the corresponding RCA fraction, with grain sizes ranging from about 2.0 mm or 1.0 mm to 0.5 mm, was eliminated and replaced with rubber waste. The particle size distribution curves of all four RCA–RTW mixtures are shown in Figure 5. It is worth noting that as the content of the fine fraction (FF), i.e., <0.063 mm, increases, the content of waste rubber decreases. The mixes created thus have the following compositions:

- M5_RCA_20R_0FF (RCA with 10% addition of 0.5–1.0 mm powder, 10% addition of granulate 1.0–2.0 mm and 0% of RCA fine fraction),
- M6_RCA_15R_10FF (RCA with 15% addition of 0.5–1.0 mm powder and 10% of RCA fine fraction),
- M7_RCA_10R_20FF (RCA with 10% addition of 0.5–1.0 mm powder and 20% of RCA fine fraction),
- M8_RCA_5R_30FF (RCA with 5% addition of 0.5–1.0 mm powder and 30% of RCA fine fraction).

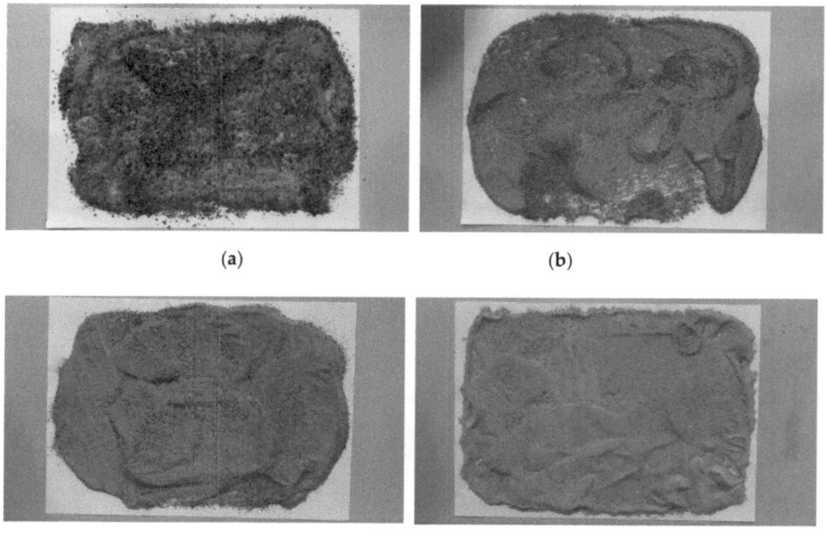

(a) (b) (c) (d)

Figure 4. A view of RCA–RTW mixtures. (**a**) M5_RCA_20R_0FF; (**b**) M6_RCA_15R_10FF; (**c**) M7_RCA_10R_20FF; (**d**) M8_RCA_5R_30FF.

These mixes could potentially be used in the construction of road embankments. The idea would be to replace the traditional natural soil with a material derived from recycling so that it performs the same functions as the traditional soil.

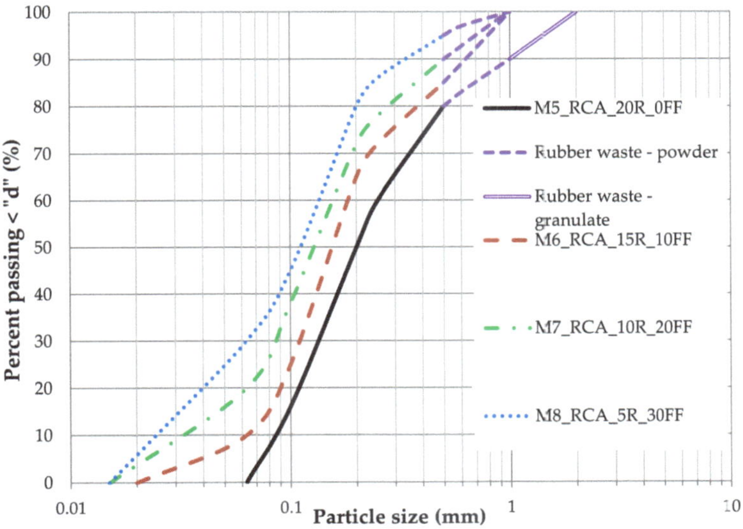

Figure 5. The particle size distribution curves of M5, M6, M7, and M8 used for this study.

2.3. Testing Methods

A series of isotropic consolidated drained (CID) triaxial tests were performed to evaluate the effect of the percentage of recycled tire waste on the strength characteristics of eight different specimens. The automated triaxial testing system adopted for this study is described in the following papers [18,25]. The specimens were prepared in the laboratory using the measurement of weight. The exact amount of RCA and RTW were mixed by hand thoroughly until a uniform mixture was achieved. Next, the mixtures were transferred to a triplicate mold, tightly coupled together, with a rubber membrane wound on the inside and secured with two O-rings. This mold was set on a pore stone additionally equipped with a filter paper, based on a triaxial apparatus. The mixtures were placed in the mold in layers and special care was taken that the rubber does not segregate. Each layer was compacted by a small hand tamper, to obtain specimens in a compacted state. When the specimens were completely prepared, standard initial steps of triaxial tests were performed, i.e., flushing, saturation, and consolidation. All the mixtures were tested at three different effective stresses, i.e., 90, 180, and 270 kPa (covering a range of pressures expected in many geotechnical applications [26]). At the end of consolidation, the axial load was increased at a constant rate of axial strain until an axial strain of 15% was reached or failure occurred. The shear rate of each specimen was set at 0.033 mm/min based on the oedometric tests performed and the experience of the researchers. During shear, data of axial deformation and axial load were recorded from the dial gauges and load cell, respectively.

3. Testing Results and Discussion

Shear strength characteristics and volumetric characteristics of the composite materials were examined concerning the percentage of rubber waste and the applied effective stress on the specimens in the CID triaxial tests. The possible influence of rubber addition on the cohesion and friction angle of RCA were analyzed considering the contents of rubber. Additionally, the influence of the fine fraction of recycled concrete aggregate on the shear properties was also studied. In this section, the results of laboratory tests are presented with a discussion highlighting the effects of the various parameters. The inclusion of all figures in this paper is cumbersome and makes the paper long. Only a limited number of them are presented here. Additionally, each composite was tested several times and the results presented are averaged.

3.1. Stress-Strain Behavior

Typical stress-strain plots from CID triaxial test for all blends created, at the mean effective stresses of 90, 180, and 270 kPa, are shown in Figure 6a,b. Blue color indicates data for 90 kPa, red for 180 kPa, and green for 270 kPa. From Figure 6, we can observe a relatively clear peak in shear resistance for all mixed specimens, regardless of rubber contents. For the mixtures not containing tire waste (see Figure 6a), the deviatoric stress is the highest for the M1 specimen, i.e., the one without a fine fraction of RCA, whereas the smallest for M4 specimen, i.e., the mixture with 30% of the fine fraction of RCA. At the same time, the smallest values of the shear strain at failure are obtained for M1 mixtures and the highest for the M4 blend. However, analysis of Figure 6b shows that for mixtures of modified recycled concrete aggregate, it is exactly the opposite as for the composites M1–M4. The addition of rubber waste does not increase the deviatoric stress. After comparing the stress-strain characteristics from Figure 6b, we found that with decreasing rubber content and increasing fine fraction content, the shear resistance of the mixtures increases by about 20% with shear strain at failure also decreasing by about 20%. The exception is the M8 mix, which has only been tested at two mean effective stresses, namely p′ = 90 and 180 kPa. Thus, it appears that fine fraction content next to rubber content influences the shear strength of the mixtures.

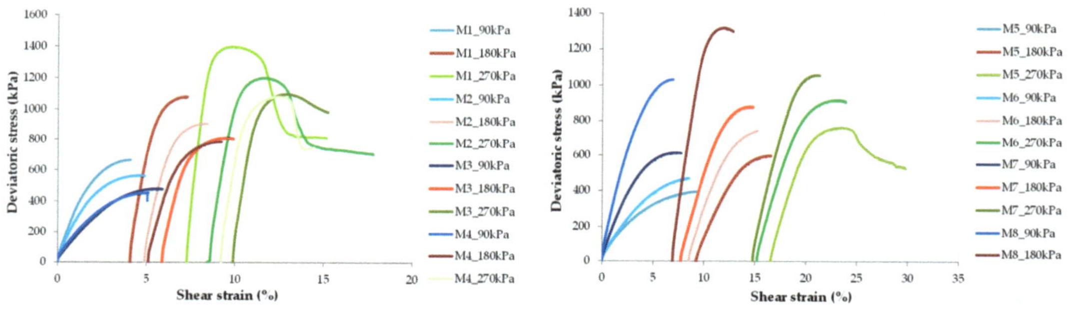

Figure 6. stress-strain characteristic for RCA and RCA–RTW mixtures. (**a**) RCA mixtures: M1, M2, M3, and M4; (**b**) RCA–RTW mixtures: M5, M6, M7, M8.

On the other hand, comparing with Figure 6a,b, it can be seen that the initial slope of the stress-strain curves of RCA–RTW mixtures at 90 kPa is greater than the stress-strain curves of RCA specimens at the same confining pressure. For the other tests, at higher pressures, no such trend is observed. This can indicate that recycled concrete aggregate modified by rubber addition will have more strength but only at small strain values.

In Figure 7, the variations of deviatoric stress versus mean effective stress (p′) for all studied blends are presented. For all tested compositions, with increasing p′ the higher deviatoric stresses are obtained. It can be observed a linear relationship between deviatoric stress and mean effective stress. Very high values of the coefficients of determination (R^2) show a very good adaptation of this model to the experimental data. In addition, the effect of strengthening is not observed in all cases in Figure 7. It was expected that the addition of rubber grains can significantly enhance the shear resistance of RCA, whereas only for two specimens, M7 and M8 with 10% and 5% of rubber insert respectively, this conclusion can be drawn. For the other two compositions, M5 and M6 with 20% and 15% of rubber content respectively, the maximum deviatoric stresses are, however, found for the blends made of recycled concrete aggregate alone (M1 and M2). Most likely, that is related to the fine fraction content of RCA and, on the other hand, to the method of specimen preparation. Higher content of the fine fraction, and a lower content of rubber waste, allow for a better, tighter arrangement of grains and soil particles. The fine fraction of RCA with dimensions <0.063 mm fills voids better than rubber waste of 0.5–2.0 mm. This results in better-compacted mixtures. However, it should be remembered that the content of the fine fraction is not high, the maximum is 30%. So, in the presented study, there are still materials called sand-like soil [27].

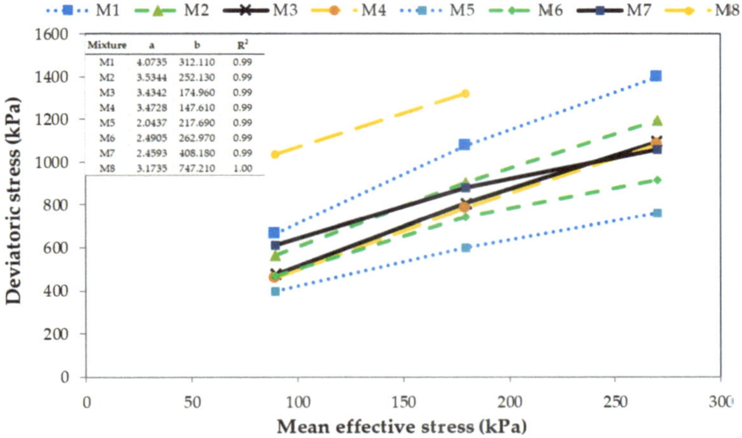

Figure 7. Deviatoric stress versus mean effective stress for all tested mixtures.

A typical plot of the variation of the shear strain corresponding to peak shear stress at different confining pressure is shown in Figure 8. Due to the high deformation of the rubber waste particles, the magnitude of the shear strain at the peak deviator stress increases with the increasing proportion of rubber in the mixture. The highest values of the shear strain, regardless of the mean effective stress values, are observed for the M5 mixture, with 20% of rubber addition. As the rubber content of the specimen decreases by 5%, the strain decreases by an average of 8%. For the compositions without RTW, M1–M4, the shear strain decreases by about 50% for specimen M1 to about 25% for specimen M4.

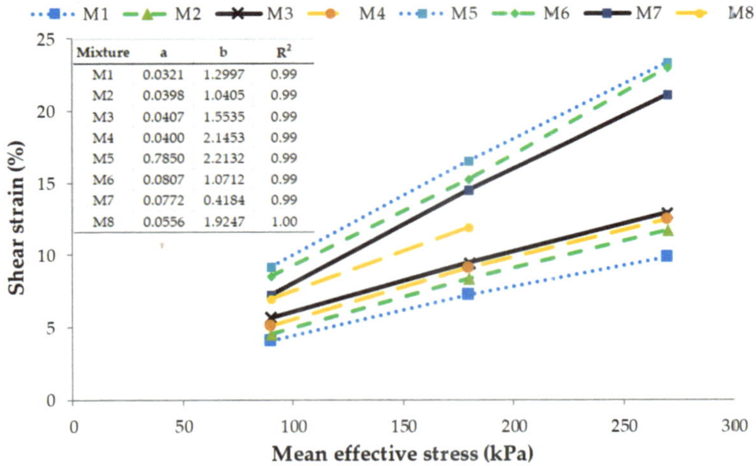

Figure 8. Variation of the shear strain corresponding to peak shear stress with change in mean effective stress.

Furthermore, it can be seen from Figure 8, that the shear strain at failure shows a linear relationship with the effective stress. In Figure 8, additionally, the summary of the linear function constants (a, b) and R^2 is included. Analyzing the slope of the shear strain–mean effective stress curves, it can be noticed that for RCA–RTW mixtures it is similar, regardless of the rubber content, and is about five degrees. Only for sample M8, it is smaller, three degrees. RCA mixtures M1–M4, have the initial slope of the shear strain–mean effective stress curves slightly smaller, by about half in comparison with other specimens. Such small differences in slope values do not allow to unequivocally state a positive effect of the rubber additive on the shear strength of the studied mixes.

The relationships between shear strain and volumetric strain are plotted in Figure 9. Figure 9a only provides the results of RCA specimens, whereas in Figure 9b the results of RCA–RTW mixtures are presented. The color coding corresponds to that in Figure 6. From Figure 9a, it can be seen that there is an initial expansion, i.e., dilation, and then compression, with an increase in shearing strain. The dilatancy characteristics of studied RCA blends are similar to the behavior of cohesionless material, having both negative and positive magnitudes of dilatancy [11,28]. In addition, with increasing confining pressure, the tendency of dilation decreases with increasing shear strain.

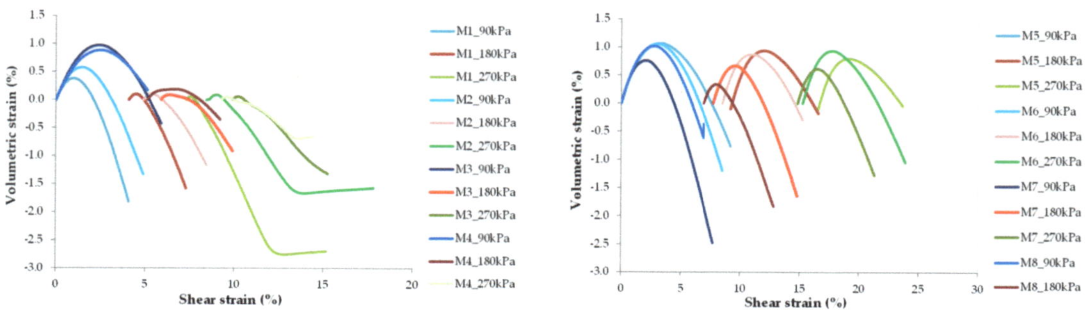

(**a**) (**b**)

Figure 9. Typical plot for variation of volumetric strain with shear strain for RCA and RCA–RTW mixtures. (**a**) RCA mixtures: M1, M2, M3, and M4; (**b**) RCA–RTW mixtures: M5, M6, M7, M8.

Analyzing Figure 9b, the deformation characteristics of rubberized RCA composites are very similar to each other, not depending on the mean effective stress. It is difficult to notice here a clear relationship between volumetric strains and varying rubber content, because of the very close strain values obtained. After a closer look at the values, it can be eventually concluded that volumetric strain decreases little with an increase in fine fraction content and a decrease in rubber content. Different behavior was observed by Saberian et al. [29]. As shown in Figure 9b, RCA–RTW mixtures, show a change in volume like RCA, but they develop more dilation than pure RCA blends. The dilatant reaction is visible and the specimens are characterized by high susceptibility to volume change. This behavior indicates that the mixtures are in a highly compacted state. There is no doubt that the addition of rubber results in higher strain values, both volumetric and shearing strains. This is due to the deformable behavior of the rubber.

3.2. Shear Strength Parameters

Shear strength can be defined as the maximum internal resistance to the applied shearing force. The main parameters used in formulating the analysis of the shear strength of soil are the angle of internal friction, the so-called friction angle, and the cohesion. Many researchers have investigated the shear strength of tire waste mixed with natural soil using triaxial tests and/or direct shear tests [30,31].

The results of Humphrey and Sandford [32] from the direct shear tests on pure tire chips samples exhibit friction angles ranging from 20 to 35°, and cohesion from 3 to 11.5 kPa. The triaxial tests from Wu [11] produced higher values of φ, greater than 40° The majority of the studies conducted found that the inclusion of rubber chips in sand increased shear strength. On the other hand, reducing the particle size of rubber waste from chips to granulated rubber adversely affected the shear strength capacity [33]. For rubber-cohesive soil mixtures, opinions are divided as to the positive effect of rubber addition on the strength characteristics. Some believe that the addition of rubber chips to clay increased the shear modulus and cohesion intercept of the matrix material [34]. While others suggest that the shear strength remains the same or reduces with increasing rubber content, thereby implying that the bond between clayey soils and waste rubber is weak [35].

In Table 7, the strength parameters of soil-rubber mixtures for the composites obtained in the present study are presented. The given values of friction angle are similar to those for compacted noncohesive soils, fine sands, and silty sands, namely between 30° and around 42°. Corresponding values for RCA mixed with two different sizes and three different percentages of crumb rubber were reported by Saberian et al. [29]. The highest value of the effective angle of internal friction, around 42°, characterizes the mixture M8, the lowest, however, equal to φ = 30.20° was the mixture M5. Generally, the large values of the φ angle are most likely due to the high value of friction between grains. All tested mixtures were in an air-dry state. With an increase in moisture content, we could expect a decrease in friction values as the soil grains begin to be covered by a layer of water. It can be noticed as well that M1–M4 mixes have on average a value of the angle of internal friction of 39°. In the case of RCA–RTW blends (M5–M8 mixes), the obtained average value of the φ angle is slightly lower at 34.6°, which is about 11% of the difference.

Table 7. Results of strength parameters and basic statistics for studied materials.

Mixture No	Material	Rubber Content (%)	Friction Angle (°)			Cohesion (kPa)		
			Average	Std. Dev.	Std. Error	Average	Std. Dev.	Std. Error
1	RCA_0FF	0	40.19	4.94	2.0	63.01	46.62	20.8
2	RCA_10FF	0	37.21	8.23	3.7	45.17	24.80	11.1
3	RCA_20FF	0	39.17	0.94	0.5	43.37	9.27	5.4
4	RCA_30FF	0	39.46	0.65	0.4	34.76	6.36	3.7
5	RCA_20R_0FF	10G [1], 10P [2]	30.20	1.28	0.7	64.74	10.20	5.9
6	RCA_15R_10FF	15P [2]	33.69	3.09	1.8	75.42	27.93	16.1
7	RCA_10R_20FF	10P [2]	33.26	2.84	1.6	114.83	27.79	16.0
8	RCA_5R_30FF	5P [2]	41.15	1.95	1.0	55.23	33.12	16.6

[1] Granulate rubber waste; [2] powder rubber waste.

In terms of cohesion, for the first four compositions, a decrease of this parameter is observed from a value of about 63 kPa for the mixture M1 to a value around half as low, i.e., 35 kPa for the mixture M4. The noticeably higher results of c are obtained for the M1 mixture, which has the highest cohesion of all four pure RCA mixes. It is an "apparent cohesion", which is not realistic for dry unbounded granular materials. Even higher values of apparent cohesion, by 10 kPa, for recycled concrete aggregate were obtained by Perera et al. [36]. The specimens M1–M4 are characterized by an average cohesion value of 47 kPa. By analyzing the obtained values of c for the next four mixtures, the larger values of cohesion are found, with an average equal to 78 kPa. The average difference in the strength parameter c between specimens M5–M8 and M1–M4 is about 40% in favor of the former.

As presented in Table 7, rubber additive causes an initial increase in the strength parameter c by approx. 60 kPa (see specimen M7 compared to M8), followed by a decrease of ca. 34% (M6 compared to M7) and 14% (M5 compared to M6). This confirmed the observation of Saberian et al. [29], although these researchers obtained significantly smaller cohesive values. It is interesting to note that for the highest rubber content, i.e., the 20% of specimen M5, an average cohesion value is obtained. This was expected to be rather higher, due to the addition of rubber waste of a larger size. According to the findings of Anbazhagan et al. [28], who stated that larger size rubber particles have a good interlocking capacity, resulting in an increase in the cohesion with an increase in rubber particle size.

Generally, the higher decline of c in the mixture (see M5–M8 blends), the bigger their grain size and percentage composition. This fact can be explained by the decreasing domination of electromagnetic force between fine particles, which results in the separation of soil particles due to the increasing content of rubber waste. The disturbance of this trend is the mixture M8 containing 5% of powder rubber waste.

In Table 7, basic statistics for the strength parameters are summarized. The following parameters are calculated: average value, standard deviation, and standard error. It is worth noticing that the standard error for the mean friction angle values for all of the presented data is low, not exceeding the value of 4°. This confirms the regularity of the conducted research. On the other hand, the standard error values for the obtained cohesion values are larger, but still do not exceed 20 kPa.

The changes observed for strength parameters, φ, and c, for RCA and RCA-RTW mixtures are compiled in Table 8. The analysis of the φ values shows that the addition of rubber waste to RCA practically does not affect this parameter. It was expected that the friction angle increases with an increase in rubber content, or at least up to some percentage. The angularity of rubber contributes to increasing friction angle by interlocking with sand particles. According to Attom [37], the increase in the angle of internal friction is due to failure in the shearing zone. In this particular zone, the rubber particles are distributed and oriented randomly at the shearing surface. As shearing starts, the rubber particles either slide or resist the shearing against cut-off, which results in an increased shearing force.

Table 8. Variations of strength parameters and results of density for studied materials.

Mixture No	Material	Rubber Content (%)	The Ratio of Improvement/Reduction in		Average Density (g/cm^3)
			Friction Angle (-)	Cohesion (-)	
1	RCA_0FF	0	-	-	1.65
2	RCA_10FF	0	0.93	0.72	1.63
3	RCA_20FF	0	1.05	0.96	1.55
4	RCA_30FF	0	1.01	0.80	1.52
5	RCA_20R_0FF	10G, 10P	-	-	1.23
6	RCA_15R_10FF	15P	1.12	1.16	1.35
7	RCA_10R_20FF	10P	0.99	1.52	1.43
8	RCA_5R_30FF	5P	1.24	0.48	1.47

Larger differences in the values are observed in the case of cohesion, though there is no consistency in the changes here. The values of cohesion are both increasing and decreasing. A noticeable decrease in apparent cohesion with the addition of recycled rubber can be explained by the fact that although rubber provided better shear strength for the RCA, due to its higher tensile strength, the increasing of the rubber content would lead to decreasing the shear parameters. A higher increase in apparent cohesion is achieved for the M7 mixture. This is a composition that contains 10% of rubber inclusion and 20% of fine fractions. This mix has therefore the highest shear strength.

In Table 8, the average densities (ρ_{avg}) values are tabulated. The obtained values of ρ_{avg} prove that the addition of rubber waste reduces the density of RCA by approx. 2%, 6%, and 9% for specimens M7, M6, and M5, respectively. Smaller differences in density, ranging from 1% to about 4%, are observed for mixes without rubber with increasing fine fraction content. Changes in the fine fraction (mixtures M1–M4) result in half the changes in RCA density.

In general, the fine content significantly affects the engineering properties of non-cohesive soils, if they consist of fine particles. In a recent study, Phan et al. [38] showed that as the fine content of sand–fine particle mixtures increased, all parameters of deviator stress, volumetric strain, shear stress, internal friction angle, and cohesion increased. Other researchers, however, have reported that as silt content increased, the steady-state strength initially decreased and subsequently increased in shear strength with further increases in the silt content to values greater than 30%.

Figure 10 provides the relationship between friction angle and the fine content of all studied compositions. It is noticed that for mixtures of pure recycled concrete aggregate (M1–M4) there is a comparable situation to the one observed in the studies of other scientists. The initial decrease in the friction angle values is stopped for the M2 blend with 10% of the fine fraction of RCA. While for M4 mixture with 30% of fine content and increase of 6% in the strength parameter φ is observed, relative to the mixture M2. In the case of RCA–RTW mixtures, a strong linear relationship (R^2 = 0.81) between friction angle and percentage of the fine fraction can be found. An increase in the fine fraction content by 10%, together with a decrease in the rubber additive by 5%, results in an increase in the value of the angle of internal friction by about 10°.

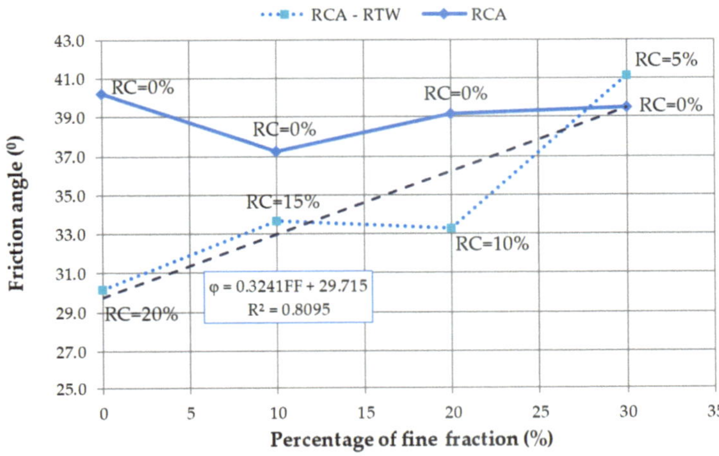

Figure 10. The relationship between friction angle and fine fraction of RCA and RCA–RTW mixtures. Note: RC—rubber content.

In Figure 11, the relationship between cohesion and the fines content of all studied compositions is shown. For RCA–RTW mixtures, a linear increase in the cohesion value with increasing fine fraction content, up to a value of FF = 20%, is observed. Mixture M7, with 20% of FF and 10% of RC has the highest cohesion, approx. 115 kPa. Mixture M8 is characterized, on the other hand, by the lowest cohesion value, ca. 55 kPa, although it has the highest percentage of the fine fraction—30%. In the case of testing M1–M4 mixtures, it can be concluded that the strength parameter c decreases with an increase in the fine fraction. For the proposed linear regression model R^2 is equal to 0.89, which means that the variation of mix cohesion is explained by the variation of fine fraction content in almost 90%.

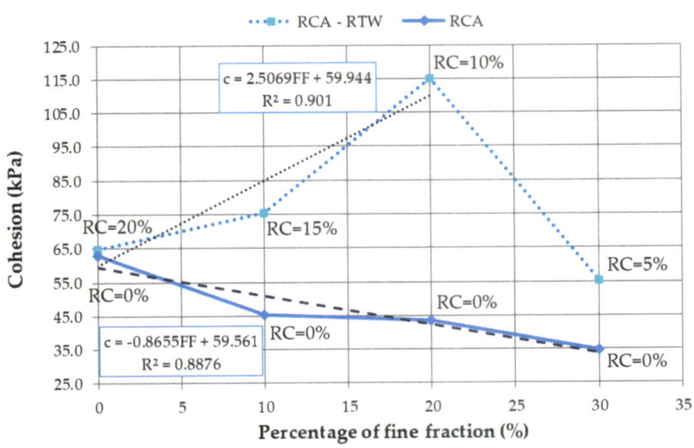

Figure 11. The relationship between cohesion and fine fraction of RCA and RCA–RTW mixtures. Note: RC—rubber content.

4. Conclusions and Recommendations for Further Research

In this study, a series of isotropic consolidation drained triaxial tests were performed to assess the selected mechanical properties of four recycled concrete aggregate mixtures with different fine fraction content and of four RCA mixed with different percentages of rubber waste. Despite the costs, it was decided to test such materials because they can effectively address the growing environmental concerns and at the same time provide solutions to geotechnical problems associated with, e.g., low soil shear strength and high dilatancy. The main conclusions of this research are:

1. All tested mixes have high mechanical properties. The values of the internal friction angles for the pure RCA compositions range from 37° to 40°, while for the RCA–RTW blends these values range from 30° to 41°. The mixes M1–M4 have on average an angle of internal friction of 39°. In the case of RCA–RTW blends (M5–M8 mixes), the obtained average value of the φ angle is slightly lower, at 34.6° (about 11% of the difference). The analysis of the φ values shows that the addition of rubber waste to RCA practically does not affect this parameter.
2. In the case of cohesion, the values ranging from 35 kPa to 63 kPa were obtained for specimens M1–M4, and from 55 kPa to 115 kPa for specimens M5–M8. The pure RCA specimens are characterized by an average cohesion value of 47 kPa. For the next four mixtures, enriched with a rubber additive, the larger values of cohesion are found, with an average equal to 78 kPa. The average difference in the strength parameter c between specimens M5–M8 and M1–M4 is about 40%, in favor of the former. The values of cohesion for RCA and RCA–RTW mixtures have various trends; they can fall or grow, initially increase and then decline, or vice versa.
3. With decreasing rubber content and increasing fine fraction content, the shear resistance of the mixtures increases by about 20% with decreasing shear strain at failure also by about 20%.
4. Based on the experimental results, it can be noticed that the inclusion of rubber in RCA can alter the shear strength, cohesion, friction angle, and volumetric strain. Generally, the improvement in RCA shear strength is observed. However, the largest improvement in shear strength occurs for the mixture M7, with 10% of recycled tire waste.
5. In this research, recycled concrete aggregate modified by rubber addition will have more strength at small strain values. The addition of rubber results in higher strain values, both volumetric and shearing strains. The magnitude of the shear strain at the peak deviator stress increases with the increasing proportion of rubber in the mixture. The highest values of the shear strain, regardless of the mean effective stress values, are observed for the M5 mixture, with 20% of rubber addition. As the rubber content of the specimen decreases by 5%, the strain decreases by an average of 8%.
6. The dilatancy characteristics of studied RCA blends are similar to the behavior of cohesionless material, having both negative and positive magnitudes of dilatancy. This is due to the deformable behavior of the rubber. Dilatant behavior is noticed because of the interlocking between the rubber and RCA particles. By increasing the mean effective stress, the tendency for dilation decreases. There is no significant reduction in RCA dilatancy due to the inclusion of rubber from recycled tires waste.
7. The addition of rubber waste reduces the density of RCA, by an average of 6%. Changes in the fine fraction (mixtures M1–M4) result in half the changes in RCA density.
8. The presented results are found to be significantly affected by the content of the fine fraction of RCA. An increase in the fine fraction content by 10%, together with a decrease in the rubber additive by 5%, results in an increase in the value of the angle of internal friction of about 10°. For RCA–RTW mixtures, a linear increase (around approx. 24%) in the cohesion value with increasing fine fraction content, up to a value of FF = 20%, is observed. For RCA mixtures the strength parameter c decreases with an increase in the fine fraction by 9 kPa on average.

Based on the results of all tests performed, including physical, geometric, chemical, and mechanical properties of the created composites, it can be stated that the samples would meet local road authority requirements for the subbase applications.

It is planned to perform further tests on RCA–RTW mixtures without FF content. On the other hand, it is worth checking the influence of other rubber dimensions (coarse rubber) on the strength characteristics of the tested mixtures. Moreover, the obtained research results can be developed as well by studying the change in the shear strength characteristics of the proposed blends influenced by such environmental impacts as rain, temperature drops, heaving, etc. This is because the road base is always exposed to these environmental effects.

Author Contributions: Conceptualization, K.G. and A.S.; methodology, K.G.; formal analysis, K.G. and A.R.; investigation, K.G.; resources, K.G.; data curation, K.G. and R.Š.; writing—original draft preparation, K.G.; writing—review and editing, K.G. and A.R.; visualization, K.G. and R.Š.; supervision, A.S. and A.R.; project administration, K.G. and A.S.; funding acquisition, K.G. All authors have read and agreed to the published version of the manuscript.

Funding: This work is the result of the implementation of research project No. DEC-2020/04/X/ST8/00732 founded by the National Science Centre of Poland.

Institutional Review Board Statement: Not applicable.

Informed Consent Statement: Not applicable.

Data Availability Statement: Data available on request due to their size properties. The data presented in this study are available on request from the corresponding author.

Acknowledgments: The authors would like to thank Emil Soból and Jan Ziółkowski for their help in organizing the materials to make the mixtures. The laboratory assistance provided by Jacek Bąkowski is greatly appreciated.

Conflicts of Interest: The authors declare no conflict of interest.

References

1. Sas, W. Application of recycled concrete aggregate in road engineering. *Acta Sci. Pol. Archit.* **2015**, *14*, 49–59.
2. Zegardło, B.; Ogrodnik, P.; Szulej, J. Ekonomiczne aspekty Ekonomiczne aspekty stosowania ceramicznych kruszyw recyklingowych do mieszanek betonowych przy budowie betonowych obiektów komunikacyjnych (in Polish) (Economic aspects of the use of recycled ceramic aggregates to the construction of concrete mixes concrete communication objects). *Logistyka* **2015**, *4*, 6993–7004.
3. Skawińska, A.; Foszcz, T. Study of rubber granules impact on selected mechanical properties of cement mortars. *Struct. Environ.* **2019**, *11*, 256–264. [CrossRef]
4. Sybilski, D. Zastosowanie odpadów gumowych w budownictwie drogowym (in Polish) (The use of waste rubber in road construction). *Przegląd Bud.* **2009**, *80*, 37–44.
5. Park, J.K.; Kim, J.Y.; Edil, T.B. Sorption capacity of shredded waste tires. In Proceedings of the Symposium on Geotechnics Related to European Environment, Bolton, UK, 28 June–1 July 1993; pp. 341–348.
6. Jones, K.P. Back to the trees. In Proceedings of the International Symposium on Recycling and Reuse of Used Tyres, Dundee, UK, 19–20 March 2001; pp. 39–49.
7. Garga, V.K.; O'Shaughnessy, V. Tire-reinforced earthfill. Part 1: Construction of a test fill, performance, and retaining wall design. *Can. Geotech. J.* **2000**, *37*, 75–96. [CrossRef]
8. Jastrzębska, M. Strength Characteristics of Clay-Rubber Waste Mixtures in UU Triaxial Tests. *Geosciences* **2019**, *9*, 352. [CrossRef]
9. Duda, A.; Sobala, D.; Siwowski, T.; Kaleta, D. Wykorzystanie materiałów z recyklingu opon samochodowych w budownictwie komunikacyjnym (In Polish) (Utilization of recycled materials from automobile tires in traffic construction). *Arch. Inst. Inżynierii Lądowej* **2016**, *21*, 97–111.
10. Pincus, H.; Edil, T.; Bosscher, P. Engineering Properties of Tire Chips and Soil Mixtures. *Geotech. Test. J.* **1994**, *17*, 453. [CrossRef]
11. Wu, Y.W.; Benda, C.C.; Cauley, R.F. Triaxial determination of shear strength of tyre chips. *J. Geotech. Geoenvironmental Eng.* **1997**, *123*, 479–482. [CrossRef]
12. Yang, S.; Lohnes, R.A.; Kjartanson, H. Mechanical properties of shredded tyres. *Geotech. Test. J.* **2002**, *25*, 44–52.
13. Bergado, D.T.; Youwai, S.; Rittirong, A. Strength and deformation characteristics of flat and cubical rubber tyre chip–sand mixtures. *Geotechnique* **2005**, *55*, 603–606. [CrossRef]
14. Akbarimehr, D.; Aflaki, E. An Experimental Study on the Effect of Tire Powder on the Geotechnical Properties of Clay Soils. *Civ. Eng. J.* **2018**, *4*, 594. [CrossRef]

15. Kowalska, M.; Chmielewski, M. Mechanical Parameters of Rubber-Sand Mixtures for Numerical Analysis of a Road Em-bankment. In Proceedings of the IOP Conference Series: Materials Science and Engineering, Beijing, China, 24–27 October 2017.
16. Bałachowski, L.; Gotteland, P. Characteristics of Tyre Chips-Sand Mixtures from Triaxial Tests. *Arch. Hydro-Eng. Environ. Mech.* **2007**, *54*, 3–14.
17. Sas, W.; Głuchowski, A.; Szymański, A. The geotechnical properties of recycled concrete aggregate with addition of rubber chips during cyclic loading. In Proceedings of the 5th International Conference—GEOMATE 2015 Géotechnique, Construction Materials and Environment, Osaka, Japan, 16–18 November 2015; pp. 193–198.
18. Gabryś, K.; Sas, W.; Soból, E.; Głuchowski, A. Application of Bender Elements Technique in Testing of Anthropogenic Soil—Recycled Concrete Aggregate and Its Mixture with Rubber Chips. *Appl. Sci.* **2017**, *7*, 741. [CrossRef]
19. Polish Committee for Standardization. PN-EN 933–11:2009. In *Tests for Geometrical Properties of Aggregates*; Classification Test the Constituents of Coarse Recycled Aggregate; BSI: Warszawa, Poland, 2009.
20. Cabrera-Covarrubias, F.G.; Gómez-Soberón, J.M.; Almaral-Sánchez, J.L.; Arredondo-Rea, S.P.; Gómez-Soberón, M.C.; Corral-Higuera, R. An Experimental Study of Mortars with Recycled Ceramic Aggregates: Deduction and Prediction of the Stress-Strain. *Materials* **2016**, *9*, 1029. [CrossRef]
21. PN-EN ISO 14688-2:2006. *Geotechnical Investigation and Testing—Determination and Classification of Soils—Part 2: Classification Rules*; PKN: Warszawa, Poland, 2006.
22. Głuchowski, A.; Gabryś, K.; Soból, E.; Šadzevičius, R.; Sas, W. Geotechnical Properties of Anthropogenic Soils in Road Engineering. *Sustainability* **2020**, *12*, 4843. [CrossRef]
23. Głuchowski, A.; Sas, W.; Dzięcioł, J.; Soból, E.; Szymański, A. Permeability and Leaching Properties of Recycled Concrete Aggregate as an Emerging Material in Civil Engineering. *Appl. Sci.* **2018**, *9*, 81. [CrossRef]
24. Kim, H.-K.; Santamarina, J.C. Sand–rubber mixtures (large rubber chips). *Can. Geotech. J.* **2008**, *45*, 1457–1466. [CrossRef]
25. Gabryś, K.; Soból, E.; Sas, W.; Šadzevičius, R.; Skominas, R. Warsaw Glacial Quartz Sand with Different Grain-Size Characteristics and Its Shear Wave Velocity from Various Interpretation Methods of BET. *Materials* **2021**, *14*, 544. [CrossRef] [PubMed]
26. Tasalloti, A.; Chiaro, G.; Banasiak, L.; Palermo, A. Experimental investigation of the mechanical behaviour of gravel-granulated tyre rubber mixtures. *Constr. Build. Mater.* **2020**, *273*, 121749. [CrossRef]
27. Ruan, B.; Miao, Y.; Cheng, K.; Yao, E.-L. Study on the small strain shear modulus of saturated sand-fines mixtures by bender element test. *Eur. J. Environ. Civ. Eng.* **2018**, *25*, 28–38. [CrossRef]
28. Anbazhagan, P.; Manohar, D.R.; Rohit, D. Influence of size of granulated rubber and tyre chips on the shear strength characteristics of sand–rubber mix. *Géoméch. Geoengin.* **2016**, *12*, 266–278. [CrossRef]
29. Saberian, M.; Li, J.; Perera, S.T.A.M.; Ren, G.; Roychand, R.; Tokhi, H. An experimental study on the shear behaviour of recycled concrete aggregate incorporating recycled tyre waste. *Constr. Build. Mater.* **2020**, *264*, 120266. [CrossRef]
30. Mohajerani, A.; Burnett, L.; Smith, J.V.; Markovski, S.; Rodwell, G.; Rahman, T.; Kurmus, H.; Mirzababaei, M.; Arulrajah, A.; Horpibulsuk, S.; et al. Recycling waste rubber tyres in construction materials and associated environmental considerations: A review. *Resour. Conserv. Recycl.* **2020**, *155*, 104679. [CrossRef]
31. Sławińska, J. The Mohr-Coulomb friction angle of granular soils under different stress conditions. *Acta Sci. Pol.-Arch. Bud.* **2018**, *17*, 51–60. [CrossRef]
32. Humphrey, D.N.; Sandford, T.C. Tire chips as lightweight subgrade fill and retaining wall backfill. In Proceedings of the Symposium on Recovery and Effective Reuse of Discarded Materials and By-Products for Construction of Highway Facilities, Denver, CO, USA, 19–22 October 1993; pp. 5–87.
33. Chaney, R.; Demars, K.; Masad, E.; Taha, R.; Ho, C.; Papagiannakis, T. Engineering Properties of Tire/Soil Mixtures as a Lightweight Fill Material. *Geotech. Test. J.* **1996**, *19*, 297–304. [CrossRef]
34. Akbulut, S.; Arasan, S.; Kalkan, E. Modification of clayey soils using scrap tire rubber and synthetic fibers. *Appl. Clay Sci.* **2007**, *38*, 23–32. [CrossRef]
35. Tatlisoz, N.; Benson, C.; Edil, T. Effect of Fines on Mechanical Properties of Soil-Tire Chip Mixtures. In *Testing Soil Mixed with Waste or Recycled Materials*; American Society for Testing and Materials: West Conshohocken, PA, USA, 1997.
36. Perera, S.; Arulrajah, A.; Wong, Y.; Maghool, F.; Horpibulsuk, S. Evaluation of shear strength properties of unbound PET plastic in blends with demolition wastes. *Constr. Build. Mater.* **2020**, *262*, 120545. [CrossRef]
37. Attom, M.F. The use of shredded waste tires to improve the geotechnical engineering properties of sands. *Environ. Earth Sci.* **2005**, *49*, 497–503. [CrossRef]
38. Phan, V.T.-A.; Hsiao, D.-H.; Nguyen, P.T.-L. Effects of Fines Contents on Engineering Properties of Sand-Fines Mixtures. *Procedia Eng.* **2016**, *142*, 213–220. [CrossRef]

materials

Article

Selected Properties of Self-Compacting Concrete with Recycled PET Aggregate

Justyna Jaskowska-Lemańska, Milena Kucharska *, Jakub Matuszak, Paweł Nowak and Wojciech Łukaszczyk

Faculty of Civil Engineering and Resource Management, AGH University of Science and Technology, 30-059 Cracow, Poland; lemanska@agh.edu.pl (J.J.-L.); jakub.matuszak70@gmail.com (J.M.); pawelsams.pn@gmail.com (P.N.); wojtek.tady@gmail.com (W.Ł.)

* Correspondence: kucharska@agh.edu.pl; Tel.: +48-12-617-20-63

Abstract: In this paper, the issue of self-compacting concrete (SCC) with the addition of polyethylene terephthalate (PET) recycled aggregate is addressed. The PET utilized was a waste fraction in the PET-bottle-recycling process. The implementation of waste in concrete mixes has a positive impact on their environmental and social profile; however, technical requirements are not necessarily met. In this investigation, PET was used as a substitute for fine aggregate in quantities ranging from 0 to 20% in increments of 5%. Both the flow properties of SCC mixes and the hardened SCC properties (compressive strength, splitting tensile strength, modulus of elasticity, and Poisson ratio) were investigated. Additionally, non-destructive tests (ultrasound and sclerometric) were performed to determine the correlation curves. The research revealed that both the flow properties and the parameters of the hardened concrete deteriorated with the PET content. Concrete with 20% PET replacement did not meet the self-compacting requirements and its compressive strength decreased by almost 50%. However, it was noted that replacing fine aggregate with PET aggregate in the amount of 5% did not significantly alter the concrete parameters and could be an attractive alternative to traditional concretes. Based on non-destructive testing, correlation curves were constructed that could be applicable to the future quality assessment of self-compacting concretes with recycled PET aggregate.

Keywords: self-compacting concrete; PET; recycled aggregate; eco-efficient concrete; fresh properties of SCC mechanical properties; non-destructive tests (NDT)

Citation: Jaskowska-Lemańska, J.; Kucharska, M.; Matuszak, J.; Nowak, P.; Łukaszczyk, W. Selected Properties of Self-Compacting Concrete with Recycled PET Aggregate. *Materials* **2022**, *15*, 2566. https://doi.org/10.3390/ma15072566

Academic Editor: Yuri Ribakov

Received: 28 February 2022
Accepted: 27 March 2022
Published: 31 March 2022

Publisher's Note: MDPI stays neutral with regard to jurisdictional claims in published maps and institutional affiliations.

1. Introduction

The rapid consumption of multiple natural resources and the production of substantial amounts of waste by the construction industry are widely recognized to contribute to a serious environmental problem. With the development of both science and technology, including building materials, this crisis should be continuously alleviated. One of the prevailing approaches is the implementation of intelligent and energy-efficient solutions in the construction sector, such as the idea of a circular economy or new generation materials with improved durability parameters.

Today, concrete is the most widely used construction material due to its versatility, good general availability, favorable mechanical properties, and relatively low price. Its production increases annually (approximately 32 billion tons worldwide [1]) and is unfortunately associated with harmful emissions of carbon dioxide (CO_2) and particulate matter (PM_{10}) to the environment, as well as the depletion of natural resources. Following the sustainable development strategy, new technologies have been introduced to the construction industry for several decades, for example, eco-efficient concrete. These measures aim, among other things, to minimize the environmental impact during production, increase the durability and resistance to environmental hazards, extend structural reliability, and reduce the life cycle costs of structures.

In the world literature, suggested strategies for the development of eco-efficient concrete involve, among others, the replacement of Portland cement's (which is responsible for approximately 8% of total CO_2 emissions [1]) mineral additives and natural aggregates with postproduction waste (e.g., recycled aggregate) or a general reduction in the contribution of the binder to the composition [2–4]. Thus, the idea of using eco-efficient concrete can help reduce harmful emissions of CO_2 and PM_{10}, limit the consumption of natural mineral resources, and reduce post-industrial waste, which can be given a "second chance of life," in line with the concept of a circular economy. In their final life cycle stage, these materials, such as traditional concrete, can be recycled by preparing crushed concrete to replace natural aggregate, as a highway substrate, or as a clean fill around buildings. Compared to other materials, concrete is relatively environmentally friendly; therefore, the eco-improvement of the performance characteristics of concrete must be multiphased and customized to the location of its implementation. The analysis of applied solutions taking into account the whole life cycle of the building object is considered crucial for concrete construction [5–7]. Therefore, the influence of design strategies on environmental and economic efficiency [8], as well as solutions favorable to the creation of an environmental profile, are being investigated.

Another global concern is the depletion of natural raw materials, including sand, which is used extensively in the construction, road, and glass industries. Although Earth's sand resources are relatively abundant, not all sand is suitable for concrete mixes. Thus, in concrete technology, the concept of substituting sand with recycled aggregates, including glass or plastic products, emerged [9].

Due to their superior properties (lightweight, relatively high strength, flexibility in transforming into different shapes, and high resistance to bacteria), plastics remain in use in many industries around the world. In the last 50 years, plastic production has increased more than 22-fold. However, given the amount of plastic consumed and the fact that not all of it can be recycled, plastic pollution is one of the fastest-growing threats in the world [10]. Polyethylene terephthalate (PET) is currently the most commonly used material for packaging (including bottles). The process of PET bottle recycling usually consists of several stages that generate different products: full quality recyclables, reduced quality recyclables, and waste [11]. After pre-cleaning, the compressed bottles are sent to grinders, where they are crushed into flakes with a target fraction of 4.0–12.0 mm, generating waste in the form of a PET powder (Figure 1a). After the grinding process, the PET flakes are washed again and the separation of fractions below 4.0 mm is also performed in this process. The PET flakes in the 4.0–12.0 mm fraction (Figure 1b) are most often recrystallized and, as a granulate, they are raw material for the production of packaging. The flakes below the fraction of 4.0 mm are not recrystallized for technical reasons, making them difficult to utilize (Figure 1c). According to data obtained from a leading recycling company in Poland, 4000 tons of PET bottles are recycled monthly, which results in 60 tons of by-product in the form of PET flakes with a fraction below 4.0 mm called "fines washed." The creation of opportunities to manage such waste is an extremely important problem faced by recycling companies in Poland and around the world.

In the literature, there are various examples of the use of PET materials in concrete technology, but most often they concern PET fibers, i.e., intentionally prepared fibers (by the author or by the industry) rather than waste materials. The volume of fiber content with respect to fiber concrete is between 0.3% and 1.5%. Therefore, the procedure reuses a small amount of PET waste, and in addition, the waste must be deliberately prepared [12]. Evaluation of the application of these fibers (in these small proportions) indicates that the decrease in compressive strength of concrete with PET fibers is not significant (up to 10%), but the parameters related to tensile strength and, therefore, also flexural strength are significantly improved [13–15]. From a waste management point of view, it seems much more beneficial to use PET fractions as aggregate substitutes. The most significant conclusions that can be drawn from the available studies on the application of various types of PET are related to the influence of the amount of this substitute and its effect on the basic physical and mechanical properties, such as bulk density, compressive strength, or splitting tensile strength [16,17]; less frequently, the values of the modulus of elasticity or flexural toughness has been determined [12,18,19]. There is also a negligible number of studies in which non-destructive tests for this type of concrete are discussed [20]. The authors are rather consistent in their basic conclusions, which can be summarized as a statement that properties such as compressive strength, modulus of elasticity, or fresh mixture properties deteriorate with a percentage increase in PET waste application. However, for small amounts of waste, the decrease may not be significant, and some properties, such as splitting tensile strength, may be improved. Considering the growing interest in this type of composite, it becomes relevant to conduct simultaneous investigations of their properties using non-destructive techniques, which would allow for an ongoing assessment of the quality of larger batches of this material.

(a) (b) (c)

Figure 1. PET bottle recycling products: (**a**) PET fines waste/powder—manufacturing waste, (**b**) washed PET fines of fraction 0.5–4.0 mm—non-quality product, and (**c**) washed PET Fines of fraction 4.0–12.0 mm—quality product.

This study investigated concrete mixes that involved the addition of waste from the PET bottle recycling process, which cannot be recycled, or the spectrum of their application is limited. The design of modern building materials with a reduced environmental impact and consume fewer natural resources while disposing of non-recyclable waste is a necessary step in light of climate change and the European Union requirements being introduced. The implementation of waste into a new generation concrete mix, such as self-compacting concrete (SCC) mix, would reduce the environmental impact while maintaining fine mechanical properties [21].

2. Materials and Methods

2.1. Constituent Materials

2.1.1. PET Recycled Aggregate

In this study, the PET fines washed material acquired from a leading PET bottle recycler in Poland was used. The physical and mechanical characteristics declared by the recycler are summarized in Table 1. Table 2 presents the maximum possible impurities of washed PET fines, although no significant impurities were observed in the material used for this study (only minor fragments of polyvinyl chloride (PVC) or polyethylene (PE)).

Given the relatively wide range of dimensions of the PET flakes declared by the manufacturer (0.5 to 4.0 mm), sieve analysis was performed according to standard [22] without the requirement of heating the aggregate at a temperature of 110 ± 5 °C to constant mass before further testing. The investigation was carried out for three samples taken from different parts of the collected test material, with each sample having a weight of 600 g.

The results of the sieve analysis are provided in the form of grading curves for washed PET fines in Figure 2.

Table 1. Physical and mechanical characteristics of washed PET fines.

Physical and Mechanical Characteristics	Value
Specific density	~1.35 g/cm^3
Bulk density	~550 kg/m^3
Flakes size	0.5 to 4.0 mm
Intrinsic viscosity	0.62 to 0.75 dL/g
Melting point	App. 250 °C
Color	Clear, light blue, blue, green

Table 2. Impurities PET fines washed.

Impurities	Value
Metal	<1%
Paper/labels	<10%
Polyolefine (PE)	<5%
PVC	<0.2%
Share in dust	<5%
Glue	Yes
Caustic soda	Yes

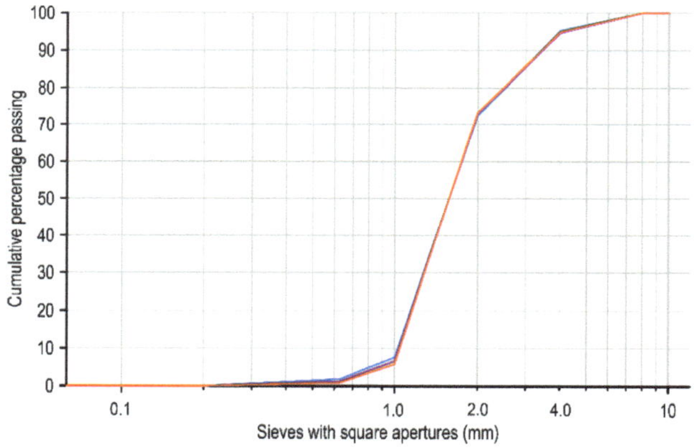

Figure 2. Grading curves for washed PET fines.

The grading curves of PET aggregate indicated that the particle size distribution for individual samples was nearly identical, and slight differences could be observed for the finest fractions. Fractions above 4.0 mm constituted approximately 5% of the material composition; fractions between 2.0 and 4.0 mm constitute 20% of the material; the largest fraction was between 1.0 and 2.0 mm, which represented almost 70% share; and the remaining 5% were fractions below 1.0 mm. The entire material was divided into three main fractions (above 4.0 mm, 2.0 to 4.0 mm, and below 2.0), as shown in Figure 3.

Figure 3. Washed PET fines sorted into fractions: (**a**) above 4.0 mm, (**b**) 2.0 to 4.0 mm, and (**c**) below 2.0 mm.

2.1.2. Concrete Mixes

In this research program, 5 concrete mixes with different contents of PET flakes (fraction below 2.0 mm) were prepared, which was used as the fine aggregate. The SCC control mix was prepared with a replacement of 5%, 10%, 15%, and 20% of the mass of fine aggregate (sand). The mixtures were based on Portland composite cement CEM II/B-M of a strength category of 42.5 and normal strength gain (symbol N). The cement consisted of Portland clinker and the addition of fly ash and ground-granulated blast furnace slag in the amount of 21–35% according to EN 197 [23]. In the mixtures, two fractions of coarse aggregate (2–8 mm and 8–16 mm) and one fraction of fine aggregate (0–2 mm) were used. Gravel constituted the coarse aggregate, while sand and an appropriate amount of PET waste flakes were considered fine aggregate. Polycarboxylate with a polyethylene condensate defoamed-based admixture was selected as a superplasticizer.

The binder content was fixed at 440 kg/m^3, which corresponded to a water-to-binder ratio of 0.36. The proportion of fine aggregate was fixed at 684 kg/m^3 for all mixtures. The superplasticizer was dosed each time until sufficient flow parameters of the SCC were achieved. Detailed compositions of the mixtures are compiled in Table 3.

Table 3. The composition of the mixes used per cubic meter.

Ingredients	Unit	SCC-P-0	SCC-P-5	SCC-P-10	SCC-P-15	SCC-P-20
CEM II/B-M 42.5 N	kg	440	440	440	440	440
Water		160	160	160	160	160
Sand (0–2 mm)		684	650	615	581	547
PET waste flakes (0–2 mm)		-	34	69	103	137
Gravel (2–8 mm)		510	510	510	510	510
Gravel (8–16 mm)		560	560	560	560	560
Superplasticizer		7.54	7.00	7.00	5.75	4.00
PET aggregate level	-	0%	5%	10%	15%	20%

A total of 21 test elements (20 cubes with 150 mm sides and 1 cube with 100 mm sides) were prepared from each concrete mix, resulting in a total of 105 test elements.

2.2. Test Procedures

2.2.1. Fresh Mixture Properties

All concrete mixtures were subjected to flow properties tests immediately after the mixing process. Flowability and plastic viscosity were evaluated using the slump flow test, during which, the time to 500 mm slump flow diameter (t_{500}) and the diameter of the slump flow after completion of the flow were measured according to [24]. To evaluate passing ability, the L-box test with 3 bars was performed [25]. The segregation of the mixture was visually graded according to the standard [26].

2.2.2. Compressive Strength

The compressive strength as the basic strength parameter of concrete was investigated according to [27] on cube samples with dimensions of 150 mm × 150 mm × 150 mm. The dimensions of the samples and the process of their preparation met the requirements of the standards [28,29]. The tests were performed after 28 days of concrete curing under laboratory conditions (temperature 20 ± 2 °C, humidity > 95%). Overall, 50 cubic samples were used to determine the compressive strength of the SCCs. Tests were performed using a universal testing machine (Walter + Bai AG, Löhningen, Switzerland).

2.2.3. Splitting Tensile Strength

Another important strength parameter of concrete is the splitting tensile strength. Investigations of this aspect were performed according to standard [30] on cube specimens of dimensions 150 mm × 150 mm × 150 mm, as allowed by Annex A of this standard. Specially dedicated loading pieces and hardboard packing strips were used for the test. The splitting tensile strength was determined using Equation (1):

$$f_{ct,split} = \frac{2F}{\pi L d}, \tag{1}$$

where F is the maximum force, L is the contact line length of the specimen, and d is the claimed cross-sectional dimension. In total, 30 cubic samples were used to determine the splitting tensile strength of the SCCs. Tests were performed using a universal testing machine (Controls Group, Atlanta, GA, USA).

2.2.4. Deformation Parameters

The determination of the deformation parameters was performed during the uniaxial compression test. These tests were performed using ISRM guidelines [31]. A hydraulic press with automatic piston feed, force registration, and longitudinal and transverse strain registration (Walter + Bai AG, Löhningen, Switzerland) was used to investigate the deformation parameters. Both the modulus of elasticity (E) and Poisson's ratio (ν) were determined over the full range of linearity of the stress–strain characteristics (the so-called average modulus) of a given specimen. The modulus of elasticity was determined as the ratio of the difference in compressive stress difference ($\sigma_2 - \sigma_1$) to the relative longitudinal strain ($\varepsilon_2 - \varepsilon_1$) over a given measurement range using Equation (2):

$$E = \frac{\sigma_2 - \sigma_1}{\varepsilon_2 - \varepsilon_1}, \tag{2}$$

Furthermore, Poisson's ratio was determined as the ratio of transverse strain ($\varepsilon_\perp$) to longitudinal strain ($\varepsilon_\parallel$) in a given compressive stress range using Equation (3):

$$\nu = \frac{\varepsilon_\perp}{\varepsilon_\parallel}, \tag{3}$$

The deformation parameters were evaluated for three types of cylindrical core specimens: cores drilled in the direction of concrete casting and cores drilled in the direction perpendicular to concreting in the upper and lower parts of the cubic specimen. In general, 80 cylindrical samples were prepared to determine the deformation parameters of the SCCs.

2.2.5. Ultrasound Test

Ultrasound examinations were performed using a PunditLab kit from the Proceq manufacturer (Schwerzenbach, Switzerland) with 54 kHz transducers with an automatic transmitter voltage. Prior to each use, the device was calibrated on a dedicated calibration rod with an ultrasound wave transit time of 25.4 μs. Measurements were performed with a coupling agent in the form of a chemically passive polyacrylate gel; both transducers' heads were covered with a layer of gel before each measurement. The specimens were divided into two types: specimens for determining the mechanical properties of 150 mm × 150 mm × 150 mm cubes and specimens for determining the mechanical properties of 50 mm × 100 mm cylinders. For samples from both groups, the testing fields were prepared via gentle grinding. For samples in group 1, tests were made only for side surfaces in the central part of the samples, while for samples in group 2, four sections were defined: two along the direction of concrete casting and one in each of the upper and lower zones of the sample.

For each test section of a given sample, 3 measurements were performed and the average values of the ultrasound wave transit velocity (v) between the transducer heads were presented as the results. The tests were performed in relation to the standards EN 12504-4 [32] and EN 13791 [33].

2.2.6. Sclerometric Tests

The sclerometric tests were conducted with a Proceq N-type Schmidt sclerometer (Switzerland). The Schmidt sclerometer allows for the surface hardness of the concrete to be measured from the indentation of an incident mass after its collision with the surface. In order to ensure proper stability of the tested samples and thus accuracy of the results, the samples were mounted in a vice. The tests were performed on the same surfaces as the ultrasound examinations. The testing fields were prepared via gentle grinding to achieve smoothness and flatness. For each test surface, nine measurements were performed and their results, properly arithmetically mediated, provided the sclerometric rebound index. The tests were performed in relation to the standards EN 12504-2 [34] and EN 13791 [33].

2.2.7. X-ray Computed Tomography

X-ray computed tomography (X-ray CT) illustrating the distribution of PET flakes in each SCC was performed for selected samples. X-ray CT was performed using a GE (Boston, MA, USA) Phoenix v-tome-x m device.

The GE Phoenix v-tome-x m device together with VG Studio Max (version 2022.1, Heidelberg, Germany) software, enables the reconstruction and analysis of the internal structure of the element tested on the basis of a series of X-ray images obtained during a 360° rotation of the sample [35]. Table 4 describes the basic parameters of the test and the values used. Cuboid samples having dimensions of 50 mm × 50 mm × 100 mm, which were obtained by cutting 100 mm cube samples, were examined.

Table 4. Parameters of the X-ray CT test.

Parameter	Value
Source voltage	160 kV
Source current	180 μA
Filter	0.2 mm copper (Cu) filter
Exposure time	250 ms
Number of X-ray pictures used to reconstruct a 3D model	3100

Beam hardening correction, automatic geometry calibration, and geometry optimization algorithms were applied during the reconstruction process. After the image reconstruction, separation of the PET flakes in a given sample from the 3D model was performed.

3. Results

3.1. Fresh Mixture Properties

The properties of the fresh SCC mixtures were determined using filling ability (slump flow) and passing ability (L-box) tests. Furthermore, the fresh visual segregation index was assessed. On the basis of the results, all mixtures were classified according to the European Guidelines for SCC [36]. As SCC with a 20% PET replacement did not meet the requirements of self-compacting concrete, but it was still flowable concrete, we refer to it as almost self-compacting concrete (ASCC). Moreover, as it was not self-compacting, the ASCC-P-20 mixture was carefully vibrated. Detailed test results are given in Table 5.

Table 5. Properties of the fresh SCC mixtures.

Fresh Mixtures Properties	SCC-P-0	SCC-P-5	SCC-P-10	SCC-P-15	ASCC-P-20
Slump flow (mm)	725	710	660	600	510
Slump flow class—SF	SF 2	SF 2	SF 2	SF 1	-
Slump flow time T_{50} (s)	2.5	4.5	5.5	7.0	3.0
Viscosity class—VS	VS 2	VS 2	VS 2	VS 2	VS 2
L-box ratio	0.88	0.87	0.86	0.81	0.62
L-box class	PL 2	PL 2	PL 2	PL2	-
Fresh visual stability index	0	1	1	1	2

The main finding of the flow tests was that regardless of the investigated parameter, deterioration increased with the addition of PET flakes. The amount of superplasticizer was controlled to avoid segregation of the SCC mixture. Thus, despite the higher proportion of superplasticizer, the slump flow decreased with an increased proportion of PET flakes; simultaneously, the viscosity of the mixture (the time of slump flow up to 50 cm diameter) increased. The lower the natural aggregate content in the mix, the more difficult it was to maintain sufficient stability of the mix; therefore, bleeding was observed in all mixes with the addition of PET flakes. Consequently, due to the high probability of significant segregation, the mixture with 20% PET in relation to the fine aggregate mass did not reach the self-compacting guidelines. A reduction in the stability of the mixture was also observed during the L-box test. The L-box ratio deteriorated with the addition of PET flake aggregate as a result of the higher viscosity of the mixture. The authors of other studies on self-compacting mixes with PET aggregates drew analogous conclusions [37–39].

3.2. Destructive Tests of Hardened Concrete Properties

The present research program involved the preparation of self-compacting concretes with recycled aggregate in the form of PET waste flakes. A series of tests were conducted on samples made from such concretes, including the determination of compressive strength, splitting tensile strength, modulus of elasticity, and Poisson ratio through the destructive test method. In addition, a series of non-destructive tests were performed, which are described in Section 3.3. The results of the destructive tests are presented graphically using box plots in Figure 4.

In whisker-box plots, the boxes are determined by quartiles (25% and 75%), the midpoint represents the mean of the results, and the minimum and maximum values of the test series are presented as whiskers. The results of the mechanical and deformation parameters are presented in relation to the type of mixture.

Compressive strength appeared to be significantly affected by the addition of PET waste flakes, starting with a 10% replacement of the fine aggregate (Figure 4a). The behavior of the control samples and those containing 5% PET aggregate may be considered equivalent, and the minor variations that appeared were a consequence of the scattering of the result. However, it should be noted that the scatter of compressive strength (COV) results for SCC-P-5 was higher than for SCC-P-0. This could have been associated with slight bleeding of a mixture. Sample analysis indicated that the results were symmetrically distributed relative to the average for concretes up to 10% PET replacement in the mix. For the SCC-P-15 and ASCC-P-20 concretes, the sample population was left-handed asymmetric, despite the overall lower values. Thus, the majority of the results were higher than the average, and the mean result decreased due to the high outliers (minimum value). The failure pattern of all samples after the compression test was consistent with the standard [27]; therefore, there was no reason to dismiss any results. Compared to the average compressive strength value for SCC-P-0, reductions of approximately 1.3%, 35.2%, 38.9%, and 47.8% occurred for concretes with 5%, 10%, 15%, and 20% PET replacements, respectively. The decrease in compressive strength may have been primarily related to the deterioration of the flow properties of SCC with recycled PET aggregate.

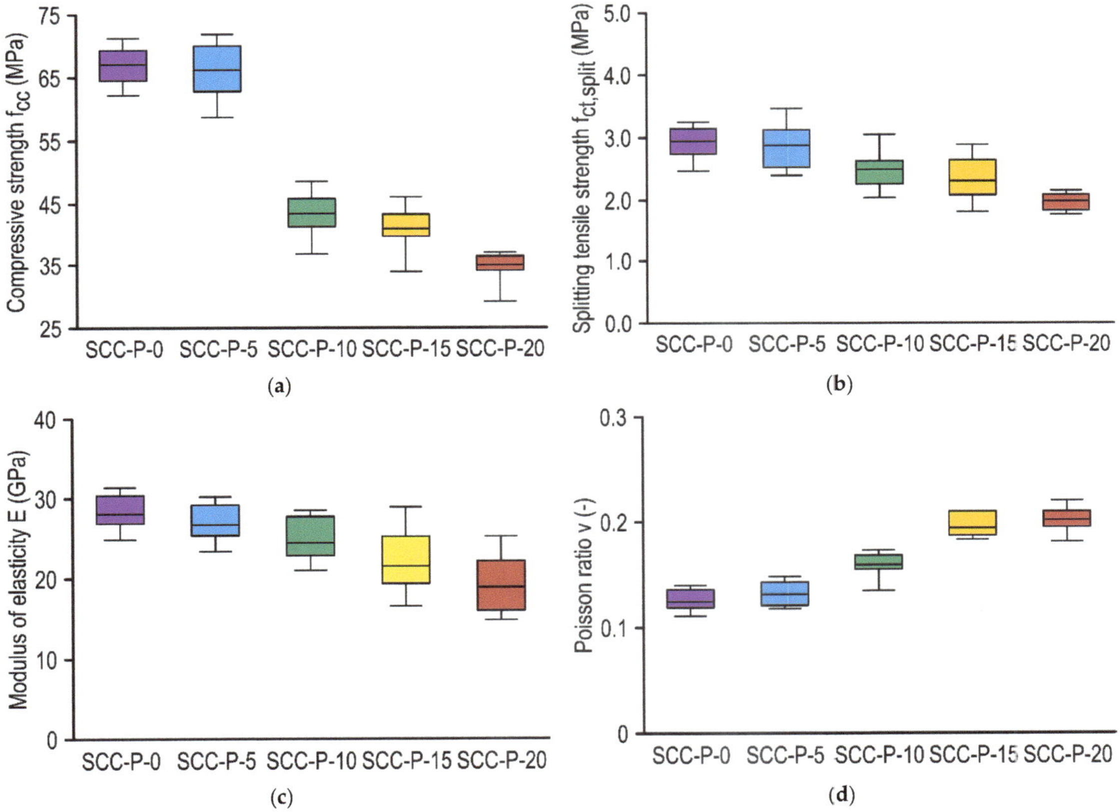

Figure 4. Results of hardened concrete parameters determined through destructive tests: (**a**) compressive strength, (**b**) splitting tensile strength, (**c**) modulus of elasticity, and (**d**) Poisson ratio.

The splitting tensile strength decreased with the addition of PET aggregates to the SCC (Figure 4b). The decrease was similar in its course to the reduction in compressive strength but was not as significant. In terms of splitting tensile strength, SCC-P-0 and SCC-P-5 exhibited insignificantly different values that resulted from the scatter of the results. Some of the SCC-P-5 samples received higher splitting tensile strength values compared to the control samples; however, in general, its population was right asymmetric. Although the splitting tensile strength of SCC-P-15 was inferior to SCC-P-10, the distribution of the results in the populations suggested that these concretes performed similarly. The results for SCC-P-20 were significantly reduced compared to the control concrete; moreover, it is worth noting that the results did not show significant scatter and were symmetrically distributed around the mean. Compared to the mean value of the tensile strength splitting of SCC-P-0, reductions of 2.4%, 15.6%, 21.7%, and 32.5% occurred for concretes with PET replacements of 5%, 10%, 15%, and 20%, respectively.

Young's modulus determined in compression decreased with the addition of PET aggregate (Figure 4c). Due to the adopted test method, the results deteriorated compared to those available when testing according to the EN procedure [40]. The reduction in the modulus of elasticity of SCC-P-5 compared to the control concrete was more significant than for the parameters mentioned above and reached almost 5%. It is also worth noting that the test results for all concretes except SCC-P-20 were left-handed asymmetric; therefore, most of the results were above the mean. The largest scatter of results was obtained for concretes with the highest PET aggregate content (15% and 20%). Compared to the average modulus of elasticity value of SCC-P-0, reductions of 5.0%, 13.2%, 23.5%, and 32.7% on average were observed for concretes with PET replacements of 5%, 10%, 15%, and 20%, respectively.

The higher the PET percentage in the mixture, the higher the Poisson ratio in compression (Figure 4d). Its change corresponded to the other parameters determined. Compared to the average Poisson ratio of SCC-P-0, reductions of 5.4%, 27.4%, 55.5%, and 61.7% were found for concretes with 5%, 10%, 15%, and 20% PET replacements, respectively.

In the literature, there is some research on destructive tests of self-compacting concretes with PET aggregate in terms of compressive strength, splitting tensile strength, and modulus of elasticity [37,41–43]. Nevertheless, the issue has been significantly better recognized in ordinary concretes or research has been conducted on concretes with the addition of PET fibers. In general, the trends shown in this study coincided with the literature; a broader comparison is presented in Section 4.

3.3. Non-Destructive Tests of Hardened Concrete Properties

Non-destructive test series were performed on the samples, including the determination of the rebound number and ultrasound wave transition velocity. As in the case of destructive tests, the results are presented graphically using box plots (Figure 5). The test results are presented in relation to the type of mixture.

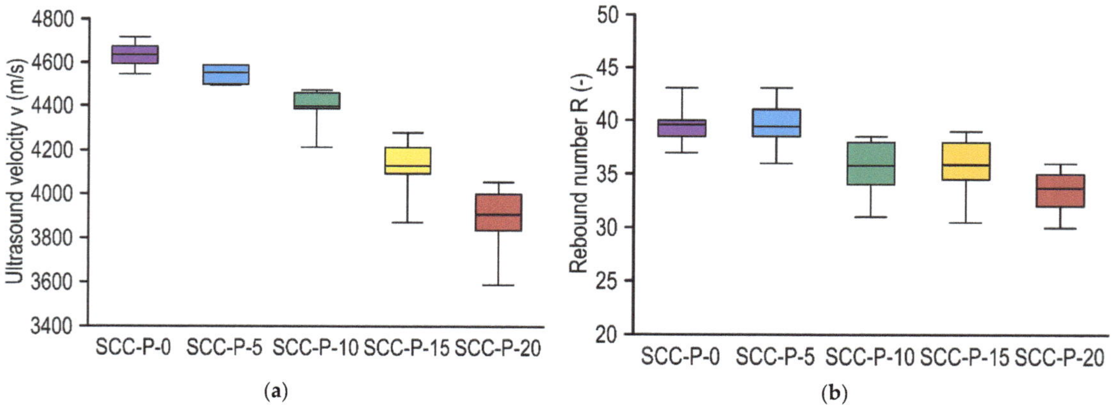

Figure 5. Non-destructive test results: (**a**) ultrasound velocity v and (**b**) rebound number R.

For both sclerometric and ultrasound tests, a decrease in the test result value was observed with increasing PET aggregate content. For ultrasound tests, the relationship of the decrease in wave transition velocity with the increase in PET flake content was more linear than for the sclerometric tests. One can also observe that the decrease in the ultrasound wave velocity was accompanied by an increase in the coefficient of variation, which was primarily caused by registering single results significantly lower than the average (especially for SCC-P-10, SCC-P-15, and SCC-P-20). The reduction in the velocity of the ultrasound wave transition for successive concretes was well correlated with the decrease in density as the proportion of PET replacement in the concretes increased. Compared to the average ultrasound wave transition velocity of SCC-P-0, reductions of 2.0%, 5.2%, 10.0%, and 15.8% on average occurred for concretes with 5%, 10%, 15%, and 20% PET replacements, respectively. In the case of the bulk density of hardened concrete, a density of 2360.3 kg/m^3 was recorded for SCC-P-0 and, on average, reductions of 2.4%, 3.8%, 5.7%, and 9.5% occurred for concretes with 5%, 10%, 15%, and 20% PET replacements, respectively.

In the sclerometric tests, the variations were not linear; for SCC specimens without and with 5% PET replacement, the rebound numbers differed minimally and the scatter was slightly less for the control specimens. For SCC-P-10 and SCC-P-15, a significant decrease was reported relative to the control samples, and the values were also close to each other. Significantly lower results were observed, which may have been due to the location of the PET flakes close to the sample surface, which may have directly affected the decrease in surface hardness, and the highest coefficient of variation was observed here. The lowest rebound number results were recorded for SCC-P-20 specimens. Compared to the average rebound number of SCC-P-0, decreases of 0.4%, 9.6%, 9.4%, and 14.9% on average were observed for concretes with 5%, 10%, 15%, and 20% PET replacements, respectively. Similar correlations were obtained for compressive strength; hence, the conclusion is that the cause may have also been due to a significant deterioration in the flow properties of the fresh mixture.

In the literature, there are individual works on non-destructive testing of self-compacting concretes [37,43], although they refer only to ultrasound tests. The issues related to non-destructive tests are also the subject of the design code for structural concretes, allowing for the development of correlation relationships between destructive and non-destructive tests, which are presented and discussed in Section 4.

3.4. *X-ray CT Scans*

X-ray CT scans provided a view of the distribution of PET flakes in the concrete specimens. To visualize them more clearly in the figures, an analysis was performed to isolate the voxels in a given grey value range corresponding to the PET material. Figure 6 presents transverse and longitudinal sections through the tomographic images in which white was used to distinguish air pores in the material and cyan was used to distinguish PET flakes. The cyan around certain pores, e.g., in the SCC-P-10 image, was related to the grey value of the cement matrix around the air bubbles. Analysis of the image allowed us to conclude that PET flakes were evenly distributed throughout the sample volume and did not clump together, which was connected with the proper stability of the mixture. In PET concretes, a reduced density material layer was observed in the top part of the samples. This phenomenon was evidence of concrete mixture bleeding, which was also observed during the determination of the fresh visual stability index. The layer increased proportionally to the observed bleeding during the mix examination (up to approximately 2 mm in the tomographic images). PET flakes also appeared in the reduced density layer; however, no increased flake distribution was observed in the higher part of the samples. Another phenomenon worth noting is that air pores tended to be trapped underneath horizontal flakes.

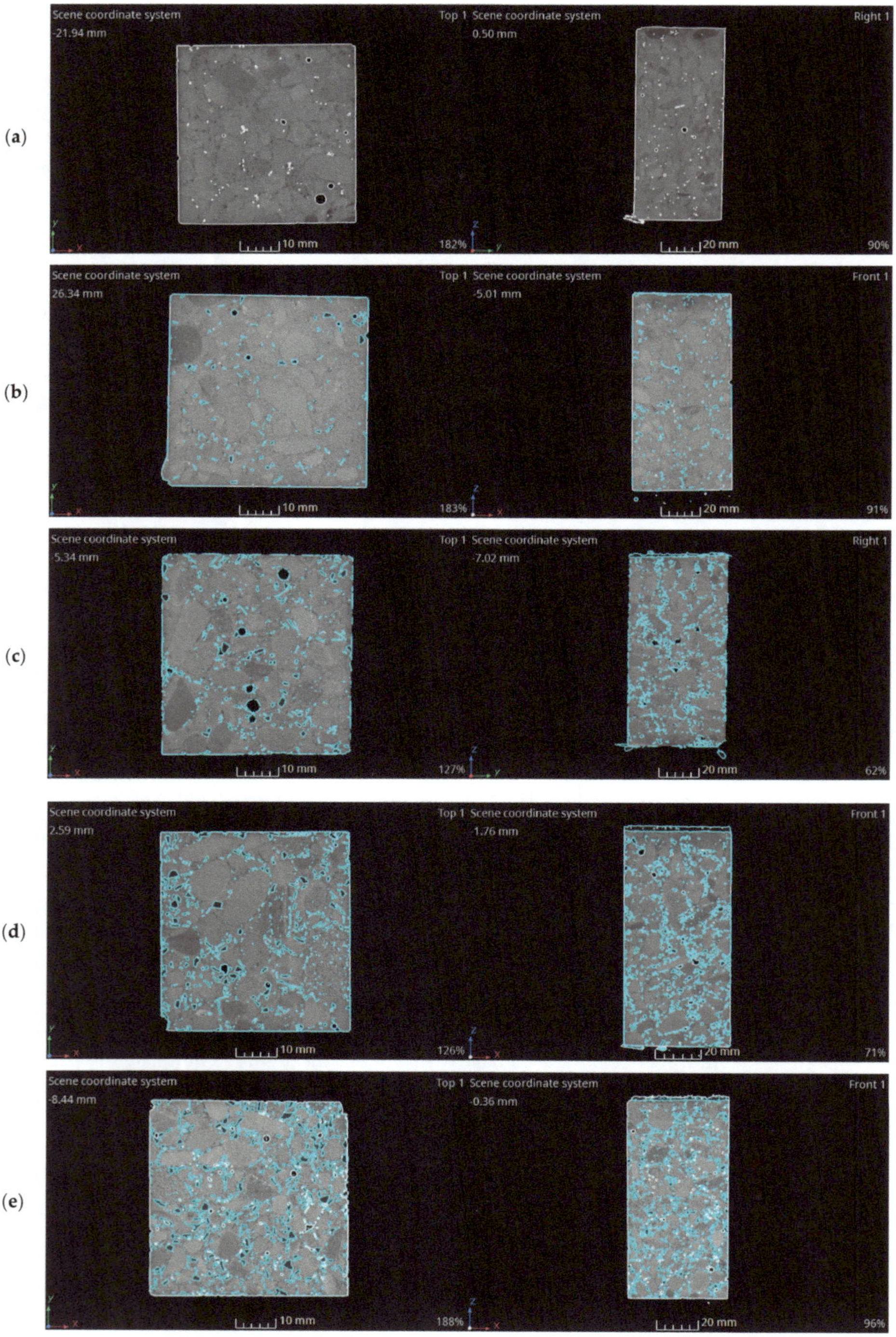

Figure 6. X-ray CT scans—identification of PET flakes in a concrete sample: (**a**) SCC-P-0, (**b**) SCC-P-5, (**c**) SCC-P-10, (**d**) SCC-P-15, and (**e**) SCC-P-20.

4. Discussion

4.1. A Discussion of the Results with the Literature Findings

In concrete materials technology, compressive strength is taken as the basic mechanical characteristic. In the present study, it was found that the modification of the fine aggregate to up to 5% of the PET aggregate had no significant effect on the compressive strength of the SCC. A considerable reduction in compressive strength was observed for mixtures containing at least 10% PET flakes relative to the mass of sand, while for subsequent percentages, the effect was not as pronounced. These results were compared with data from the available literature and presented in a summary graph in Figure 7. The following types of concretes were analyzed: traditional concrete with PET flakes (red tone lines) [12,18,42,43], self-compacting mixtures with PET flakes (blue tone lines) [37,39], and self-compacting mixture with PET granules (green line) [44]. The results of the present study are given in black. In the case of the literature studies that assumed volumetric substitution of fine aggregate, the proportion of substitution was converted to mass replacement relative to sand mass in order to properly compare the results of the study.

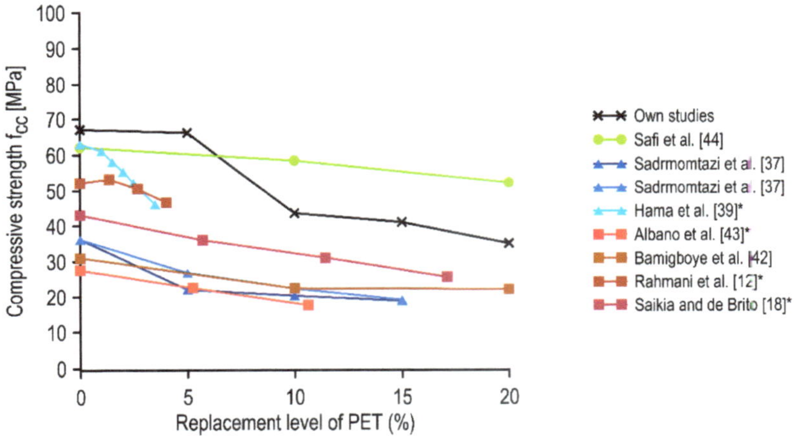

Figure 7. Variations of 28-day compressive strength of different concretes with respect to the replacement level of PET. * Converted from volume replacement.

The mixture containing PET granules did not show a significant decrease in compressive strength up to 10% PET granule content. This may have been related to the favorable shape of this PET aggregate, as well as the generally higher compressive strength of concrete (high strength concretes are generally characterized by better quality).

In general, SCC mixes with PET flakes exhibited a reduction in compressive strength, even in the case of a small replacement percentage. It should be noted that the control concretes varied in compressive strength between experiments; therefore, the comparison was made on the variations rather than the values themselves. Self-compacting concrete appeared to be more prone to a decrease in compressive strength as the proportion of PET aggregate increased. This was related to the deterioration of flow properties and the increased sensitivity of self-compacting mixtures to segregation when adequate technological rigor is not maintained. In the case of the highest replacement percentage of fine aggregate in Hama et al. [39], which was an approximately 5% mass replacement, there was a 27% decrease in compressive strength compared to the control concrete. Sadrmomatzi et al. [37] observed a reduction of approximately 40% at a sand mass replacement of 10%. In the present study, for concrete with a 10% replacement of fine aggregate with PET aggregate, a decrease of 35% was observed. The greater reduction in compressive strength found in [37] compared to the findings of our study could have been the result of the larger PET fraction used in the mixtures (up to 4.75 mm), which corresponded to the replacement of coarse aggregate. In the analyzed experiments, a further increase in PET content did not affect the reduction of compressive strength as significantly.

The compressive strength of traditional concretes also tended to decrease as the proportion of PET in the mix increased. For lower-strength normal concretes, these reductions were more significant, although not as pronounced for self-compacting lower-strength concretes. In the case of the study by Rahmani et al. [12], the total reduction achieved was up to 11% (comparing concrete with approximately 5% replacement by sand mass to the control one). However, it is worth noting that for a minor sand mass substitution (approximately 1.5%), the compressive strength of concrete increased. In the cases of the other researchers analyzing normal concrete with PET aggregate, a contribution of 10% by weight of sand resulted in an average reduction in compressive strength of 31%, which was slightly lower than the result obtained in our studies.

A noteworthy hypothesis is that the decrease in strength of concrete with PET recycled aggregate is related to the increased softness of the material compared to natural aggregate, which was proposed by Faraj et al. [17]. The review authors stated that during loading, PET aggregate behaves as air voids inside the cement matrix, resulting in the initiation of cracks around the particles. This claim was verified using a tomographic image of the SCC-P-15 fragment subsequent to compressive strength tests, which illustrated the propagating cracks surrounding the particles (Figure 8).

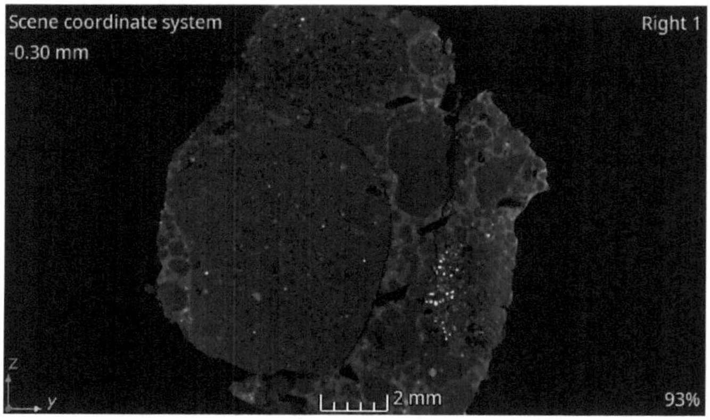

Figure 8. X-ray CT scan—section of a fragment of an SCC-P-15 sample after compression.

Furthermore, on the basis of SEM studies, it was concluded by other authors that as the proportion of PET aggregate increases, the porosity of concrete and the width of the poor quality interfacial transition zone (ITZ) increases. Kangavar et al. [19] reported that the structure of concrete incorporated with PET granules appeared to be more cavernous with larger air bubbles when the substituted volume exceeded 30%. In comparison to other types of plastic, PET aggregate generates the smallest ITZ with lower quality due to its surface roughness resulting from the processing of this material [45]. Based on the literature, the lower compressive strength of concretes with higher proportions of PET is related to the lower bond between the cement matrix and PET and the increased porosity of concrete, which is also reflected in the lower density of concrete.

The second characteristic that is of great importance for concrete composites is tensile strength. In the present study, analogous to compressive strength, it was found that substitution of fine aggregate to PET aggregate up to 5% did not significantly affect this property. The subsequent decrease in the splitting tensile strength with increasing PET content can be considered to be relatively linear. These results were compared with available literature data and presented in a summary graph in Figure 9, where the amount of data to be compared was less than for compressive strength due to the significantly different research approaches of other authors. Comparisons were made between traditional concrete mixes with PET flakes (red tone lines) [12,18,42,43] and self-compacting mixes with PET flakes (blue tone lines) [37]. The results of the present study are given in black. In the case of the studies that assumed volumetric substitution of fine aggregate, the proportion of substitution was converted to mass replacement relative to sand mass in order to properly compare the results of the study.

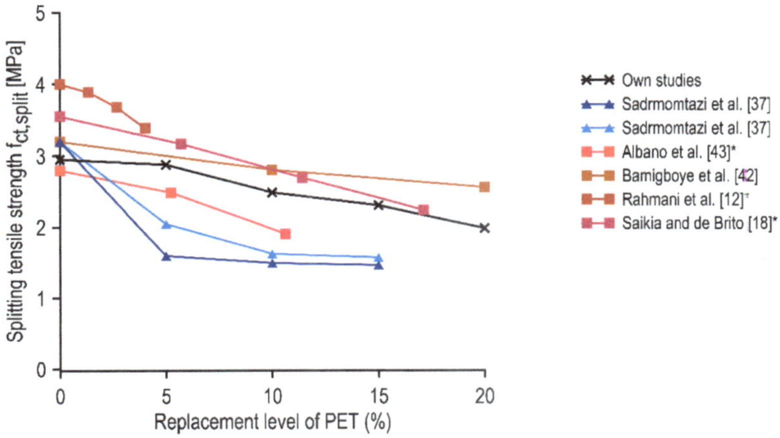

Figure 9. Variation of 28-day splitting tensile strength of different concrete with respect to the replacement level of PET. * Converted from volume replacement.

All traditional concrete mixes with PET aggregate exhibited a similar linear decrease in splitting tensile strength with an increased proportion of PET aggregate. In the case of the replacement of 5% fine aggregate with PET aggregate, both Albano et al. [43] and Saikia et al. [18] obtained a decrease in splitting tensile strength of about 11%, while Rahmani et al. [12] reported a greater decrease of 17% compared to the control concrete. In [37], for SCC, the highest decrease in splitting tensile strength (for a 5% replacement of sand with PET aggregate) was recorded, up to 36% for samples with fly ash content and 50% in the case of samples containing only cement as a binder. In all the analyzed studies, the further decrease in splitting tensile strength for higher PET aggregate content was not as significant. It should be noted that in the case of concretes with PET aggregate, the increase in splitting tensile strength that is visible in the application of concrete with PET fibers, e.g., in [13], is not usually observed. This is a direct consequence of the different, more spherical, and irregular shapes of the PET aggregate used.

The ultrasound results obtained were similar to those reported in the literature. Albano et al. [43] tested normal concretes with PET aggregate and found wave transition velocities of 3000 m/s to 4200 m/s, which was lower than in this study. However, they also reported a decrease in wave transition velocity with increasing PET flake content and the magnitude of these variations did not change with concrete aging (tests conducted after 7 to 60 days of curing). For self-compacting concrete, Sadrmomtazi et al. [37] found that the control SCC had a wave transition velocity of 4700 m/s, while for concretes containing a different proportion of PET additives, they were 3830 m/s to 4580 m/s; these findings are very similar to those obtained herein.

The empty spaces created by the air pores and PET particles attenuate the ultrasound wave due to the acoustic impedance. When the ultrasound wave passes through different media (materials), it is partially reflected and transmitted; therefore, its velocity decreases. First of all, the decrease in the ultrasound wave velocity is caused by the reduction in concrete density [46]. Moreover, as noted by Albano et al. [43], due to the fact that the ultrasonic wave velocity is a function of the elastic properties and volume concentrations of the constituents when replacing natural fine aggregate with PET, the ultrasound wave velocity must decrease.

4.2. Comparison of NDT and DT Testing

The tests performed on the mixtures with different PET replacements indicated significantly varying results of ultrasonic wave transit velocity tests. Therefore, it was possible to correlate the compressive strength with the velocity of the ultrasound wave and relate these results to the standard basic curve presented in Figure 10a. The correlation of these results with a polynomial function can be described as strong (R^2 = 0.845). The basic curve, on the other hand, seemed to be strongly underestimated relative to the results obtained. However, it should be considered that the basic curve presented in EN 13791 [33] is the lower envelope and should, in principle, be placed below the results obtained from destructive tests. Furthermore, a standard procedure for the correction of the basic curve was performed, which involved the change in the basic curve by Δf (Equation (4)), which consists of the average difference obtained between the destructive tests results of the samples and the results obtained from the basic curve $\delta f_{m(n)}$ reduced by the product of the standard deviation s and a factor k_1 that depends on the number of pairs of results (in this research, k_1 = 1.48):

$$\Delta f = \delta f_{m(n)} - k_1 \cdot s, \tag{4}$$

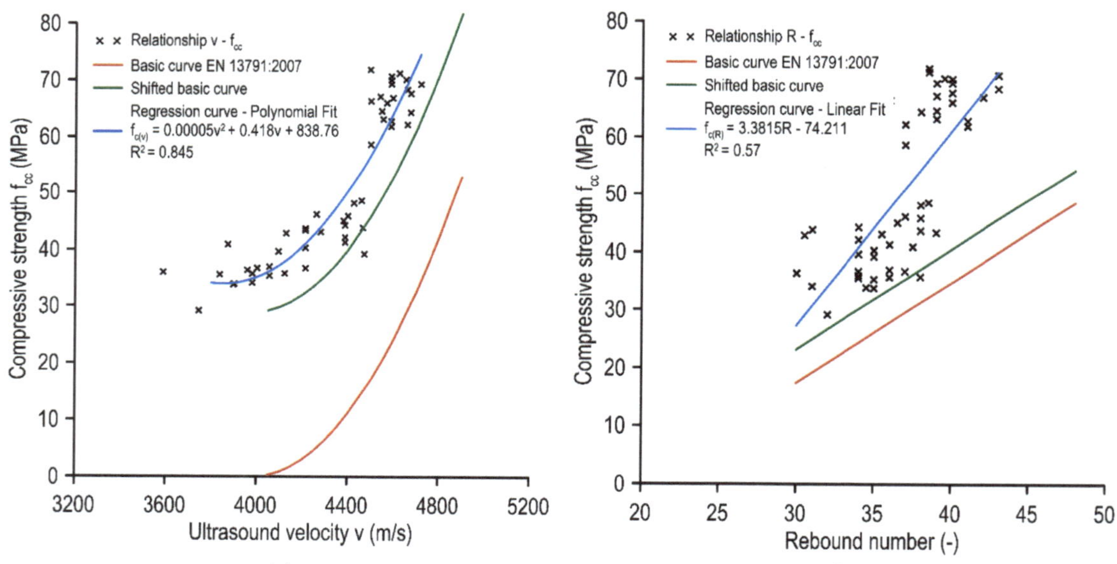

Figure 10. A comparison of non-destructive tests results to the compressive strength with normative basic curves: (**a**) ultrasound vs. compressive strength test results and (**b**) sclerometric vs. compressive strength test results.

The shifted basic curve provides a better representation of the relationships described. There is only a single point located under the curve (thus the estimation using this function would be highly secure). The analyses carried out indicated that it was necessary to adopt a limitation of the curve application (both shifted and correlation) similarly to the basic curve. For the shifted basic curve, it was 3900 m/s to 4800 m/s, and for the acquired correlation curve, it was 3700 m/s to 4600 m/s. Crucially, the correlation function exhibited a pattern very similar to that of the basic and shifted functions in terms of direction.

For sclerometric tests, analogous considerations were made with respect to the standard [9]. First, a correlation analysis of the NDT test results with DT was performed. A moderate-strength linear correlation was obtained (R^2 = 0.57). As in the case of ultrasound examinations, these results were summarized in a combined graph with a basic curve and a shifted basic curve (Figure 10b). It can be observed that the basic and shifted basic curves had a decreased slope compared to the curve derived from the linear regression obtained. For minor values of the rebound number, the difference in compressive strength gained from the standard basic curves compared to the correlation curve was not significant and it grew as the rebound number increased. The basic curve was found to be below all the measurements obtained and for the shifted basic curve, as for ultrasound, there was only a single result below the curve. Therefore, it might be applicable with a high level of safety (with the remark that the higher the rebound number, the more underestimated the result). It should be emphasized that sclerometric tests are conducted on the surface and do not comprehensively describe the strength properties of the tested materials. However, their simplicity and promptness favor the application of this method for regular in situ concrete quality control.

4.3. Implementation of PET Aggregate in New Generation Concretes

The use of PET aggregate in the form of flakes, which is a by-product of the recycling process (it should be noted that the fraction used in this study is not recyclable into granulate, which constitutes a valuable raw material), is far more beneficial from an environmental point of view than the use of granulates or specially prepared fibers. One can assume that the acquisition of PET flakes in fractions above 4.0 mm (used in the production of PET granulate) is inevitably connected with the production of waste in the form of smaller-fraction flakes and PET powder. Therefore, the costs associated with PET bottle treatment need to be borne anyway, which improves the economic profile of SCCs with PET aggregates compared to SCCs with PET fibers. However, as demonstrated in numerous studies, the use of PET fibers in small quantities (less than 2% by volume) has the potential to improve the mechanical properties of concrete, especially in terms of compressive strength and flexural strength, including SCCs [47,48]. Nevertheless, it should be emphasized that due to the hydrophobic nature of PET, the flow properties of SCC mixtures deteriorate in the case of both PET fibers addition and sand replacement with PET aggregate. Therefore, caution is required when implementing larger amounts of plastic into self-compacting mixes in order to ensure that segregation phenomena do not occur.

Given that the world's annual production of concrete is steadily increasing and currently reaches approximately 32 billion tons per year [1], and its production requires non-renewable natural resources, the exploration of substitutes, even if partial, is highly valuable. On average, about 300 kg of sand is used to produce one ton of concrete; hence, the annual consumption of this raw material is about 9.6 billion tons. Considering the replacement of 5% of natural sand with waste PET aggregate of the fraction below 2.0 mm, material savings of 480 million tons per year of this resource could be reached.

Nowadays, SCCs are considered to be a special type of concrete used in structures with dense reinforcement and complex geometries. However, given the further aspects of environmental and social assessment, it seems inevitable that the utilization of mixtures that do not require additional technological processes (such as vibration and troweling) will increase in demand. It is essential to investigate the incorporation of waste materials into new generation concretes in order to effectively implement these solutions in the future.

5. Conclusions

This paper discusses the possibility of using PET waste flakes of the fraction below 2.0 mm in self-compacting concrete mixtures. The presented results and their juxtaposition with data from the literature provide the following conclusions:

- PET aggregates in self-compacting concrete were uniformly distributed regardless of their content in the mix.
- The higher the PET content in the mix, the lower the density of the concrete, and therefore, the velocity of the ultrasonic wave and the rebound number of the sclerometer. The non-destructive test results acquired relatively good correlations with the compressive strength test results.

- The replacement of fine aggregate with PET recycled aggregate negatively affected the parameters of both fresh mixes and hardened concretes. In the case of the concrete with the highest aggregate replacement, the decreases compared to the control concrete were 9.5%, 47.8%, 32.5%, 32.7%, and 61.7% in terms of bulk density, compressive strength, splitting tensile strength, modulus of elasticity, and Poisson ratio, respectively. The ASCC-P-20 also did not meet the requirements of self-compaction.
- However, the replacement of 5% of natural aggregate with PET aggregate of the appropriately selected fraction was possible for SCC and did not cause significant variations in the physical and mechanical properties: 2.4%, 1.3%, 2.4%, 5.0%, and 5.4% in terms of bulk density, compressive strength, splitting tensile strength, modulus of elasticity, and Poisson ratio, respectively.

Research on the properties of SCC mixtures with PET aggregates in the form of flakes should be further developed to determine properties such as corrosion resistance, frost resistance, water penetration resistance, the durability of concrete or bond to reinforcing steel, and the bond to subsequent layers of concrete. In addition to the differences in the obtained basic research results relative to the literature data, as important steps forward in this research, one can indicate the determination of correlation curves for non-destructive testing and the determination of the Poisson's ratio, which is not commonly referred to, but can be useful in the future structural design using SCC with PET (especially for FEM calculations).

Author Contributions: Conceptualization, J.J.-L., M.K., J.M., P.N. and W.Ł.; methodology, J.J.-L. and M.K.; validation, J.J.-L. and M.K.; formal analysis, J.J.-L. and M.K.; investigation, J.J.-L., M.K., J.M., P.N. and W.Ł.; resources, J.J.-L., M.K., J.M., P.N. and W.Ł.; data curation, J.J.-L. and M.K.; writing—original draft preparation, J.J.-L. and M.K.; writing—review and editing, J.J.-L. and M.K.; visualization, J.J.-L. and M.K.; supervision, J.J.-L. and M.K.; project administration J.J.-L. and M.K. All authors have read and agreed to the published version of the manuscript.

Funding: This research received no external funding.

Institutional Review Board Statement: Not applicable.

Informed Consent Statement: Not applicable.

Data Availability Statement: The data that support the findings of this study are available from the corresponding author upon request.

Acknowledgments: The authors would like to acknowledge PRT Radomsko Sp. z o.o (PET Recycling Team) for providing the PET waste material for this research.

Conflicts of Interest: The authors declare no conflict of interest.

References

1. *ISO/TC 71*; Strategic Business Plan for International Organization for Standardization Technical Committee 71. International Organization for Standardization: Geneva, Switzerland, 2021.
2. Hüsken, G.; Brouwers, H.J.H. Eco-SCC: From Theory to Practical Application. In Proceedings of the 1st International Conference on Sustainable Construction Materials: Design, Performance and Application, Ancone, Italy, 28–30 June 2010; UWM Center for By-Products Utlilization: Milwaukee, WI, USA, 2010; Volume 1, pp. 720–732.
3. Pacheco-Torgal, F.; Jalali, S.; Labrincha, J.A.; John, V.M. *Eco-Efficient Concrete*; Woodhead Publishing Limited: Cambridge, UK, 2013; ISBN 9780857094247.
4. Dybeł, P.; Kucharska, M. Multilayer Casting of Eco-Efficient Self-Compacting Concrete with Reduced Binder Content. *Materials* **2021**, *14*, 5685. [CrossRef]
5. Hájek, P.; Fiala, C.; Kynčlová, M. Life Cycle Assessments of Concrete Structures—A Step towards Environmental Savings. *Struct. Concr.* **2011**, *12*, 13–22. [CrossRef]
6. Marinkovic, S.B. Life Cycle Assessment (LCA) Aspects of Concrete. In *Eco-Efficient Concrete*; Woodhead Publishing: Sawston, UK, 2013; pp. 45–80. ISBN 9780857094247.
7. Wałach, D.; Dybeł, P.; Sagan, J.; Gicala, M. Environmental Performance of Ordinary and New Generation Concrete Structures—A Comparative Analysis. *Environ. Sci. Pollut. Res.* **2019**, *26*, 3980–3990. [CrossRef] [PubMed]
8. Rohden, A.B.; Garcez, M.R. Increasing the Sustainability Potential of a Reinforced Concrete Building through Design Strategies: Case Study. *Case Stud. Constr. Mater.* **2018**, *9*, e00174. [CrossRef]

9. Batayneh, M.; Marie, I.; Asi, I. Use of Selected Waste Materials in Concrete Mixes. *Waste Manag.* **2007**, *27*, 1870–1876. [CrossRef]
10. Greater Hyderabad Municipal Corporation. *Enhancing Circular Economy Perspectives—Plastic Waste Management Strategy and Action Plan for Greater Hyderabad Municipal Corporation;* Greater Hyderabad Municipal Corporation: Telangana, India, 2020.
11. Welle, F. Twenty Years of PET Bottle to Bottle Recycling—An Overview. *Resour. Conserv. Recycl.* **2011**, *55*, 865–875. [CrossRef]
12. Rahmani, E.; Dehestani, M.; Beygi, M.H.A.; Allahyari, H.; Nikbin, I.M. On the Mechanical Properties of Concrete Containing Waste PET Particles On the Mechanical Properties of Concrete Containing Waste PET Particles. *Constr. Build. Mater.* **2013**, *47*, 1302–1308. [CrossRef]
13. Kim, S.B.; Yi, N.H.; Kim, H.Y.; Kim, J.H.J.; Song, Y.C. Material and Structural Performance Evaluation of Recycled PET Fiber Reinforced Concrete. *Cem. Concr. Compos.* **2010**, *32*, 232–240. [CrossRef]
14. Ochi, T.; Okubo, S.; Fukui, K. Development of Recycled PET Fiber and Its Application as Concrete-Reinforcing Fiber. *Cem. Concr. Compos.* **2007**, *29*, 448–455. [CrossRef]
15. Thomas, L.M.; Moosvi, S.A. Hardened Properties of Binary Cement Concrete with Recycled PET Bottle Fiber: An Experimental Study. *Mater. Today: Proc.* **2020**, *32*, 632–637. [CrossRef]
16. Thorneycroft, J.; Orr, J.; Savoikar, P.; Ball, R.J. Performance of Structural Concrete with Recycled Plastic Waste as a Partial Replacement for Sand. *Constr. Build. Mater.* **2018**, *161*, 63–69. [CrossRef]
17. Faraj, R.H.; Hama, H.F.; Far, A.; Sherwani, H.; Hassan, B.R.; Karim, H. Use of Recycled Plastic in Self-Compacting Concrete: A Comprehensive Review on Fresh and Mechanical Properties. *J. Build. Eng.* **2020**, *30*, 101283. [CrossRef]
18. Saikia, N.; De Brito, J. Mechanical Properties and Abrasion Behaviour of Concrete Containing Shredded PET Bottle Waste as a Partial Substitution of Natural Aggregate. *Constr. Build. Mater.* **2014**, *52*, 236–244. [CrossRef]
19. Kangavar, M.E.; Lokuge, W.; Manalo, A.; Karunasena, W.; Frigione, M. Investigation on the Properties of Concrete with Recycled Polyethylene Terephthalate (PET) Granules as Fine Aggregate Replacement. *Case Stud. Constr. Mater.* **2022**, *16*, e00934. [CrossRef]
20. Irwan, J.M.; Othman, N.; Koh, H.B.; Aeslina, A.K.; Annas, M.M.K.; Asyraf, R.M. Overview of an Environmental Friendly Technique of Using Fine Aggregate in Concrete from Polyethylene Terephthalate Bottles. In Proceedings of the Regional Symposium on Engineering and Technology 2011 Kuching, Sarawak, Malaysia, 21–23 November 2011.
21. Wałach, D. Analysis of Factors Affecting the Environmental Impact of Concrete Structures. *Sustainability* **2021**, *13*, 204. [CrossRef]
22. *EN 933-1:2012*; Tests for Geometrical Properties of Aggregates—Part 1: Determination of Particle Size Distribution—Sieving Method. European Committee for Standarization: Brussels, Belgium, 2012.
23. *EN 197-1*; Cement—Part 1: Composition, Specifications and Conformity Criteria for Common Cements. European Committee for Standarization: Brussels, Belgium, 2012.
24. *EN 12350-8*; Testing Fresh Concrete—Part 8: Self-Compacting Concrete—Slump-Flow Test. European Committee for Standarization: Brussels, Belgium, 2010.
25. *EN 12350-10*; Testing Fresh Concrete—Part 10: Self-Compacting Concrete—L-Box Test. European Committee for Standarization: Brussels, Belgium, 2010.
26. *ASTM C1611*; Standard Test Method for Slump Flow of Self-Consolidating Concrete. ASTM International: West Conshohockedn, PA, USA, 2018.
27. *EN 12390-3*; Testing Hardened Concrete. Compressive Strength of Test Specimens. European Committee for Standarization: Brussels, Belgium, 2009.
28. *EN 12390-1*; Testing Hardened Concrete—Part 1: Shape, Dimensions and Other Requirements for Specimens and Moulds. European Committee for Standarization: Brussels, Belgium, 2013.
29. *EN 12390-2*; Testing Hardened Concrete—Part 2: Making and Curing Specimens for Strength Tests. European Committee for Standardization: Brussels, Belgium, 2019.
30. *EN 12390-6*; Testing Hardened Concrete—Part 6: Tensile Splitting Strength of Test Specimens. European Committee for Standarization: Brussels, Belgium, 2009.
31. ISRM Turkish National Group. *The Complete ISRM Suggested Methods for Rock Characterization, Testing and Monitoring: 1974–2006;* Ulusay, R., Hudson, J.A., Eds.; Commission on Testing Methods, International Society of Rock Mechanics: Ankara, Turkey, 2007; ISBN 9789759367541 9759367548.
32. *EN 12504-4*; Testing Concrete in Structures—Part 4: Determination of Ultrasonic Pulse Velocity. European Committee for Standarization: Brussels, Belgium, 2021.
33. *EN 13791*; Assessment of In-Situ Compressive Strength in Structures and Precast Concrete Components. European Committee for Standardization: Brussels, Belgium, 2007.
34. *EN 12504-2*; Testing Concrete in Structures—Part 2: Non-Destructive Testing—Determination of Rebound Number. European Committee for Standarization: Brussels, Belgium, 2021.
35. Kaczmarczyk, G.P.; Kinasz, R.; Bilozir, V.; Bidenko, I. Application of X-ray Computed Tomography to Verify Bond Failures Mechanism of Fiber—Reinforced Fine—Grain Concrete. *Materials* **2022**, *15*, 2193. [CrossRef]
36. The Self-Compacting Concrete European Project Group. *The European Guidelines for Self Compacting Concrete: Specification, Production and Use;* Association House: Surrey, UK, 2005.
37. Sadrmomtazi, A.; Dolati-Milehsara, S.; Lotfi-Omran, O.; Sadeghi-Nik, A. The Combined Effects of Waste Polyethylene Terephthalate (PET) Particles and Pozzolanic Materials on the Properties of Selfcompacting Concrete. *J. Clean. Prod.* **2016**, *112*, 2363–2373. [CrossRef]

38. Mohammed, M.K.; Al-Hadithi, A.I.; Mohammed, M.H. Production and Optimization of Eco-Efficient Self Compacting Concrete SCC with Limestone and PET. *Constr. Build. Mater.* **2019**, *197*, 734–746. [CrossRef]
39. Hama, S.M.; Hilal, N.N. Fresh Properties of Self-Compacting Concrete with Plastic Waste as Partial Replacement of Sand. *Int. J. Sustain. Built Environ.* **2017**, *6*, 299–308. [CrossRef]
40. *EN 12390-13*; Testing Hardened Concrete—Part 13: Determination of Secant Modulus of Elasticity in Compression. European Committee for Standarization: Brussels, Belgium, 2021.
41. Faraj, R.H.; Sherwani, A.F.H.; Daraei, A. Mechanical, Fracture and Durability Properties of Self-Compacting High Strength Concrete Containing Recycled Polypropylene Plastic Particles. *J. Build. Eng.* **2019**, *25*, 100808. [CrossRef]
42. Bamigboye, G.O.; Tarverdi, K.; Umoren, A.; Bassey, D.E.; Okorie, U.; Adediran, J. Evaluation of Eco-Friendly Concrete Having Waste PET as Fine Aggregates. *Clean. Mater.* **2021**, *2*, 100026. [CrossRef]
43. Albano, C.; Camacho, N.; Hernández, M.; Matheus, A.; Gutiérrez, A. Influence of Content and Particle Size of Waste Pet Bottles on Concrete Behavior at Different w/c Ratios. *Waste Manag.* **2009**, *29*, 2707–2716. [CrossRef] [PubMed]
44. Safi, B.; Saidi, M.; Aboutaleb, D.; Maallem, M. The Use of Plastic Waste as Fine Aggregate in the Self-Compacting Mortars: Effect on Physical and Mechanical Properties. *Constr. Build. Mater.* **2013**, *43*, 436–442. [CrossRef]
45. Abu-Saleem, M.; Zhuge, Y.; Hassanli, R.; Ellis, M.; Rahman, M.; Levett, P. Evaluation of Concrete Performance with Different Types of Recycled Plastic Waste for Kerb Application. *Constr. Build. Mater.* **2021**, *293*, 123477. [CrossRef]
46. Tomczak, K.; Jakubowski, J.; Kotwica, Ł. Self-Sealing Process Evaluation Method Using Ultrasound Technique in Cement Composites with Mineral Additives. *Materials* **2020**, *13*, 3336. [CrossRef]
47. Al-Hadithi, A.I.; Hilal, N.N. The Possibility of Enhancing Some Properties of Self-Compacting Concrete by Adding Waste Plastic Fibers. *J. Build. Eng.* **2016**, *8*, 20–28. [CrossRef]
48. Al-Hadithi, A.I.; Noaman, A.T.; Mosleh, W.K. Mechanical Properties and Impact Behavior of PET Fiber Reinforced Self-Compacting Concrete (SCC). *Compos. Struct.* **2019**, *224*, 111021. [CrossRef]

materials

Article

Orthogonal Experimental Study on Concrete Properties of Machine-Made Tuff Sand

Dunwen Liu, Wanmao Zhang, Yu Tang *, Yinghua Jian and Yongchao Lai

School of Resources and Safety Engineering, Central South University, Changsha 410083, China; dunwen@csu.edu.cn (D.L.); 205512136@csu.edu.cn (W.Z.); jyh__0412@csu.edu.cn (Y.J.); 215512141@csu.edu.cn (Y.L.)
* Correspondence: tangyu12@csu.edu.cn

Abstract: Machine-made sand instead of natural sand has become an inevitable choice for the sustainable development of the concrete industry. Orthogonal experiment and grey correlation analysis were used to investigate the performance of machine-made tuff sand concrete. The optimal concrete mix ratio of machine-made sand was obtained by orthogonal test and its working performance was verified. Grey correlation analysis was applied to compare the factors affecting the mechanical properties of the machine-made sand concrete. The test results show that the sand rate has the greatest degree of influence on slump and slump expansion. The mineral admixture has the greatest effect on the 7-day compressive strength of the concrete. Additionally, the water–cement ratio has the greatest influence on the 28-day compressive strength. The mechanical and working properties of the machine-made sand concrete reach the optimum condition when the mineral admixture is 20%, the sand rate is 46%, the stone powder content is 10% and the water–cement ratio is 0.30. Comparing different fine aggregate concretes of similar quality, we conclude that the mechanical and working properties of tuff sand concrete and limestone sand concrete and river sand concrete are similar. The compressive strengths of the mechanism concrete show the greatest correlation with roughness and the least correlation with stone powder content. The stone powder content has almost no effect on the compressive strength of concrete when the stone powder content does not exceed a certain range. The results of the study point out the direction for the quality control of concrete with machine-made sand.

Keywords: machine-made sand; concrete; orthogonal experimental design; grey correlation analysis; compressive strength

Citation: Liu, D.; Zhang, W.; Tang, Y.; Jian, Y.; Lai, Y. Orthogonal Experimental Study on Concrete Properties of Machine-Made Tuff Sand. *Materials* **2022**, *15*, 3516. https://doi.org/10.3390/ma15103516

Academic Editor: Yuri Ribakov

Received: 23 April 2022
Accepted: 12 May 2022
Published: 13 May 2022

1. Introduction

Concrete is the most widely used material for building structures today. As China invests more in infrastructure construction, concrete consumption is still on the rise. Fine aggregates account for about one-third of the raw materials for concrete. Fine aggregates can be divided into natural sand and artificial sand [1]. River sand is the most widely utilized natural sand. In recent years, the production of natural sand is no longer able to meet the demand for sand in the construction industry. There is an urgent need to find other materials to replace natural sand [2]. In the middle of the last century, the United States and other developed countries began to apply machine-made sand in actual engineering. China started to use machine-made sand in an orderly and standardized way in the late 1970s. In China, the most-used artificial sand is machine-made sand. Compared with ordinary river sand, it has many advantages such as controlled raw material, stable performance, and low cost. It is suitable for promotion and application in the construction industry [3,4]. The raw materials of machine-made sand mainly include mine waste stone, tunnel excavation waste stone, mountain stone and pebbles. It is obtained from these raw materials by crushing, screening and dedusting.

There are many methods for designing the proportioning of machine-made sand concrete, and different proportioning methods have their own focus [5,6]. They are designed to meet specific performance requirements [7,8]. On the other hand, it is due to practical conditions such as raw materials and environmental conditions in the area [9]. The excellent quality of fine aggregates largely determines the concrete performance. Shen et al. [10] invented a method for the design of the mix ratio based on gradation optimization and volumetric analysis. Additionally, the ultra-high strength concrete of C120 grade was successfully prepared with sandstone machine-made sand. Ji et al. [11] found an extreme point where the compressive strength of concrete as the content of fines in the cementitious material increases from 0% to 30%. Li et al. [12] proposed that the stone powder in the machine-made sand has the effect of filling the pores of concrete and improving the gradation of aggregate particles. Menadi, Topcu and Eren et al. [13–15] proposed that the compressive and flexural strengths of the concrete with mechanism sand reached the maximum at 10% of stone powder. Above or below 10%, the machined sand concrete does not perform its best mechanical properties. Munoz et al. [16] compared the effect of various clay powders on strength. The effect of sodium montmorillonite and calcium montmorillonite on strength was significantly greater than that of kaolinite and illite. Chen et al. [17] formulated machanism sand concrete by replacing cement slurry with limestone rock powder in proportion to its volume. It was found that the incorporation of stone powder could significantly increase the strength of concrete at a fixed volume of slurry with different water–cement ratios. Felekoglu [18] compared washed river sand, washed machine-made limestone sand, machine-made limestone sand mixed with clay powder and machine-made limestone sand mixed with clay. Chouhan et al. [19] studied the effect of limestone waste on mechanical properties and durability of concrete. Partial or total replacement of concrete fine aggregates with waste materials resulted in increase in compressive and tensile strength of the colluvial sand specimens with the rise in waste material admixture. The durability and bonding properties of the specimens were also improved. Bravo et al. [20] carried out an in-depth analysis of the properties of natural and recycled aggregates. Compressive strength, splitting tensile strength, modulus of elasticity and abrasion resistance tests were also carried out. Wang et al. [21,22] studied the effect of different methylene blue (MB) values on the performance of low and high strength machine-made sand concrete. It was found that under fixed cement dosage, water–cement ratio, and water-reducing agent admixture, the compressive strength at each age increased and then decreased with increasing MB value when MB value > 1.8 g/kg. The impermeability performance showed a different pattern.

Today, the complexity of civil engineering has prompted the industry to develop a variety of new materials. In line with the concept of sustainability, the construction industry is trying to replace ordinary silicate cement concrete using geopolymer concrete [23–25]. Similarly, the formulation of high-performance concrete from machine-made sand is one of the concrete material sustainability studies [26,27]. Due to the wide distribution of limestone machine-made sand, a large number of studies focus on machine-made limestone sand [28–31], and less research on machine-made sand of other lithologies. River sand resources are scarce in the region of eastern China, which restricts the development of local engineering construction. However, tuff is abundant and the raw materials for sand making are widely available, which also includes tunnel rejects. The use of tuffaceous sand concrete in engineering practice is relatively small [32]. Additionally, it's working state and mechanical properties in engineering structural components are worth exploring in depth [33,34]. In addition, due to the large quality fluctuations in the production process of machine-made sand, there are fewer studies on the effect of machine-made sand quality on the performance of concrete [35]. Therefore, further research is needed on the effect of quality variation of machine-made sand of different lithologies on performance.

Therefore, in this study, the research on machine-made tuff sand concrete is carried out through the following aspects. Firstly, the factors influencing the high-performance machine-made sand concrete mix ratio are analyzed. These factors to be optimized are used as influencing factors for orthogonal tests. The test results are analyzed to find the optimum level of concrete performance under each influencing factor. The optimal machine-made sand concrete ratio is obtained, and the performance of concrete under the optimal ratio is verified. We used the grey correlation analysis to calculate the correlation between fine aggregate stone powder content, clay lump content, roughness and compressive strength. The main factors affecting the mechanical properties of the machine-made sand concrete were analyzed. Then, we compare the correlation matrices of tuff and limestone machine-made sand. The magnitude and laws of different factors on the concrete properties are analyzed to show the direction for the quality control of the machine-made sand concrete.

2. Materials and Methods

2.1. Materials

In this study, P.O 42.5 grade ordinary silicate cement was used for the cement. Its 28-day flexural strength was 7.5 MPa and compressive strength was 46.8 MPa, specific surface area was 348 m^2/kg. Fly ash used Grade II fly ash with a fineness of 45 μm sieve margin of 15.2%, a burn-off of 1.78%, and a water demand ratio of 101%. Slag powder using S95 grade slag powder, specific surface area 411 m^2/kg, 7-day activity index 83%, water content 0.85%, flow rate ratio of 98%. Water reducing agent is selected from a high efficiency water reducing agent. Polycarboxylic acid water reducing agent has the characteristics of low dosing and high water-reduction rate. The water reduction rate is 25%. In this study, the dosing range is 1.2%~1.8% of the cementitious material. The experimental mixing water is tap water, which meets the requirements of Chinese national standards. The fine aggregates are selected from machine-made tuff sand (number NH), machine-made limestone sand (number SH) and river sand (number HS). The physical index of fine aggregate is shown in Table 1. The coarse aggregate is tuffaceous gravel with nominal particle size of 5~20 mm, continuous gradation, apparent density of 2.63 g/cm^3 and bulk density of 1.52 g/cm^3. The gradation curve of fine aggregate is shown in Figure 1.

Table 1. Fine aggregate index properties.

Type	Fineness Modulus	Apparent Density/(g/cm³)	Bulk Density/(g/cm³)	Bulk Void Ratio/%	Stone Powder Content/%	Methylene Blue Value/(g/kg)	Roughness/(s)
Machine-made Tuff Sand (NH)	2.8	2.65	1.53	42.3	7.5	1.36	16.5
Machine-made Limestone Sand (SH)	2.7	2.71	1.67	38.4	6.4	1.34	13.4
River Sand (HS)	2.6	2.61	1.53	41.4	1.2	1.12	9.60

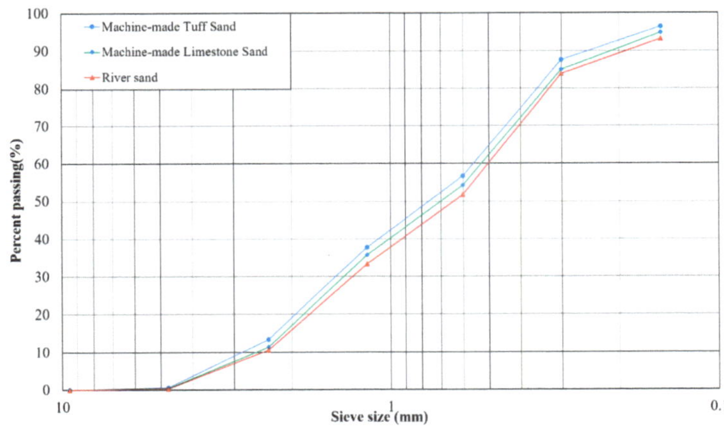

Figure 1. Fine aggregate grading curve.

2.2. Orthogonal Experimental Design

Orthogonal experimental design is a quantitative design analysis method for studying multiple factors and levels. This experiment method is commonly used to optimize mortar and concrete properties [36,37]. There are many factors affecting the performance of the machine-made sand concrete, and the number of experiments is large if all the factors are experimented. The factors that have a greater influence on the mix ratio are derived from the analysis and evaluation. The proportion of mineral admixture, stone powder content, sand rate, and water–cement ratio are selected as orthogonal test design factors, which are denoted by the symbols A, B, C, and D. The mineral admixture type is also a factor to be optimized. The only mineral admixture in the calculated ratio is fly ash. Single mineral admixture is not conducive to the development of concrete properties, and the subsequent tests used fly ash and slag powder compounding [38,39]. Machine-made sand was first removed by dry sieving and then mixed with tuff stone powder to adjust the stone powder content of fine aggregates.

We used the full calculation method to perform preliminary tests on the properties of the machine-made sand concrete. The mix ratio was optimized by considering the level of factors and the number of tests. Four levels were set for each factor, and a 4-factor, 4-level orthogonal test table was selected for the experiments. The parameter values under each level are shown in Table 2, the experimental scheme is shown in Table 3, and the amounts of raw materials under each test scheme are shown in Table 4.

Table 2. $L_{16}(4^4)$ orthogonal experimental design table.

Factor / Level	Mineral Admixture/%	Stone Powder Content/%	Sand Rate/%	Water Cement Ratio
1	10	3.0.	44	0.30
2	15	6.5	45	0.31
3	20	10.0	46	0.32
4	30	13.0	47	0.33

Table 3. $L_{16}(4^4)$ orthogonal experimental scheme.

Factor / Test Number	Mineral Admixture A/%	Stone Powder Content B/%	Sand Rate C/%	Water Cement Ratio D
1	10	3.0	44	0.30
2	10	6.5	45	0.31
3	10	10.0	46	0.32
4	10	13.0	47	0.33
5	15	3.0	45	0.32
6	15	6.5	44	0.33
7	15	10.0	47	0.30
8	15	13.0	46	0.31
9	20	3.0	46	0.33
10	20	6.5	47	0.32
11	20	10.0	44	0.31
12	20	13.0	45	0.30
13	30	3.0	47	0.32
14	30	6.5	46	0.31
15	30	10.0	45	0.33
16	30	13.0	44	0.32

Table 4. Amount of each raw material for the test of machine-made sand concrete (kg/m³).

Test Number	Cement	Mineral Admixture	Machine-Made Sand	Stone	Water	Additive/%
1	471	52	779	975	157	1.5
2	471	52	794	971	162	1.5
3	471	52	810	950	167	1.5
4	471	52	824	930	173	1.5
5	444	79	792	968	167	1.5
6	444	79	772	982	173	1.5
7	444	79	832	938	157	1.5
8	444	79	812	953	162	1.5
9	419	104	824	930	173	1.5
10	419	104	827	933	167	1.5
11	419	104	777	988	162	1.5
12	419	104	797	973	157	1.5
13	366	157	830	935	162	1.5
14	366	157	814	956	157	1.5
15	366	157	789	965	173	1.5
16	366	157	774	986	167	1.5

2.3. Comparison of Fine Aggregate Performance Test

Three different types of fine aggregates, machine-made tuff sand, machine-made limestone sand and river sand (number NH, SH, HS), were selected and tested in different batches of the same fine aggregate. Additionally, the fineness modulus, bulk density, apparent density, clay lump content and roughness of each batch were measured. Four batches of machine-made tuff sand (number NH1, NH2, NH3, NH4), four batches of machine-made limestone sand (number SH1, SH2, SH3, SH4), and two batches of river sand (number HS1, HS2) were sieved to obtain two types of stone powder to adjust the stone powder content of the sand. The stone powder content of each batch of the two types of sand was set to 3.0%, 6.5%, 10.0% and 13.0%. The clay lump content of the batches of the two types of machine-made sand fluctuated widely from 0.2% to 0.9%. The clay lump content of river sand was higher at 0.7% and 0.9%. The performance indicators of each batch of fine aggregate meet the corresponding national standards, and the main performance indicators are shown in Table 5.

Table 5. Main performance indicators of fine aggregates.

Test Number	Apparent Density/(g·cm^{-3})	Bulk Density/(g·cm^{-3})	Fineness Modulus	Stone Powder Content/%	Clay Lump Content/%	Roughness/S
NH1	2.65	1.52	2.8	3.0	0.2	17.5
NH2	2.64	1.53	2.7	6.5	0.4	16.4
NH3	2.65	1.52	2.6	10.0	0.8	18.6
NH4	2.67	1.53	2.6	13.0	0.9	19.7
SH1	2.71	1.65	2.9	3.0	0.2	14.6
SH2	2.70	1.64	3.0	6.5	0.4	15.7
SH3	2.71	1.68	2.8	10.0	0.7	17.1
SH4	2.72	1.67	2.6	13.0	0.8	16.3
HS1	2.65	1.49	2.7	1.6	0.7	11.9
HS2	2.66	1.52	2.6	1.3	0.9	10.6

The degree of influence of index properties such as stone powder content, clay lump content and roughness of the machine-made sand on the mechanical properties of concrete is different. The compressive strengths of concrete with different fine aggregates were tested and obtained through comparative studies of concrete with different batches of tuff and machine-made limestone sand and river sand. The correlation between the stone powder content, clay lump content, roughness of the machine-made sand and compressive strength of the concrete was also calculated by using the grey correlation analysis. The correlation matrix was analyzed to derive the magnitude and influence law of each factor on the degree of performance.

2.4. Test Methods

The concrete strength grade in this study is C50, which is mainly applied to the T-beam in the bridge structure. The T-beam is the main component of the main structure of the bridge and has a design service life of 100 years. The concrete is required to have good working properties and mechanical properties, and the trial strength reaches 1.15 times of the design strength.

The working performance test indexes are slump and slump expansion. Additionally, the specific operation is carried out in accordance with the Chinese standard GB/T 50080-2016 "Standard for Test Methods of Properties of Ordinary Concrete Mixes". The mechanical performance test indexes are compressive strength at 7 and 28 days. Additionally, the specimen size is 150 mm × 150 mm × 150 mm concrete test block, the test method is conducted according to the Chinese standard GB/T 50081-2002 "Standard on Mechanical Properties of General Concrete". The process of this study is shown in Figure 2. The field test process is shown in Figure 3.

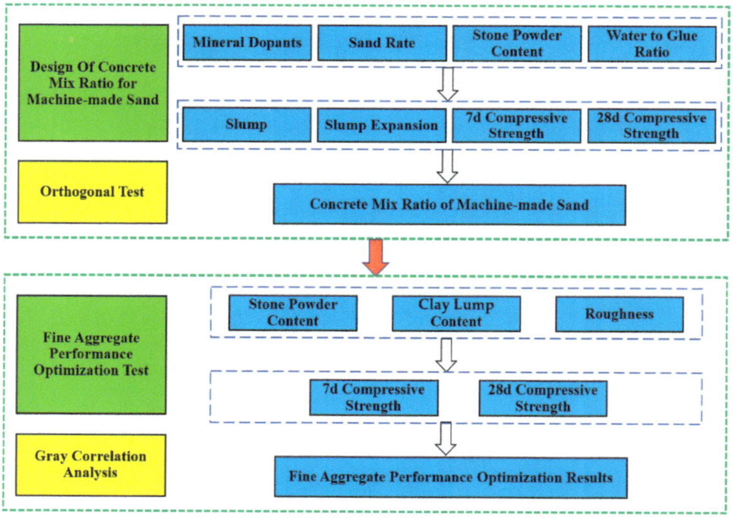

Figure 2. Flow chart of this study.

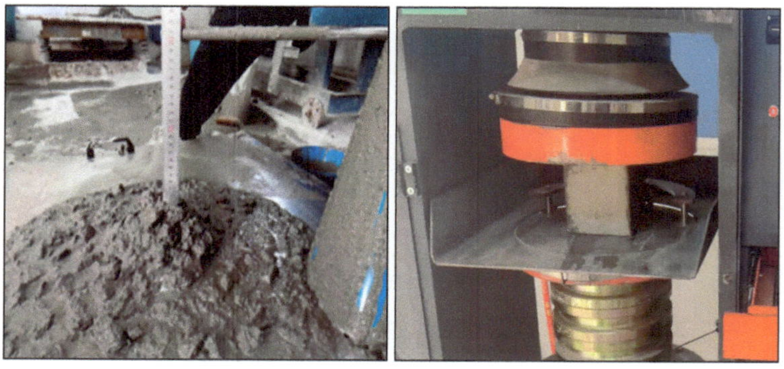

Figure 3. Field test chart of this study.

3. Results and Discussion

3.1. Test Results

Sixteen sets of concrete test blocks designed for orthogonal experiments were tested for working properties and mechanical properties. The results of the concrete slump and compressive strength tests are shown in Table 6.

Table 6. Machine-made sand concrete test results.

Test Number	Slump/mm	Slump Expansion/mm	7-Day Compressive Strength/MPa	28-Day Compressive Strength/MPa
1	165	450	42.5	62.5
2	170	465	43.3	62.4
3	180	480	43.0	62.0
4	165	455	40.8	61.2
5	175	470	41.9	61.9
6	170	470	41.5	60.8
7	165	460	42.5	62.9
8	175	475	42.8	62.6
9	185	490	40.5	62.1
10	170	470	40.6	61.8
11	175	480	41.9	63.2
12	175	470	42.2	62.4
13	165	460	39.7	61.5
14	180	480	40.6	63.4
15	170	470	39.5	62.3
16	165	460	38.9	62.5

3.2. Analysis of Extreme Differences

The orthogonal experimental design software Minitab 20 (State College, PA, USA) was used to perform a polar difference analysis on the experimental results. The extreme difference refers to the difference between the maximum value and the minimum value of the mean value of the test index corresponding to each level in each column. The effect of each level on the test indexes was ranked from largest to smallest by the extreme difference analysis. The results of slump and slump expansion analysis of concrete are shown in Table 7.

Table 7. Slump and slump expansion degree extreme difference analysis table (mm).

Type	Level	Mineral Admixture A/%	Stone Powder Content B/%	Sand Rate C/%	Water Cement Ratio D
Slump	$K_1/4$	170.0	172.5	168.8	171.3
	$K_2/4$	171.3	172.5	172.5	171.3
	$K_3/4$	176.3	172.5	180.0	172.5
	$K_4/4$	170.0	170.0	166.3	172.5
	Polar Difference D	6.3	2.5	13.8	1.3
	Ranking	2	3	1	4
Slump Expansion	$K_1/4$	462.5	467.5	465.0	465.0
	$K_2/4$	468.8	471.3	468.8	470.0
	$K_3/4$	477.5	472.5	481.3	470.0
	$K_4/4$	467.5	465.0	461.3	471.3
	Polar Difference D	15.0	7.5	20.0	6.3
	Ranking	2	3	1	4

From Table 7, the influence of each factor on slump and slump expansion is ranked as sand rate > mineral admixture > stone powder content > water–cement ratio. With the increase in sand rate, the slump of fresh concrete tends to increase. There is an optimum ratio of coarse and fine aggregates for fresh concrete. In a certain amount of admixture, as the amount of mineral admixture increases, the concrete slump increases. This is due to the lubricating effect of fine particles of admixtures, so that the fluidity of fresh concrete increases. When the amount of admixture is too large, the amount of water needed to wrap the surface increases, resulting in a decrease in the fluidity of fresh concrete and a decrease in slump. With the increase in stone powder content, the concrete resistance to segregation is enhanced. The specific surface area of stone powder particles is large, and the water requirement for wrapping the powder surface is large. This leads to a decrease in concrete fluidity, but the cohesiveness and water retention of fresh concrete is good. With the increase in the water–cement ratio, the water consumption gradually increases, which to some extent promotes the fluidity and cohesiveness of fresh concrete.

The results of the 7-day and 28-day compressive strength extreme difference analysis of the machine-made sand concrete are shown in Table 8.

Table 8. Table of extreme difference analysis of compressive strength (MPa).

Type	Level	Mineral Admixture A/%	Stone Powder Content B/%	Sand Rate C/%	Water Cement Ratio D
7-day Compressive Strength	$K_1/4$	42.40	41.15	41.20	41.95
	$K_2/4$	42.17	41.50	41.72	41.92
	$K_3/4$	41.30	41.73	41.73	41.10
	$K_4/4$	39.68	41.17	40.90	40.58
	Polar Difference D	2.73	0.57	0.82	1.38
	Ranking	1	4	3	2
28-day Compressive Strength	$K_1/4$	62.03	62.00	62.25	62.80
	$K_2/4$	62.05	62.10	62.25	62.42
	$K_3/4$	62.38	62.60	62.52	62.05
	$K_4/4$	62.42	62.18	61.85	61.60
	Polar Difference D	0.40	0.60	0.68	1.20
	Ranking	4	3	2	1

From Table 8, the degree of influence of each factor on 7-day compressive strength is as follows: mineral admixture > water–cement ratio > sand rate > stone powder content. The degree of influence of each factor on 28-day compressive strength is as follows: water cement ratio > sand rate > stone powder content > mineral admixture. The activity of fly ash in the mineral admixture is very low in the early stage, which does not participate in the reaction and has a weakening effect on the strength of concrete. With the growth of the age of concrete, the mineral admixture activity gradually emerges. However, some studies [38–41] pointed out that adding the right amount of mineral admixture can play its filling effect. It reduces the internal voids of concrete and increases the strength of concrete. When the mineral admixture is 20%, the strength of concrete has been significantly enhanced, and further increase the admixture has less effect on the improvement of concrete strength. With the increase in the water–cement ratio, the compressive strengths of concrete showed a gradual decrease. The increase in water consumption of concrete leads to an increase in the probability of excess water forming voids inside the concrete, thus gradually increasing the possibility of reducing the strength of concrete. To ensure the excellent mechanical properties of the machine-made sand concrete, a lower water–cement ratio should be used. The increase in fine aggregate content makes the cement paste more densely filled in the coarse and fine aggregates, which has the effect of enhancing the strength. However, when the sand rate exceeds a certain range, more slurry is required for the bonding surface of the concrete internal aggregate interface with the slurry, making the void rate between fine aggregates not completely filled by the slurry. The particle size of stone powder is between the cementitious material and the machine-made sand. Additionally, it can effectively fill in between the aggregate particles, forming a tightly packed body. However, when the stone powder content exceeds a certain range, it will reduce the fluidity of fresh concrete. Due to its lower activity, it will reduce the strength of concrete.

3.3. Single Factor Analysis

Single factor analysis refers to the situation where the test results vary with a single factor when other factors are constant. The analysis of the effect of single factor on the slump and slump expansion of the machine-made sand concrete is shown in Figures 4 and 5. The analysis of the effect of a single factor on the compressive strengths of the machine-made sand concrete is shown in Figures 6 and 7.

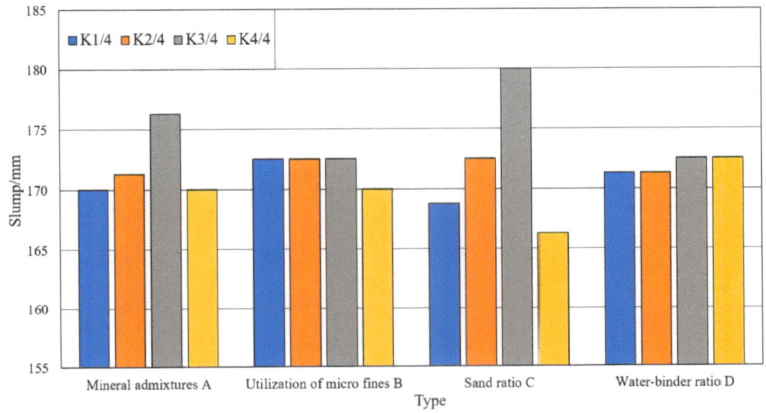

Figure 4. Trend of the influence of factors on slump test results.

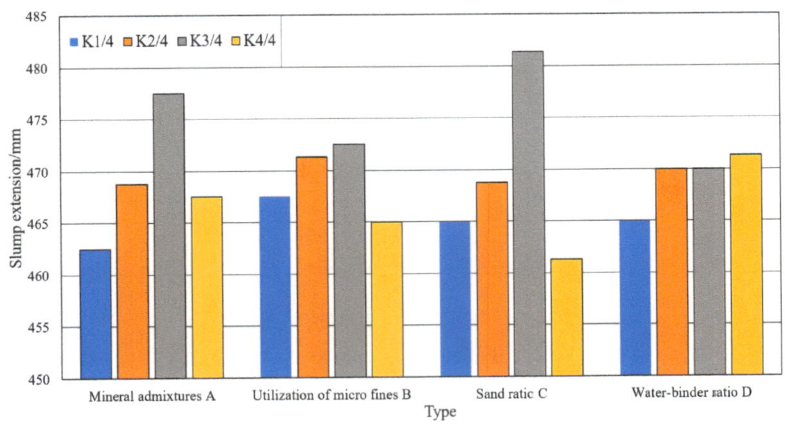

Figure 5. Trend of the influence of factors on the results of slump expansion test.

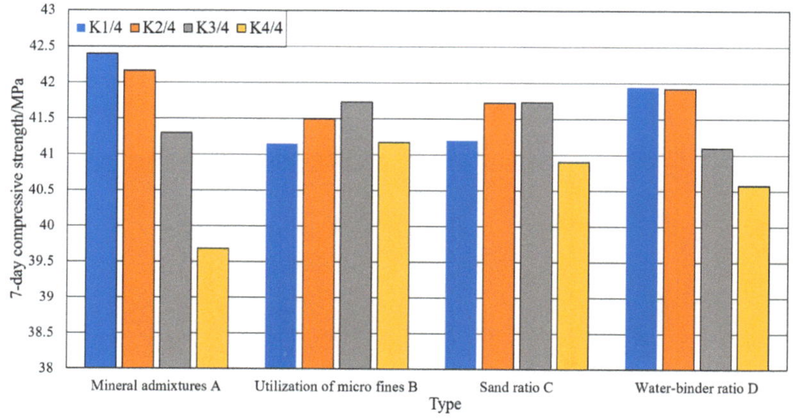

Figure 6. Trend of the influence of factors on the results of 7-day compressive strength test.

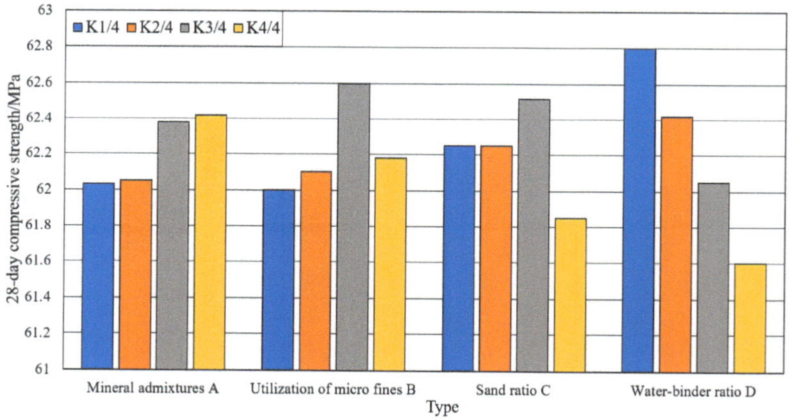

Figure 7. Trend of the influence of factors on the results of 28-day compressive strength test.

(1) The amount of mineral admixture

As shown in Figures 4 and 5, with the gradual increase in mineral admixture, the concrete slump and slump expansion increase first and then decrease. The slump and slump expansion have great values at three levels. As shown in Figures 6 and 7, the 7-day compressive strength of concrete gradually decreases with the increase in mineral admixture. The 28-day compressive strength is the opposite, gradually increasing with the increase in mineral admixture. However, the changes in 7-day compressive strength in the four levels are significantly greater than that of 28-day compressive strength. The growth of 28-day compressive strength is not obvious when the mineral admixture is mixed from 20% to 30%.

(2) Stone powder content

As shown in Figures 4 and 5, the slump of the concrete keeps constant and then decreases with the increase in stone powder content. The slump expansion of concrete shows a trend of increasing and then decreasing. Due to the small particle size of stone powder, it acts as a lubricant in the fresh concrete system. Therefore, when the content of stone powder in the fine aggregate does not exceed a certain range, the fluidity and water retention of concrete has a certain role in promoting. As shown in Figures 6 and 7, the compressive strengths show a trend of increasing and then decreasing with increasing stone powder content, with a great value at the third level. After the stone powder content exceeds 10%, the strength value decreases. However, the magnitude of the decrease is small.

(3) Sand rate

As shown in Figures 4 and 5, the effect of sand rate on slump and slump expansion is similar to that of mineral admixture. With the increase in sand rate, it shows a trend of increasing and then decreasing. There are great values at all three levels. Additionally, the average value of different levels differs greatly. The maximum difference of slump and slump expansion are 13.8 mm and 20.0 mm. As shown in Figures 6 and 7, the compressive strengths of concrete reached great values at all three levels. The difference of maximum difference of compressive strength between levels is small.

(4) Water–cement ratio

As shown in Figures 4 and 5, the slump and slump expansion of concrete show an increasing trend as the water–cement ratio increases. Both of them have maximum values at level 4. As shown in Figures 6 and 7, the compressive strengths of concrete change in the opposite pattern to slump and slump expansion. As the water–cement ratio increases, the compressive strengths show a decreasing trend. Unlike factors such as stone powder content, the increase in the water–cement ratio leads to a larger decrease in the 28-day compressive strength of concrete.

In summary, the optimal working performance of the machine-made sand concrete is achieved at a mineral admixture of 20%. Although the corresponding compressive strengths of concrete are not the respective maximum values, they differ from the maximum values by a small margin. The concrete slump and slump expansion are at maximum, and the compressive strengths of concrete are maximum when the sand percentage is 46%. When the stone powder content is 10%, the working performance is better, and it is in the state of imminent deterioration of performance. At this time, the compressive strengths have great values. With the increase in the water–cement ratio, the concrete slump and slump expansion show a gradual increase in trend, which has a certain improvement effect on the work performance. The water–cement ratio has a significant effect on the compressive strengths of concrete, and the strength decreases significantly as the water–cement ratio increases. Therefore, in order to ensure that the concrete has excellent mechanical properties, a lower water–cement ratio should be used as much as possible.

3.4. Experimental Validation

From the above points, it can be seen that the mineral admixture, sand rate and stone powder content are all at level 3 and the water–cement ratio is at level 1, i.e., $A_3B_3C_3D_1$. This is the optimal combination ratio of the machine-made sand concrete obtained from the orthogonal test. However, this group of tests was not available in the orthogonal tests. To further verify the accuracy of the results, three sets of tests were set $A_3B_3C_3D_1$, $A_4B_3C_3D_1$ and $A_3B_3C_3D_2$. The numbers are Y-1, Y-2 and Y-3. The working properties and mechanical properties of the machine-made sand concrete were tested. The matching ratio and test results of the verification experiments are shown in Table 9.

Table 9. Test verification ratio parameters and test results.

Test Number	Proportion/(kg·m^{-3})						Test Result		
	Cement	Mineral Admixture	Machine-Made Sand	Stone	Water	Additive	Slump/mm	7-Day Compressive Strength/MPa	28-Day Compressive Strength/MPa
Y-1	419	104	814	956	157	1.5%	180	42.8	63.4
Y-2	366	157	814	956	157	1.5%	190	39.5	62.8
Y-3	419	104	812	953	162	1.5%	190	41.2	61.9

From Table 9, it can be seen that for the working performance, all three groups have better compatibility and are free from water secretion. The slump of both groups Y-2 and Y-3 is greater than that of group Y-1. Compared with the remaining two groups, the concrete of Y-1 group was slightly less fluid. The compressive strength test results showed that the 7-day compressive strength of concrete in Y-2 group was significantly smaller than that in the other two groups. 28-day compressive strength in Y-2 group was between the two. Compressive strengths of concrete in Y-1 group were the largest. 28-day compressive strength in Y-3 group was the smallest, but still reached the design strength requirement. In summary, the overall performance of the concrete of the three groups of the concrete with the machine-made sand is good.

3.5. Fine Aggregate Performance Test Results

From Table 5, it can be obtained that the difference in bulk density and apparent density between the same fine aggregates is small. The loose bulk density and apparent density of machine-made tuff sand are close to that of river sand, but smaller than that of machine-made limestone sand. The fineness modulus of all batches of fine aggregates ranged from 2.6 to 3.0, which was in the range of medium sands. Roughness is a measure of the fluidity and angularity of fine aggregates. The experimental results showed that the roughness index values of the same fine aggregates differed less. Compared with the other two fine aggregates, the machine-made tuff sand has a rougher grain shape. Concrete was prepared using multiple batches of different types of machine-made sand and river sand based on the optimal combination of ratios. The performance of the concrete with different fine aggregates was compared. The specific mix parameters and test results are shown in Table 10. The variation of compressive strength of concrete with different fine aggregates is shown in Figure 8. (1_{NH1} denotes the first set of experiments, using fine aggregate of batch number NH1, and so on.)

Table 10. Test mix ratio and test results.

Test Number	Proportion/(kg·m^{-3})						Test Result		
	Cement	Mineral Admixture	Machine-Made Sand	Stone	Water	Additive	Slump/mm	7-Day Compressive Strength/MPa	28-Day Compressive Strength/MPa
1_{NH1}	419	104	814	956	157	1.6	180	43.1	60.2
2_{NH2}	419	104	814	956	157	1.6	180	44.5	63.8
3_{NH3}	419	104	814	956	157	1.7	170	47.3	65.3
4_{NH4}	419	104	814	956	157	1.7	165	43.3	61.6
5_{SH1}	419	104	814	956	157	1.6	195	42.5	60.6
6_{SH2}	419	104	814	956	157	1.6	190	44.2	62.8
7_{SH3}	419	104	814	956	157	1.7	180	45.3	64.3
8_{SH4}	419	104	814	956	157	1.7	170	44.1	62.5
9_{HS1}	419	104	814	956	157	1.6	200	46.8	62.3
10_{HS2}	419	104	814	956	157	1.7	180	45.8	63.3

Combining Table 10 and Figure 8, it can be obtained that the 28-day compressive strength of concrete of both tuff and machine-made limestone sand is greater than 60 MPa. The 7-day and 28-day compressive strength of concrete of both types of machine-made sand increases and then decreases with the increase in stone powder content and clay lump content. The compressive strengths of concrete were maximum for the 10% stone powder content of the machine-made sand. The compressive strengths of machine-made tuff sand fluctuated more. For the magnitude of concrete compressive strength variation, machine-made tuff sand concrete was significantly larger than machine-made limestone sand concrete. The maximum strengths of river sand concrete were slightly lower than those of the mechanism sand concrete, but still had good mechanical properties.

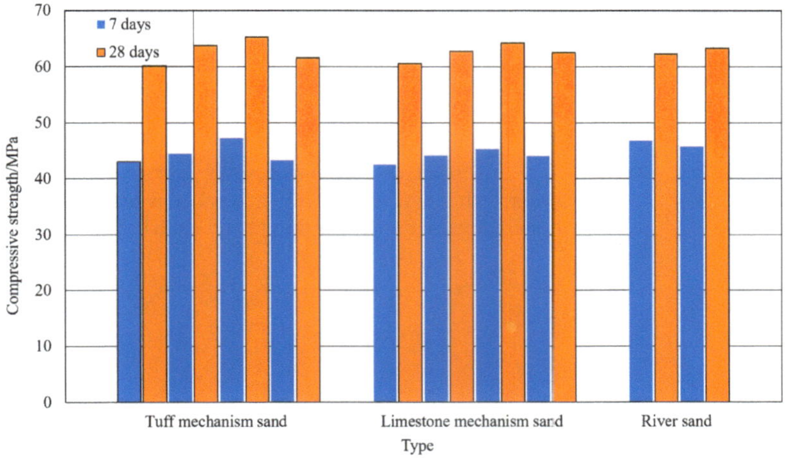

Figure 8. Compressive strength of concrete with different fine aggregate machine-made sand.

3.6. Grey Correlation Analysis

Grey correlation analysis is based on the similarity of the geometry of the series curves to determine whether they are closely related or not [42–44]. Grey correlation analysis is often used in concrete performance research. In exploring the analysis of the effects of stone powder content, clay lump content, roughness and mechanical properties of concrete of machine-made sand, the three subfactors are stone powder content, clay lump content and roughness. Two parent factors are 7-day and 28-day compressive strength of concrete, respectively. Each parent factor has three correlations to the three subfactors. The specific calculation steps are as follows.

1. Data column representation.

Specify the reference data column, which is often noted as x_0, i.e., the parent factor. The first case is denoted as $x_0(1)$, and the kth case is denoted as $x_0(k)$. The reference parent series can be expressed as $x_0 = (x_0(1), x_0(2), \ldots, x_0(k))$.

2. Initialization process.

Before calculating the correlation coefficient, the series needs to be primed. Each number in the series can be divided in turn by the maximum value in the series to obtain a dimensionless data series for easy comparison.

3. Calculation formula of correlation coefficient and correlation degree.

Referring to the parent sequence x_0, with k comparison subsequences $x_1, x_2, \ldots, x_k$, and the correlation coefficients are calculated as follows.

$$\xi_i = \frac{\min_i \min_k |x_0(k) - x_i(k)| + \rho \max_i \max_k |x_0(k) - x_i(k)|}{|x_0(k) - x_i(k)| + \rho \max_i \max_k |x_0(k) - x_i(k)|}, \tag{1}$$

where ρ is the resolution, which is used to reduce the error and improve the significance of the difference between the correlation coefficients. The value of 0.5 is taken in the paper. $\min_i \min_k |x_0(k) - x_i(k)|$ is the absolute value of the second-degree minimum difference between the $x_0(k)$ series and the $x_i(k)$ series at point k. $\max_i \max_k |x_0(k) - x_i(k)|$ is the absolute value of the secondary maximum difference between the $x_0(k)$ series and the $x_i(k)$ series at point k.

The degree of correlation is calculated by the following equation.

$$r_i = \frac{1}{N} \sum_{k=1}^{N} \xi_i(k), \quad (2)$$

4. Construction of the strengths analysis matrix.

The correlations of the m parent factors and the corresponding n child factors are ordered in a row to obtain the correlation matrix R. The correlation matrix allows the dominance analysis of the parent and child series.

$$R = \begin{pmatrix} r_{11} & r_{12} & \cdots & r_{1n} \\ r_{21} & r_{22} & \cdots & r_{2n} \\ \cdots & \cdots & \ddots & \cdots \\ r_{m1} & r_{m2} & \cdots & r_{mn} \end{pmatrix}, \quad (3)$$

Based on the experimental results in Table 10, the values of the subseries factors and the parent series factors for tuff sand and limestone sand concrete were primed. The results are presented in Table 11. The calculations were performed according to the steps in the grey correlation analysis to obtain the machine-made tuff sand correlation matrix $R_{(1)}$ and the machine-made limestone sand correlation matrix $R_{(2)}$.

$$R_{(1)} = \begin{pmatrix} 0.5803 & 0.6137 & 0.8963 \\ 0.5731 & 0.6065 & 0.8845 \end{pmatrix}, \quad (4)$$

$$R_{(2)} = \begin{pmatrix} 0.5738 & 0.6088 & 0.9033 \\ 0.5736 & 0.6085 & 0.9023 \end{pmatrix}, \quad (5)$$

Table 11. Initialization results.

Test Number	Stone Powder Content	Clay Lump Content	Roughness	7-Day Compressive Strength	28-Day Compressive Strength
1_{NH1}	0.2308	0.2222	0.8883	0.9112	0.9053
2_{NH2}	0.5000	0.4444	0.8325	0.9408	0.9549
3_{NH3}	0.7692	0.8889	0.9442	1.0000	1.0000
4_{NH4}	1.0000	1.0000	1.0000	0.9154	0.9263
5_{SH1}	0.2308	0.2500	0.8538	0.9382	0.9425
6_{SH2}	0.5000	0.5000	0.9181	0.9757	0.9767
7_{SH3}	0.7692	0.8750	1.0000	1.0000	1.0000
8_{SH4}	1.0000	1.0000	0.9532	0.9735	0.9720

A combined comparison of the machine-made tuff sand concrete correlation matrix $R_{(1)}$ and the limestone sand concrete correlation matrix $R_{(2)}$,$r_{1i3} > r_{1i2} > r_{1i1}$ and $r_{2i3} > r_{2i2} > r_{2i1}$ (i = 1, 2) indicate that the correlation between compressive strength and roughness at different ages is the largest. Additionally, the stone powder content has the least correlation. Among the three factors, roughness is the main influencing factor on the strength of the concrete of the machine-made sand. The greater the roughness of the concrete aggregate, the stronger the bond between the aggregates and between the aggregates and the cement paste. Concrete is about less prone to damage when compressed, and thus exhibits higher compressive strength. $r_{11j} > r_{12j}$ and $r_{21j} > r_{22j}$ (j = 1, 2, 3) indicate that the correlation between the 7-day compressive strength of concrete and roughness, stone powder content, clay lump content is greater compared to the 28-day compressive strength. As the early concrete fly ash has not been fully hydrated, the strength is low, and the enhancing effect of mineral admixture has not been fully developed. Therefore, the early strength of the machine-made sand concrete is more influenced by the roughness and mud lump content. r_{1i1}, $r_{2i1} < 0.6$ (I = 1, 2) indicate that there is no significant correlation between compressive strengths of concrete and stone powder content. When the stone powder t content does not exceed a certain range (about 10%), the stone powder content has little effect on the compressive strength of concrete. For ordinary concrete, the aggregate and cement paste interface is the high incidence of force damage. Additionally, it is dominated by damage to the bonding surface. The effect of stone powder on the bonding effect of aggregate and cement paste is small. Therefore, the stone powder does not improve the compressive strength of concrete significantly.

The performance of concrete specimens was determined by reducing the stone powder content and clay lump content of the machine-made tuff sand and blending the improved quality fine aggregate, Number NH5. Additionally, we compare it with NH2 batch of machine-made sand under the optimal combination of matching ratio. The results were also compared with the test results of groups 2 and 3. The apparent and bulk densities of the sand in groups 11 and 12 were not significantly different, and the fineness modulus was 2.7. The fine aggregates in groups 11 and 2 were from the same batch, and the performance index values were not significantly different. Other properties of fine aggregates and experimental results are shown in Table 12. (12_{NH2} indicates the 12th group of experiments, using NH2 batch of fine aggregates, others as well.)

Table 12. Performance indicators and experimental results of high-quality fine aggregates.

Test Number	Stone Powder Content/%	Clay Lump Content/%	Roughness/s	7-Day Compressive Strength/MPa	28-Day Compressive Strength/MPa
11_{NH2}	6.5	0.4	15.5	43.1	62.3
12_{NH5}	3.0	0.1	18.3	46.2	65.8
2_{NH2}	6.5	0.4	16.4	44.5	63.8
3_{NH3}	10.0	0.8	18.6	47.3	65.3

From the test results, it can be seen that the compressive strengths of the concrete tested in group 12 are greater than the corresponding values in group 11. The index value of fine aggregate roughness in group 12 is greater than that in group 11 and slightly less than that in group 3. In addition, the difference between the compressive strength values of concrete in group 12 and group 3 is small. The experimental results are consistent with the findings of the study.

3.7. Application of Research Results

According to the concept of sustainable development, new environmentally friendly construction materials can protect the environment and save construction costs [24,25]. Qin et al. [44] used coral reef and coral sand as coarse and fine aggregates to prepare concrete samples in the context of offshore engineering. Additionally, they studied the uniaxial compressive strength of coral reef sand concrete. Shi et al. [31] studied the effect of lime powder in machine-made sand on the performance of concrete, and aimed at different stone powder content on the performance of concrete. A large number of studies have also focused on the development of new construction materials. This study is based on the background of a highway construction in eastern China, where river sand resources are scarce. However, tuff resources are abundant and raw materials for sand production are widely available. The use of tunnel rejects as sand making base material solves the problem of sand shortage in local projects and protects the local ecological environment. The concrete proportioning of tuff-derived sand was optimized by orthogonal test method. The optimal combination of concrete proportions was obtained. It was also verified that the concrete in the T-beam part of the bridge needs to have excellent performance. Only fine aggregates in concrete were investigated in this study, and environmentally friendly cementitious materials were not involved. In addition, the application of machine-made tuff sand concrete is also influenced by the local raw materials for sand production.

4. Conclusions

With the scarcity of natural sand resources, replacing natural sand with machine-made sand has become an inevitable choice for the sustainable development of concrete industry. In this study, the concrete performance of machine-made tuff sand is studied by orthogonal test and grey correlation analysis. From the experimental and analytical results, the following conclusions can be drawn.

1. The results of orthogonal tests were analyzed to study the degree of influence of each level on the test index. Among them, the sand rate has the greatest degree of influence on slump and slump expansion. The influence of mineral admixture dosing on 7-day compressive strength was the greatest. The water–cement ratio has the greatest influence on the 28-day compressive strength.
2. The optimal mix combination was obtained from the orthogonal test results. The optimal working performance of the machine-made sand concrete was achieved at the mineral admixture of 20%, sand rate of 46%, stone powder content of 10% and a water–cement ratio of 0.30. The performance of the concrete under the optimal mix combination was also verified.
3. We compared different fine aggregate concretes of similar quality. It is found that the mechanical properties of machine-made tuff sand concrete are similar to those of machine-made limestone sand concrete and river sand concrete.
4. Grey correlation analysis was applied to compare the factors influencing the mechanical properties of tuff and machine-made limestone sands. The correlation between compressive strength of concrete and roughness was found to be the highest. Additionally, the correlation with the stone powder content is the smallest. This points the way to quality control of high-performance machine-made tuff sand concrete.

Equal amounts of fly ash and slag powder were mainly considered as mineral admixtures in this study. We can analyze the effect of their different amounts on the properties of machine-made sand concrete in future studies. For the research of high-performance machine-made sand concrete, other environmentally friendly cementitious materials can also be considered.

Author Contributions: Methodology, D.L. and Y.T.; software, W.Z.; validation, W.Z. and Y.J.; formal analysis, W.Z. and Y.L.; data curation, W.Z. and Y.L.; writing—original draft preparation, W.Z. and Y.J.; writing—review and editing, D.L. and Y.T.; project administration, D.L. and Y.T.; funding acquisition, D.L. and Y.T. All authors have read and agreed to the published version of the manuscript.

Funding: The research was funded by the Initiation fund for postdoctoral research of Central South University, grant number 228697.

Institutional Review Board Statement: Not applicable.

Informed Consent Statement: Not applicable.

Data Availability Statement: The data that support the findings of this study are available upon request from the authors.

Acknowledgments: The authors would like to thank Shanghai Tunnel Engineering Co., Ltd. for their assistance with conducting the experiments.

Conflicts of Interest: The authors declare no conflict of interest.

References

1. Kirthika, S.K.; Singh, S.K.; Chourasia, A. Alternative fine aggregates in production of sustainable concrete—A review. *J. Clean. Prod.* **2020**, *268*, 122089.
2. Wang, J.; Yang, Z.; Liu, Y. Effects of the lithologic character of manufactured sand on properties of concrete. *J. Wuhan Univ. Technol.* **2014**, *29*, 1213–1218. [CrossRef]
3. Goncalves, J.P.; Tavares, L.M.; Toledo Filho, R.D.; Fairbain, E.M.R.; Cunha, E.R. Comparison of natural and manufactured fine aggregates in cement mortars. *Cem. Concr. Res.* **2007**, *37*, 924–932. [CrossRef]

4. Zhong, C.; Liu, M.; Zhang, Y.; Wang, J. Study on mix proportion optimization of manufactured sand RPC and design method of steel fiber content under different curing methods. *Materials* **2019**, *12*, 184511. [CrossRef]
5. Zhou, M.; Wu, Z.; Ouyang, X.; Hu, X.; Shi, C. Mixture design methods for Ultra-High-Performance concrete—A review. *Cem. Concr. Comp.* **2021**, *124*, 104242. [CrossRef]
6. Suttaphakdee, P.; Dulsang, N.; Lorwanishpaisarn, N.; Kasemsiri, P.; Posi, P.; Chindaprasirt, P. Optimizing mix proportion and properties of lightweight concrete incorporated phase change material paraffin/recycled concrete block composite *Constr. Build. Mater.* **2016**, *127*, 475–483. [CrossRef]
7. Tang, C.; Cheng, C.; Tsai, C. Mix design and mechanical properties of high-performance pervious concrete. *Materials* **2019**, *12*, 257716. [CrossRef]
8. Lee, J.; Yoon, Y.; Kim, J. A new heuristic algorithm for mix design of high-performance concrete. *KSCE J. Civ. Eng.* **2012**, *16*, 974–979. [CrossRef]
9. Yan, W.; Wu, G.; Dong, Z. Optimization of the mix proportion for desert sand concrete based on a statistical model *Constr. Build. Mater.* **2019**, *226*, 469–482. [CrossRef]
10. Shen, W.; Liu, Y.; Cao, L.; Huo, X.; Yang, Z.; Zhou, C.; He, P.; Lu, Z. Mixing design and microstructure of ultra high strength concrete with manufactured sand. *Constr. Build. Mater.* **2017**, *143*, 312–321. [CrossRef]
11. Ji, T.; Chen, C.; Zhuang, Y.; Chen, J. A mix proportion design method of manufactured sand concrete based on minimum paste theory. *Constr. Build. Mater.* **2013**, *44*, 422–426. [CrossRef]
12. Li, H.; Huang, F.; Cheng, G.; Xie, Y.; Tan, Y.; Li, L.; Yi, Z. Effect of granite dust on mechanical and some durability properties of manufactured sand concrete. *Constr. Build. Mater.* **2016**, *109*, 41–46. [CrossRef]
13. Menadi, B.; Kenai, S.; Khatib, J.; Ait-Mokhtar, A. Strength and durability of concrete incorporating crushed limestone sand. *Constr. Build. Mater.* **2009**, *23*, 625–633. [CrossRef]
14. Topcu, I.B.; Ugurlu, A. Effect of the use of mineral filler on the properties of concrete. *Cem. Concr. Res.* **2003**, *33*, 1071–1075. [CrossRef]
15. Eren, O.; Marar, K. Effects of limestone crusher dust and steel fibers on concrete. *Constr. Build. Mater.* **2009**, *23*, 981–988. [CrossRef]
16. Munoz, J.F.; Tejedor, M.I.; Anderson, M.A.; Cramer, S.M. Detection of aggregate clay coatings and impacts on concrete. *ACI Mater. J.* **2010**, *107*, 387–395.
17. Chen, J.J.; Kwan, A.K.H.; Jiang, Y. Adding limestone fines as cement paste replacement to reduce water permeability and sorptivity of concrete. *Constr. Build. Mater.* **2014**, *56*, 87–93. [CrossRef]
18. Felekoglu, B. A Comparative study on the performance of sands rich and poor in fines in self-compacting concrete. *Constr. Build. Mater.* **2008**, *22*, 646–654. [CrossRef]
19. Chouhan, H.S.; Kalla, P.; Nagar, R.; Gautam, P.K. Influence of dimensional stone waste on mechanical and durability properties of mortar: A review. *Constr. Build. Mater.* **2019**, *227*, 116662. [CrossRef]
20. Bravo, M.; de Brito, J.; Pontes, J.; Evangelista, L. Mechanical performance of concrete made with aggregates from construction and demolition waste recycling plants. *J. Clean. Prod.* **2015**, *99*, 59–74. [CrossRef]
21. Wang, J.; Niu, K.; Tian, B.; Sun, L. Effect of methylene blue (MB)-value of manufactured sand on the durability of concretes. *J. Wuhan Univ. Technol.* **2012**, *27*, 1160–1164. [CrossRef]
22. Wang, J.; Yang, Z.; Niu, K.; Ke, G.; Zhou, M. Influence of MB-value of Manufactured sand on the shrinkage and cracking of high strength concrete. *J. Wuhan Univ. Technol.* **2009**, *24*, 321–325. [CrossRef]
23. Che, W.; Liu, J.; Li, H.; He, C.; Bai, Y.; Song, Z.; Wang, Y.; Qian, W.; Chen, Z. Experimental investigation and discrete element method simulation on mechanical properties and failure mechanism of polymer-stabilized sand. *B Eng. Geol. Environ.* **2022**, *81*, 1–22. [CrossRef]
24. Shehata, N.; Mohamed, O.A.; Sayed, E.T.; Abdelkareem, M.A.; Olabi, A.G. Geopolymer concrete as green building materials: Recent applications, sustainable development and circular economy potentials. *Sci. Total Environ.* **2022**, *836*, 155577. [CrossRef] [PubMed]
25. Li, W.; Shumuye, E.D.; Shiying, T.; Wang, Z.; Zerfu, K. Eco-Friendly Fibre reinforced geopolymer concrete: A critical review on the microstructure and long-term durability properties. *Case Stud. Constr. Mater.* **2022**, *16*, e00894. [CrossRef]
26. Shi-Cong, K.; Chi-Sun, P. Properties of concrete prepared with crushed fine stone, furnace bottom ash and fine recycled aggregate as fine aggregates. *Constr. Build. Mater.* **2009**, *23*, 2877–2886.
27. Lopez Boadella, I.; Lopez Gayarre, F.; Suarez Gonzalez, J.; Gomez-Soberon, J.M.; Lopez-Colina Perez, C.; Serrano Lopez, M.; de Brito, J. The influence of granite cutting waste on the properties of ultra-high performance concrete. *Materials* **2019**, *12*, 6344. [CrossRef]
28. Joudi-Bahri, I.; Lecomte, A.; Ben Ouezdou, M.; Achour, T. Use of limestone sands and fillers in concrete without superplasticizer. *Cem. Concr. Comp.* **2012**, *34*, 771–780. [CrossRef]
29. Wang, D.; Shi, C.; Farzadnia, N.; Shi, Z.; Jia, H.; Ou, Z. A review on use of limestone powder in cement-based materials: Mechanism, hydration and microstructures. *Constr. Build. Mater.* **2018**, *181*, 659–672. [CrossRef]
30. Litfi, M.; Zafar, I. Effect of total substitution of crushed limestone sand on concrete durability. *Eur. J. Environ. Civ. Eng.* **2022**, *26*, 121–143. [CrossRef]
31. Shi, L.; Ma, H. Experimental study on effect of limestone powder on concrete properties in machine-made sand. *Integr. Ferroelectr.* **2021**, *215*, 38–46. [CrossRef]

32. Cherrak, M.; Bali, A.; Silhadi, K. Concrete mix design containing calcareous tuffs as a partial sand substitution. *Constr. Build. Mater.* **2013**, *47*, 318–323. [CrossRef]
33. Mohseni, E.; Tang, W.; Cui, H. Chloride diffusion and acid resistance of concrete containing zeolite and tuff as partial replacements of cement and sand. *Materials* **2017**, *10*, 3724. [CrossRef] [PubMed]
34. Wen, X.; Zhang, Z.; Cai, Y.; Feng, L.; Qiu, T. Impact and improvement of crushed tuff sand on sulfate resistance of cement concrete at low temperature. *J. Mater. Civil. Eng.* **2018**, *30*, 0501800410. [CrossRef]
35. He, H.; Wang, Y.; Wang, J. Effects of Aggregate Micro Fines (AMF), Aluminum Sulfate and Polypropylene Fiber (PPF) on Properties of Machine-Made Sand Concrete. *Appl. Sci.* **2019**, *9*, 2250. [CrossRef]
36. Liu, Y.; Li, Y.; Jiang, G. Orthogonal experiment on performance of mortar made with dune sand. *Constr. Build. Mater.* **2020**, *264*, 120254. [CrossRef]
37. Chen, H.; Xie, C.; Fu, C.; Liu, J.; Wei, X.; Wu, D. Orthogonal analysis on mechanical properties of basalt-polypropylene fiber mortar. *Materials* **2020**, *13*, 2937. [CrossRef]
38. Li, F.; Chen, Y.; Long, S.; Chen, J. Properties and microstructure of marine concrete with composite mineral admixture. *J. Wuhan Univ. Technol.* **2009**, *24*, 497–501. [CrossRef]
39. Liu, S.; Wang, Z.; Li, X. Long-Term properties of concrete containing ground granulated blast furnace slag and steel slag. *Mag. Concrete Res.* **2014**, *66*, 1095–1103. [CrossRef]
40. Jain, N. Effect of Nonpozzolanic and pozzolanic mineral admixtures on the hydration behavior of ordinary portland cement. *Constr. Build. Mater.* **2012**, *27*, 39–44. [CrossRef]
41. Wang, Q.; Yan, P.; Mi, G. Effect of blended steel slag-GBFS mineral admixture on hydration and strength of cement. *Constr. Build. Mater.* **2012**, *35*, 8–14. [CrossRef]
42. Kuo, Y.; Yang, T.; Huang, G. The use of grey relational analysis in solving multiple attribute decision-making problems. *Comput. Ind. Eng.* **2008**, *55*, 80–93. [CrossRef]
43. Arici, E.; Kelestemur, O. Optimization of mortars containing steel scale using taguchi based grey relational analysis method. *Constr. Build. Mater.* **2019**, *214*, 232–241. [CrossRef]
44. Qin, Q.; Meng, Q.; Yang, H.; Wu, W. Study of the anti-abrasion performance and mechanism of coral reef sand concrete. *Constr. Build. Mater.* **2021**, *291*, 123263. [CrossRef]

materials

Article

New Types and Dosages for the Manufacture of Low-Energy Cements from Raw Materials and Industrial Waste under the Principles of the Circular Economy and Low-Carbon Economy

Sergio Martínez-Martínez [1,2,*], Luis Pérez-Villarejo [1,*], Dolores Eliche-Quesada [1] and Pedro J. Sánchez-Soto [2]

1 Department of Chemical, Environmental and Materials Engineering, University of Jaén, Campus Las Lagunillas s/n, 23071 Jaén, Spain
2 Institute of Materials Science of Sevilla (ICMS), Joint Center of the Spanish National Research Council (CSIC), University of Sevilla, 41092 Sevilla, Spain
* Correspondence: smartine@ujaen.es (S.M.-M.); lperezvi@ujaen.es (L.P.-V.)

Abstract: The cement manufacturing industry is one of the main greenhouse gas emission producers and also consumes a large quantity of raw materials. It is essential to reduce these emissions in order to comply with the Paris Agreement and the principles of the circular economy. The objective of this research was to develop different types of cement clinker blends using industrial waste and innovative design to produce low-energy cement. Several types of waste have been studied as alternative raw materials. Their main characteristics have been analyzed via X-ray fluorescence (XRF), X-ray diffraction (XRD), Attenuated total reflectance Fourier trans-form infrared spectroscopy (ATR-FTIR), thermal analysis (TG-DTG-DSC) and scanning electron microscopy and energy dispersive X-ray spectroscopy analysis (SEM-EDS). The results obtained from the experimental work carried out in this research focused on the study of crude blends for low-energy cement created from industrial waste. The effect of the addition of different industrial waste types, as a substitution for raw materials, in the production of low-energy cement with high dicalcium silicate content has been investigated. Thus, the dosage design has been performed using modified Bogue equations and quality indexes (LSF, AM, and SM). The calculations of both the modified Bogue equations and quality indexes necessitate knowledge of the weight percentages of CaO, SiO_2, Al_2O_3, and Fe_2O_3, determined via XRF. In this theoretical design of the different blends, it has been established that a dicalcium silicate ratio of 60–65 wt % and an LSF of 78–83% as the limit are values common to all of them. The calculation basis for the crude blends has been based on calcined materials. Therefore, the chemical composition was established, following this premise. Thus, it was possible to develop cement clinker blends with compositions of 50 wt % and 100 wt % using industrial wastes. This research has shown that the clinkerization process is one of the main options for the valorization of waste and its consideration for inclusion as a raw material within the circularity of the cement industry's production process. Thus, waste is used as a raw material for the production of a more useful substance, taking into account the fundamental principles of the circular economy.

Keywords: low-energy cements; waste products; belite; clinker; circular economy; construction materials; CO_2 emissions; low-carbon economy

Citation: Martínez-Martínez, S.; Pérez-Villarejo, L.; Eliche-Quesada, D.; Sánchez-Soto, P.J. New Types and Dosages for the Manufacture of Low-Energy Cements from Raw Materials and Industrial Waste under the Principles of the Circular Economy and Low-Carbon Economy. *Materials* **2023**, *16*, 802. https://doi.org/10.3390/ma16020802

Academic Editor: Yuri Ribakov

Received: 15 December 2022
Revised: 6 January 2023
Accepted: 10 January 2023
Published: 13 January 2023

1. Introduction

Humanity is at a critical juncture in terms of successfully tackling the greatest challenge of our time, climate change. Since the outbreak of the Industrial Revolution, and most markedly in the last century and a half, humankind has been responsible for a dramatic increase in greenhouse gas (GHG) emissions (Appendix A, Table A1), produced largely by industrialization and the burning of fossil fuels, as well as by deforestation and large-scale agriculture. Among the greenhouse gases, the most abundant is carbon dioxide (CO_2), accounting for about two-thirds of all greenhouse gases [1]. This GHG is mainly produced by burning fossil fuels.

In 2015, after four years of negotiations, the Paris Agreement was adopted at the twenty-first session of the Conference of the Parties to the United Nations Framework Convention on Climate Change (COP21) and the eleventh session of the Conference of the Parties, serving as the meeting of the Parties to the Kyoto Protocol (COP-MOP11) in Paris. This global, legally binding climate change agreement obliges all countries to participate in global GHG reductions [2].

Taking the above into account, in December 2019, the "European Green Deal" was launched within the European Union [3]. This European Green Pact sets out an action plan to boost resource efficiency by moving to a clean and circular economy, as well as restoring biodiversity and reducing pollution.

In addition to the increase in GHG emissions, the expansion of industrial activity also increases the amount of waste generated. Currently, most of the waste generated in the construction or mining industry is disposed of in landfills, which, in turn, generates numerous environmental problems and higher costs for companies. Therefore, it is essential to find new methods to absorb the waste generated by the production process.

One author describes the concept of the circular economy, as follows: "A circular economy aims to maintain the value of products, materials, and resources for as long as possible by returning them to the product cycle at the end of their use, while minimizing the generation of waste. The fewer products we discard, the less material we extract, the better for our environment. This process starts at the very beginning of a product's lifecycle: smart product design and production processes can help save resources, avoid inefficient waste management, and create new business opportunities" [4]. In 2015, the European Commission adopted the ambitious EU Action Plan for the Circular Economy, which sets out a program of 54 concrete actions, covering the entire product life cycle from production and consumption to waste management, as well as the secondary raw materials market [4].

Cement is one of the main building materials that is manufactured and used worldwide, with a global production rate in 2021 of 4270 Mt. China and India are the world's largest producers, accounting for approximately 55% and 8% of the world's total cement production, respectively [5]. Good durability performance and compatibility with other construction elements allow the production of composite materials with improved properties [6]. Conversely, the cement manufacturing industry is one of the main industries in terms of GHG emissions, taking as a reference the fact that approximately 0.59 tonnes of CO_2 are generated for each tonne of cement produced, according to data from the year 2021 [5]. The total CO_2 emissions produced by the cement industry account for approximately 8% of global CO_2 emissions [7]. These CO_2 emissions are released in two main ways: 60% of emissions is due to the thermal decomposition of $CaCO_3$ into CaO and CO_2, while 40% is due to the consumption of thermal energy required for the production process [8–11]. Depending on the raw materials and the production process, 60 to 130 kg of fuel oil, or its equivalent, is required for each tonne of cement produced, as well as about 110 kWh of electricity [12].

In addition to the above, the cement manufacturing industry consumes a large number of raw materials that are rich in calcium oxides, silica, alumina, and iron oxide, which can be of mineral origin, mainly derived from clayey, limestone, or loamy materials, or from waste products [13]. The extraction and processing of these raw materials for the cement manufacturing industry are also associated with numerous costs and GHG emissions, as well as the corresponding deterioration of the environment through the generation of degraded areas.

Considering all this, it is essential to reduce GHG emissions from the cement manufacturing industry, in order to comply with the Paris Agreement, as well as to use it as a lever for change in the new concept of the circular economy, bearing in mind that the cement industry possesses a great capacity to absorb industrial wastes that could be used as raw material in the cement manufacturing process.

Clinker is the main ingredient in cement, with the CO_2 emissions generated in cement manufacture being directly proportional to the amount of clinker used. Between 2015 and 2020, the clinker–cement ratio was estimated to increase by approximately 1.6%/year, on average, reaching 0.72 in 2020 [5]. Thus, direct CO_2 emissions from cement production increased from 2203.0 Mt CO_2 in 2015 to 2526.8 Mt CO_2 in 2021 [5]. Ordinary Portland cement (OPC) clinker is generally composed, on average, of 67% CaO, 22% SiO_2, 5% Al_2O_3, and 3% Fe_2O_3, in addition to ~3% of other minor oxides. Depending on the temperature of the process and the proportion of the oxides, a series of solid-state reactions take place that give rise to the formation of the phases that make up the cement clinker [14,15]. In OPC, the proportions of each of these phases range from 50–70 wt % of C_3S (alite or tricalcium silicate), 15–30 wt % of C_2S (belite or dicalcium silicate), 5–10 wt % of C_3A (tricalcium aluminate) and 5–15 wt % of C_4AF (tetracalcium alumino-ferrite), with the characteristics and properties of the cement varying according to the variability of its composition [14,15].

Belitic Portland cement (BPC) or belitic cement, compared to OPC, has a higher proportion of C_2S compared to C_3S in their composition. In addition, C_2S has lower energy requirements in its formation process. This inversion of the proportions of the predominant silicate ratio in the BPC clinker leads to a reduction in CO_2 emissions, due to two main factors: the lower clinkering temperature requirement, which is approximately 100–150 °C lower than OPC [16], implying a reduction in energy consumption and a reduction in the need for limestone as a raw material. This, in turn, reduces the CO_2 emissions due to the decarbonation reaction during the clinkerization process in the kiln. The reduction in the energy demand of BPCs compared to OPCs can be accounted for in savings of approximately 150 kJ/kg [17]. In addition to this, the energy factor accounts for 30–40% of the total cost of production [18].

In contrast, cements rich in C_2S have lower initial strengths than OPC, mainly because belite has slower hydration kinetics than alite (Figure 1). However, the hydration kinetics in belite will depend on the polymorph obtained during the manufacturing process, which in belite can be of five different forms, known as α, α'_H, α'_L, β, and γ.

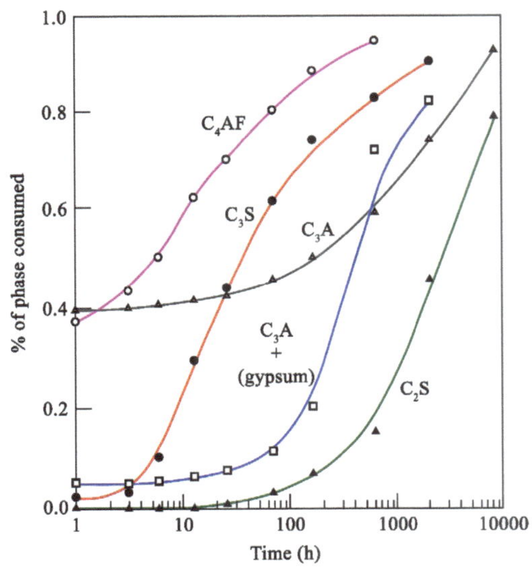

Figure 1. Typical hydration kinetics of the pure phases of cement clinker at room temperature (adapted from reference [19]).

Thus, the objective of this research is the development of different types of cement clinker blends with a high belite content, the main characteristic of which is the use of industrial wastes in their composition as raw materials, as well as their design for the production of low-energy cement clinkers, with a reduction in the manufacturing temperature of OPCs of approximately 100–150 °C, thereby improving the circularity of cement production and CO_2 emissions. The motivation for this research is the possibility of manufacturing cement clinker from industrial wastes, either at a high percentage of 50 wt % or the total substitution of raw materials, i.e., a cement clinker composed entirely of industrial waste.

2. Materials and Techniques

2.1. Samples and Sample Preparation

A total of eight raw materials and three industrial waste types have been selected, which come from different geographical areas (Figure 2). As indicated in the Introduction, the main materials to be used for clinker manufacture with a high dicalcium silicate (belite, C_2S) content are rich in calcium oxide (CaO) and silicon dioxide (SiO_2), in terms of their chemical composition. For clinker design, the necessary limit to the composition of aluminum oxide (Al_2O_3), ferric oxide (Fe_2O_3) and magnesium oxide (MgO) must be considered. This composition will directly influence the clinker quality indices, calculated by means of modified Bogue equations, as well as the expansivity of the cement.

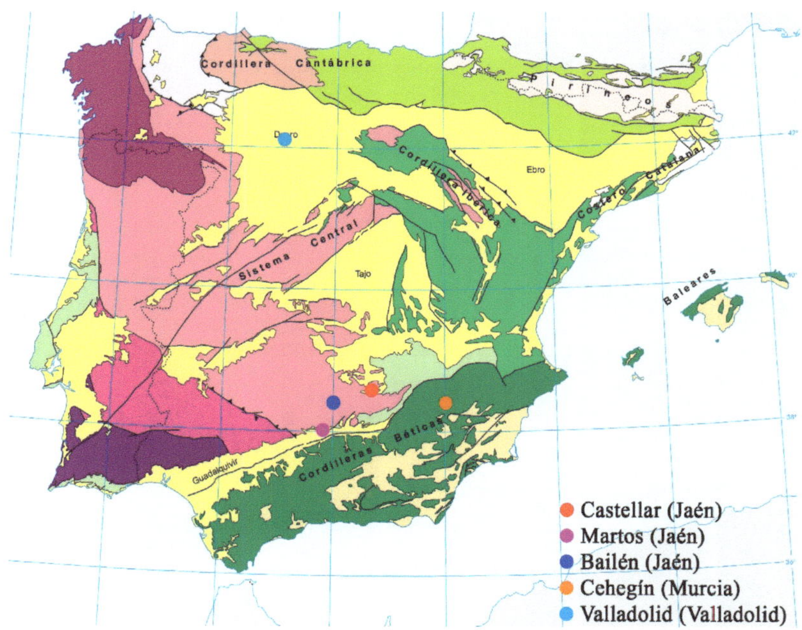

Figure 2. Map of the Iberian Peninsula and the Balearic Islands, showing the main geological units and the location of the points of origin of Spain's raw materials and industrial waste [20].

In this way, a series of raw materials and industrial wastes have been initially selected, considering both their initial chemical composition and their production capacity. In all the cases, materials are incorporated that, due to their versatility and widespread use, can be found on all five continents. Thus, the main source of CaO will be limestone and marble-cutting sludge. The main source of SiO_2 will be selected clays (yellow, black, and red clays) and the crushed waste material, chamotte. Salt slags from secondary aluminum production (PAVAL) will be used as the Al_2O_3 content corrector. Finally, gypsum will be used as a setting regulator for the cement paste.

ANAYA limestone (designated as C_A): This limestone comes from Castellar (Jaén, Spain) (Figure 2). From the geological point of view, the area of exploitation is located in the outer reaches of two large geological units: the Iberian Massif and the Baetic Mountain Range, this also being the northeastern apex of the Guadalquivir basin [21]. The sample was supplied by the trading company, ÁRIDOS ANAYA, S.L. The extraction works are carried out in open-cast mines by blasting benches of approximately 15–17 m in height using explosives.

SANTO NICASIO limestones 1 and 2 (designated as C_SN_1 and C_SN_2, respectively): The samples come from Martos (Jaén, Spain) (Figure 2). From the geological point of view, the exploitation area is located in the northern sector of the Baetic mountain range, at the boundary of the Outer Zones with the Guadalquivir Depression [22]. The limestone was supplied by the trading company, ÁRIDOS SANTO NICASIO, S.L. The extraction works are carried out in open-cast mines by blasting benches of approximately 20 m high with explosives.

TITAN limestone (designated C_T): This limestone comes from Patras (Greece). The sample was supplied by the trading company TITAN GREECE. This limestone is extracted in open-cast mines by blasting with explosives.

The three clays used in this research, called yellow clay, black clay, and red clay (designated as A_A, A_N and A_R respectively), come from Bailén (Jaén, Spain) (Figure 2). From the geological point of view, the area where these clays are sourced is situated on the territory where two of the main morphological units of the main range where the Iberian Peninsula can be divided: the Iberian Plateau and the Guadalquivir Depression [23]. The clays were supplied by the trading company, ARCILLAS BAILÉN, S.L. The extraction work is carried out via open-cast mines using mechanical means (a backhoe or bulldozer).

The industrial wastes or by-products used in this research are:

- Chamotte (designated as CHM), obtained by grinding or crushing fragments of fired ceramic materials (produced at high temperature) that are defective, either because the kiln has not reached the necessary temperature in the sintering process, because the material has spent less time than necessary in the kiln for sintering, or because of the appearance of cracks, defects, etc., during the manufacturing process. In this case, the defective ceramic pieces were supplied by the trading company LADRILLOS BAILÉN S.A., located in Bailén (Jaén, Spain) (Figure 2).
- Cutting and polishing marble sludges (designated as MRM) sourced from Cehegín (Murcia, Spain) (Figure 2). From a geological point of view, the study area is mainly located within the Subbetic zone [24]. In terms of the extraction of the marble blocks from which the sludge was sourced, the main process used was the cutting of marble blocks with a wire cutter (diamond wire), which generates sludge with a grain size of less than 0.15 mm [25]. Likewise, sludge from marble block-cutting processes using looms or block cutters, as well as from the marble slab-polishing process, have also been used.
- Secondary aluminum production salt slag (designated as PVL), also known as PAVAL, comes from the factory owned by the trading company, BEFESA MEDIO AMBIENTE, S.L., located in Valladolid (Spain) (Figure 2). This sample has an aluminum oxide (Al_2O_3) content of around 70%, so it will be used as a corrector of this oxide in clinker obtained from industrial waste.

All the materials, except the gypsum (designated as Y), were received in the laboratory and then subjected to a moisture control process. For the calculation of the moisture percentage, carried out in triplicate, Equation (1) is used, where part of the initial material sample (m_i) is weighed with an analytical balance and placed in a laboratory oven at 105 °C for 24 h. After 24 h, the material (m_f) is weighed again, repeating this process until a constant value is reached between two weighings. The moisture percentage is determined according to the formula:

$$\%\ \text{moisture} = ((m_i - m_f)/m_i) \times 100 \quad (1)$$

The moisture contents can be seen in Table 1, although these values will depend to a large extent on the period of year when the materials are received, interfering with the humidity of the environment and the humidity caused by rainfall in the stockpile areas, especially in mining operations.

Table 1. Moisture content (wt %) of raw materials, as received in the laboratory.

Raw Materials	Moisture Content (wt %)
C_A	14.5 ± 1.90
C_SN_1	21.7 ± 3.51
C_SN_2	19.3 ± 1.97
C_T	13.1 ± 0.93
A_A	25.3 ± 1.76
A_N	24.7 ± 0.92
A_R	25.4 ± 0.98
CHM	7.1 ± 0.93
MRM	13.3 ± 2.20
PVL	14.6 ± 1.08

Once the moisture content of the raw materials has been determined, the same drying process is carried out with the entire volume of material received, in order to facilitate the subsequent grinding and sieving of all the samples. It is very important to obtain a homogeneous particle size that allows all the reactions in the clinkerization process to be carried out correctly, as the control of the particle size influences the physical and mechanical characteristics of the final product [26]. In this investigation, all materials were sieved to a particle size of 100 μm or less.

2.2. *Techniques*

The experimental techniques used for the characterization of the raw materials and industrial wastes included establishing their chemical composition, determined by X-ray fluorescence (XRF, PANalytical® model Zetium, Malvern, UK), mineralogical composition, determined by X-ray diffraction (XRD, PANalytical® model Empyrean, Malvern, UK), thermogravimetric analysis, derivative thermogravimetric analysis, and differential scanning calorimetry (TG-DTG-DSC, NETZSCH® model STA 449 F3 Jupiter, Selb, Germany), a morphological study via scanning electron microscopy and chemical analysis by energy-dispersive X-ray spectroscopy analysis (SEM, Carl Zeiss® model MERLIN, Jena, Germany, EDS, Oxford Inca Energy 350X-MAX 50, Abingdon, UK), as well as functional group identification by attenuated total reflectance Fourier transform infrared spectroscopy (ATR-FTIR, Bruker® model Tensor 27, Billerica, MA, USA).

2.3. *Dosage Design, Modified Bogue Equations and Quality Indexes*

As discussed above, the main difference in the design of ordinary Portland cement (OPC) and cement with a high content of C_2S is the proportion of tricalcium silicate (C_3S) and dicalcium silicate (C_2S) in the cement clinker. This, in turn, is associated with the ratio of one of the most important quality indices, the lime saturation factor (LSF). The LSF marks the ratio between the lime contained in the clinker and the maximum amount of lime that can be set by the components at their maximum basicity; in the case of values lower than 100, they have a composition that can withstand subsequent lime additions. The opposite scenario with values higher than 100 indicates a composition with a theoretical excess of lime [27].

In addition to the LSF, there are two other quality indices to be taken into account in the clinker design. These two quality indices are the aluminum modulus (AM) and the silica modulus (SM). The AM establishes a ratio between the Al_2O_3 and Fe_2O_3 contents, which is equivalent to relating the "fluxing" components, i.e., C_3A and C_4AF [27]. The SM characterizes the ratio between the solid phase and the liquid phase; its value indicates the specific weight of the silicates in the clinker, compared to the fluxes. The higher the SM, the lower the percentage of liquid phase, and the worse the characteristics of the raw blend during the firing process of lime production [27].

For the design of the raw materials, in addition to theoretically calculating the quality indices that the cement clinker will have, the theoretical proportions of the mineralogical phases of the manufactured clinker are calculated, which are C_3S, C_2S, C_3A, and C_4AF. The approximate calculation of these theoretical phases of the cement clinker is performed using modified Bogue equations (Equations (2)–(5)) [6,28,29]. It is important to note that for the calculation of the crude blends and, therefore, the calculation of the addition of the different industrial wastes, the proportions of the main oxides (CaO, SiO_2, Al_2O_3, and Fe_2O_3) must be kept constant.

$$C_3S = 4.0710{\cdot}CaO - 7.6024{\cdot}SiO_2 - 1.4297{\cdot}Fe_2O_3 - 6.7187{\cdot}Al_2O_3 \quad (2)$$

$$C_2S = 8.6024{\cdot}SiO_2 + 1.0785{\cdot}Fe_2O_3 + 5.0683{\cdot}Al_2O_3 - 3.0710{\cdot}CaO \quad (3)$$

$$C_3A = 2.6504{\cdot}Al_2O_3 - 1.6920{\cdot}Fe_2O_3 \quad (4)$$

$$C_4AF = 3.0432{\cdot}Fe_2O_3 \quad (5)$$

The equations used for the calculation of LSF (Equation (6)), AM (Equation (7)), and SM (Equation (8)) are as follows [6,27,30]:

$$LSF = (100{\cdot}CaO)/(2.8{\cdot}SiO_2 + 1.18{\cdot}Al_2O_3 + 0.65{\cdot}Fe_2O_3) \quad (6)$$

$$AM = Al_2O_3/Fe_2O_3 \quad (7)$$

$$SM = SiO_2/(Al_2O_3 + Fe_2O_3). \quad (8)$$

The calculations of both modified Bogue equations and the quality index imply the percentages by weight of each oxide (CaO, SiO_2, Al_2O_3, and Fe_2O_3).

A Microsoft Excel® spreadsheet designed by researchers at the METLAB laboratory of the Department of Chemical Engineering of the University of Patras (Patras, Greece) was used for the design of the crude blends [6]. For the purposes of this theoretical design, a C_2S ratio of 60–65% and an LSF of 78–83% are used as limiting values that are common to all of them [31,32].

3. Results and Discussion

3.1. Characterization of Raw Materials and Industrial Wastes

3.1.1. Chemical Results

The results of the chemical composition of the raw materials and industrial wastes, expressed in oxides, as determined by the X-ray fluorescence technique (XRF), as well as the loss on ignition (LOI) or calcination at 1000 °C, are presented in Tables 2 and 3.

The limestones (Table 2) are mainly composed of CaO, with values varying between 34.54 wt % by weight of the C_A and 46.54 wt % of the C_T, noting that between the two fractions obtained from the SANTO NICASIO limestone (C_SN_1 and C_SN_2), there are no significant differences (41.86 wt % and 42.77 wt %). The main difference between the limestones is fundamental and is found in their magnesium oxide content, which is a determining factor in the calculation of the raw material dosages. This will, to a great extent, be responsible for the future expansivity of the cement. In this sense, it is possible to observe the great variation between C_A (19.12 wt %), C_SN_1 (4.59 wt %), C_SN_2 (4.63 wt %), and C_T (0.63 wt %). It should be borne in mind that the calculation basis for the crude blends is based on calcined materials, i.e., eliminating the corresponding losses on ignition (LOI), so that the percentage of magnesium oxide, as is the case with the other chemical elements, will be higher. Furthermore, C_A has higher ignition losses than the rest (46.31 wt %), the lowest being those of C_T (38.45 wt %). In the case of C_T, the lower content of magnesium oxide and the loss on ignition is compensated for by a higher percentage of silica (10.51 wt %).

Table 2. Chemical composition (XRF, wt %) of the raw materials. N.D.: not detected.

	C_A	C_SN_1	C_SN_2	C_T	A_A	A_N	A_R
Al_2O_3	0.00	2.44	0.96	1.17	10.53	10.96	19.81
CaO	34.54	41.86	42.77	46.54	17.30	17.41	3.89
SiO_2	0.00	4.56	2.31	10.51	48.60	44.29	48.69
Fe_2O_3	0.26	0.47	0.22	0.86	5.33	5.56	7.71
Na_2O	0.00	0.17	0.15	0.11	0.40	0.40	0.22
K_2O	0.01	0.16	0.05	0.29	2.30	2.92	5.47
MgO	19.12	4.59	4.63	0.63	1.86	1.79	3.43
SO_3	0.02	0.00	0.00	0.00	0.09	8.26	0.05
TiO_2	0.02	0.05	0.00	0.00	0.64	0.65	0.80
P_2O_5	0.00	0.02	0.01	0.00	0.08	0.07	0.10
MnO	0.03	0.00	0.00	0.00	0.03	0.03	0.17
LOI	46.31	43.90	44.20	38.45	12.23	10.70	9.07
Zr (ppm)	6.80	0.00	0.00	0.00	314.20	304.10	136.90
Cl (ppm)	0.06	0.00	0.00	0.00	0.01	0.02	0.02
Cr (ppm)	2.89	0.00	0.00	0.00	0.00	0.00	0.01
Sr (ppm)	N.D.	N.D.	N.D.	N.D.	0.04	0.05	0.02
Ce (ppm)	N.D.	N.D.	N.D.	N.D.	0.00	0.01	0.01
Pb (ppm)	N.D.	N.D.	N.D.	N.D.	0.00	0.00	0.01
Ni (ppm)	N.D.	N.D.	N.D.	N.D.	0.01	0.01	0.01
Cu (ppm)	N.D.	N.D.	N.D.	N.D.	0.01	0.00	0.00
Zn (ppm)	N.D.	N.D.	N.D.	N.D.	0.01	0.01	0.01
Ba (ppm)	N.D.	N.D.	N.D.	N.D.	0.07	0.06	0.07
Rb (ppm)	N.D.	N.D.	N.D.	N.D.	0.01	0.01	0.03

Table 3. Chemical composition (XRF, wt %) of the gypsum and the sample wastes.

	Y	CHM	MRM	PVL
Al_2O_3	0.18	13.75	0.00	65.40
CaO	24.25	14.75	52.14	1.73
SiO_2	12.51	51.23	0.00	5.07
Fe_2O_3	0.04	6.92	0.06	1.94
Na_2O	0.00	0.43	0.00	1.24
K_2O	0.00	3.11	0.00	0.49
MgO	1.81	2.38	0.12	8.28
SO_3	38.53	3.63	0.20	0.41
TiO_2	0.00	0.72	0.01	0.67
P_2O_5	0.00	0.09	0.00	0.02
MnO	0.00	0.04	0.00	0.18
LOI	22.69	4.00	43.31	12.29
Zr (ppm)	0.00	278.10	5.50	129.70
Cl (ppm)	0.00	0.01	0.01	0.65
Cr (ppm)	0.00	0.01	0.00	0.00
Sr (ppm)	0.00	0.05	0.01	0.04
Pb (ppm)	0.00	0.01	0.00	0.02
Bi (ppm)	0.00	0.00	0.00	0.01
F (ppm)	0.00	0.00	0.00	0.98
Ni (ppm)	0.00	0.01	0.00	0.05
Cu (ppm)	0.00	0.01	0.00	0.45
Zn (ppm)	0.00	0.02	0.00	0.19
Sn (ppm)	0.00	0.00	0.00	0.01
Ba (ppm)	0.00	0.07	0.00	0.20
Rb (ppm)	0.00	0.02	0.00	0.00

The clays (Table 2) are mainly composed of SiO_2, with a content of more than 44 wt % in all of them. A_A and A_N have very similar percentages of Al_2O_3 (~10 wt %), CaO (~17 wt %), Fe_2O_3 (~5 wt %), K_2O (~2 wt %), and MgO (~1.8 wt %). However, A_N has a higher percentage of SO_3 (8.26 wt %), it being practically nil in A_A. A_R has a higher content of Al_2O_3 (19.81 wt %), Fe_2O_3 (7.71 wt %), K_2O (5.47 wt %), and MgO (3.43 wt %) than the other clays, although its CaO content is lower (3.89 wt %) and its SO_3 content is almost zero (0.05 wt %). The ignition losses are generally associated with the physically retained water and structural OH of silicates, as well as the decomposition process of the carbonates. The values obtained are higher for the A_A sample (12.23 wt %) compared with the lower values for the A_R (9.07 wt %).

Gypsum (Y) (Table 3), which will be used as a setting regulator in the hydrated cement pastes, is mainly composed of SO_3 (38.53 wt %) and CaO (24.25 wt %). The sample contains smaller amounts of SiO_2 and MgO, as well as ignition losses of less than 23 wt %.

The CHM waste (Table 3) is mainly characterized by its high SiO_2 content (51.23 wt %), with significant additional contents of CaO and Al_2O_3 that are always above 10 wt %, as well as Fe_2O_3 (6.92 wt %). It can be seen, in general terms, that chamotte waste has a chemical composition similar to that of the clays.

The MRM waste (Table 3) is basically composed of CaO (52.14 wt %), although it presents very high ignition losses (43.31 wt %), due, to a great extent, to the decomposition of the calcium carbonates in the form of CO_2 as evolved gas. They also contain very low quantities of impurities in the form of Fe_2O_3, MgO, or SO_3, always in percentages of less than 0.20 wt %.

The PVL waste (Table 3) is mainly composed of aluminum-rich oxide and hydroxide phases [33,34]. Their Al_2O_3 composition value is high (65.40 wt %), a finding that coincides with the data provided by the producer [35]. The MgO (8.28 wt %) and SiO_2 (5.07 wt %), as well as the Na_2O and F (~1 wt %) contents, are also noteworthy, albeit to a lesser extent. As already mentioned in the case of limestones, the MgO value must be taken into account when calculating the different raw material dosages because of its effects on the future expansivity of the cement.

3.1.2. Mineralogical Composition

The crystalline phases identified for sample C_A (Figure 3a) are calcite ($CaCO_3$) and dolomite [$CaMg(CO_3)_2$]. More intense diffraction peaks are observed for dolomite, which is in agreement with the high percentage of magnesium oxide determined in its chemical composition (Table 2). This limestone can therefore be considered a dolomitic limestone. The crystalline phases identified for samples C_SN_1 (Figure 3b) and C_SN_2 (Figure 3c) are calcite and dolomite. It is clear that the two limestones have a similar XRD pattern, according to their chemical composition results. Compared to the results obtained for C_A, a lower intensity of the diffraction peak corresponding to dolomite is observed in the two diffractograms of C_SN_1 and C_SN_2. This is due to the lower amounts of magnesium oxide in their chemical composition (Table 2). The crystalline phases identified for C_T (Figure 3d) are calcite and quartz (SiO_2). These results agree with the chemical composition of the limestone (Table 2), with CaO accounting for 46.54 wt % and SiO_2 for 10.51 wt %. Unlike the diffractograms of C_A, C_SN_1, and C_SN_2, dolomite is not identified in C_T. This is due to the almost scarce presence of MgO in the limestone composition (Table 2); therefore, the conditions for the substitution of calcium ions by magnesium in calcite do not occur.

The crystalline phases identified for sample A_A (Figure 3e) are quartz, calcite and dolomite, muscovite (illite mica) [$KAl_2(AlSi_3O_{10})(OH)_2$], and feldspars such as orthoclase ($KAlSi_3O_8$) and albite ($NaAlSi_3O_8$), as well as montmorillonite (magnesium aluminum silicate: $MgOAl_2O_{35}SiO_2xH_2O$). In the case of sample A_N (Figure 3f), the crystalline phases that were identified are quartz, calcite and dolomite, bassanite or gypsum hemihydrate ($CaSO_4.1/2H_2O$), chlorite [$(Al,Fe,Mg)_{4-6}(Al,Si,Fe)_4O_{10}(OH,O)_8$], muscovite (illite mica) and rutile (TiO_2). For sample A_R (Figure 3g), the crystalline phases identified are quartz, calcite, and dolomite, muscovite (illite mica), and hematite (Fe_2O_3), the latter crystalline phase being responsible for the red color of the clay.

The crystalline phases identified for the sample of gypsum (Y) (Figure 3h), the setting regulator, are calcium sulfate dihydrate or pure phase mineral gypsum ($CaSO_4$–$2H_2O$), quartz, and anhydrite ($CaSO_4$). Quartz has been identified due to the high percentage of SiO_2 in the chemical analysis (Table 3), this not being associated with silicates in the sample.

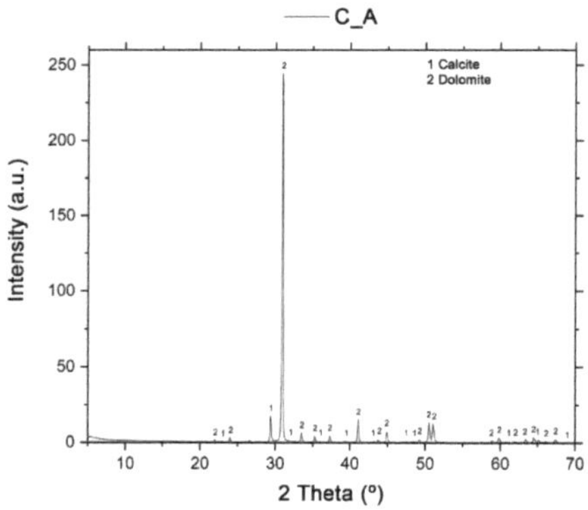

(a)

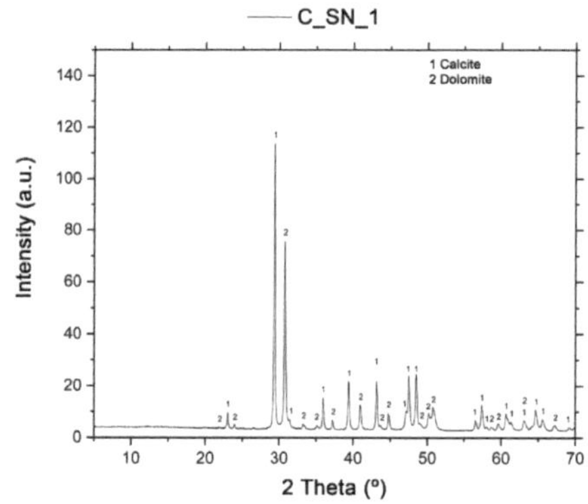

(b)

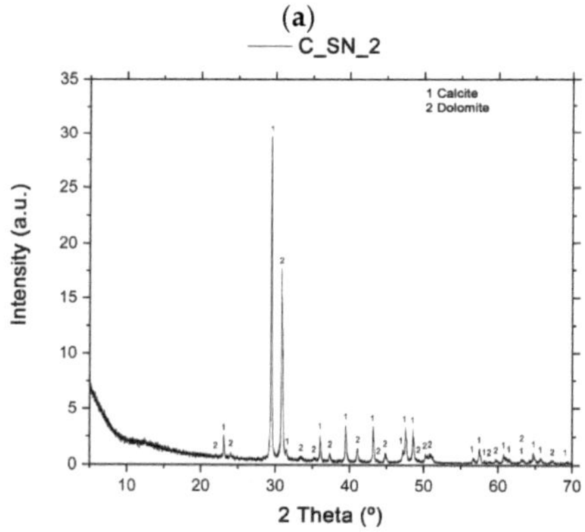

(c)

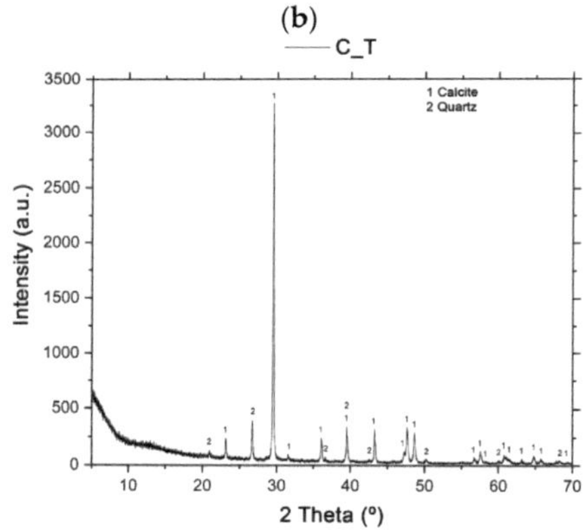

(d)

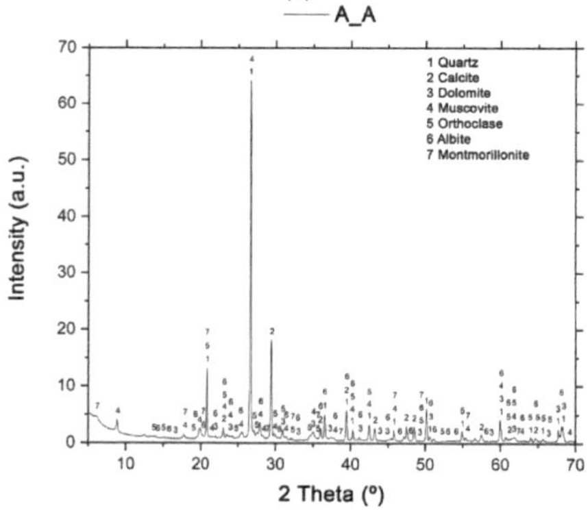

(e)

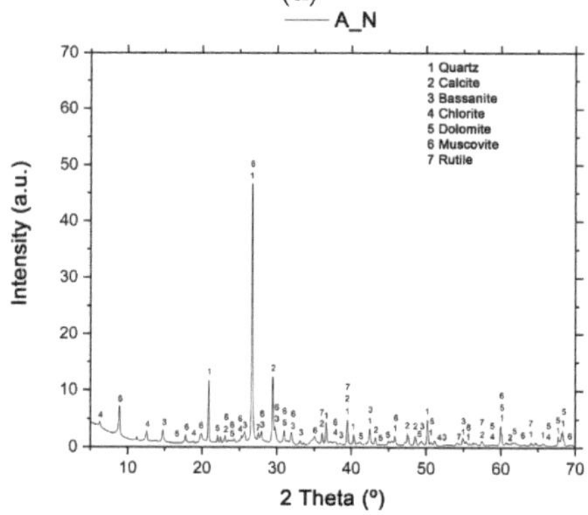

(f)

Figure 3. *Cont.*

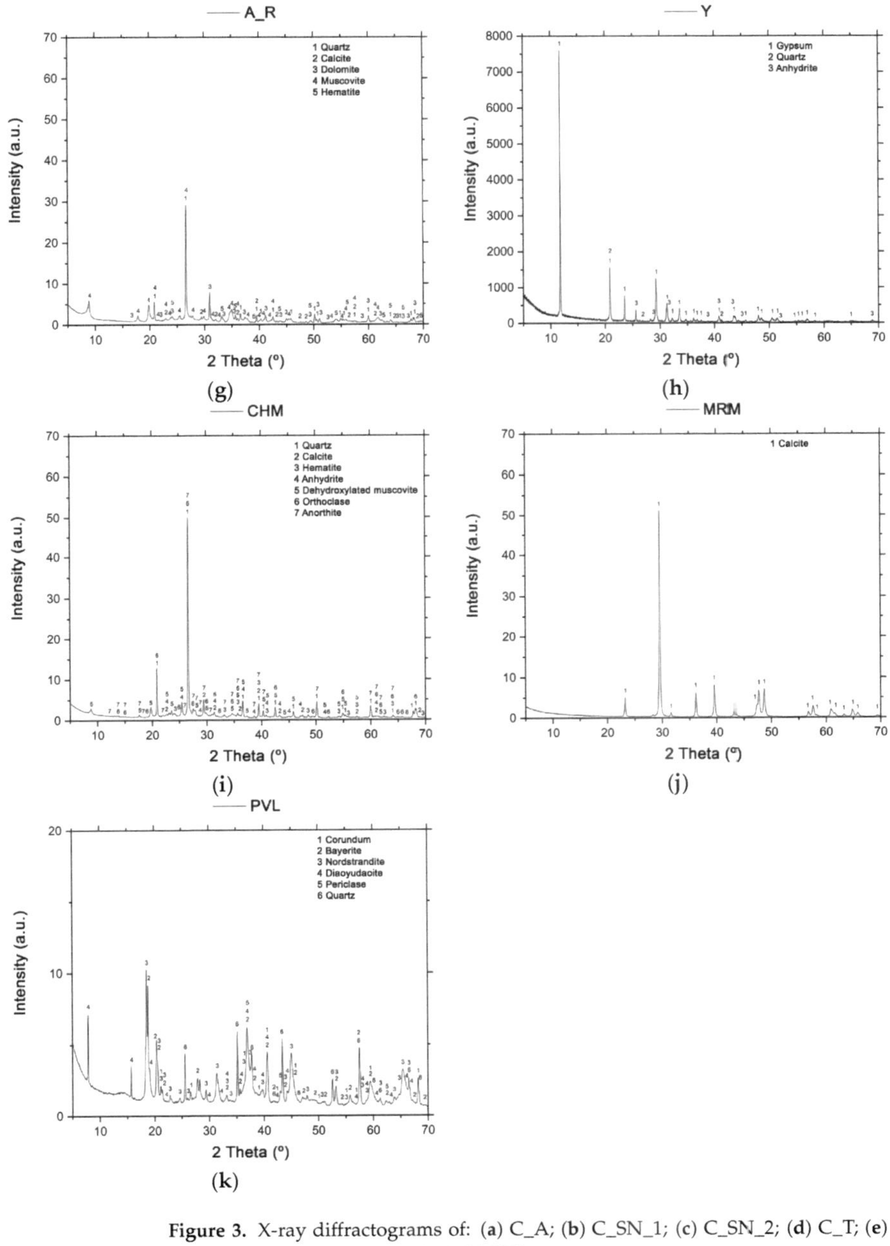

Figure 3. X-ray diffractograms of: (**a**) C_A; (**b**) C_SN_1; (**c**) C_SN_2; (**d**) C_T; (**e**) A_A; (**f**) A_N; (**g**) A_R; (**h**) Y; (**i**) CHM; (**j**) MRM; (**k**) PVL.

The crystalline phases identified for the CHM waste (Figure 3i) are quartz, calcite, hematite, anhydrite, dehydroxylated muscovite (illite mica), feldspars such as orthoclase and calcium, and an aluminum feldspar called anorthite ($CaAl_2Si_2O_8$), which is a product of the heat treatment [6].

In the MRM diffractogram (Figure 3j), only calcite can be identified, which fits perfectly with the standard $CaCO_3$ pattern [6].

The crystalline phases identified for the PVL waste (Figure 3k) are corundum (α-Al_2O_3), which is a mineral belonging to the hematite group, bayerite [γ-$Al(OH)_3$], which is an aluminum hydroxide crystallizing in the monoclinic system, nordstrandite [$Al(OH)_3$], which is an aluminum hydroxide crystallizing in the triclinic system, diaoyudaoite ($NaAl_{11}O_{17}$), which is an aluminum oxide containing sodium in solid solution, periclase (MgO), and quartz. According to the characteristics provided by the producer and reported in [35], PVL also has fluorite (CaF_2) in its mineralogical composition, in a percentage of 1–2%, as well as halite (NaCl), in a percentage of 0–1%, both mineralogical phases not being identified by XRD. A large number of aluminum-based mineral phases coincide with the origin of the waste obtained in the recycled aluminum process. In addition, the presence of compounds such as fluorite is due to their use as scorification salts during the recycling process.

3.1.3. Thermal Analysis (TG-DTG-DSC)

The analyses were carried out on C_A (Figure 4a), C_SN_1 (Figure 4b), and C_SN_2 (Figure 4c), which are mineralogically composed of calcite and dolomite, according to the XRD patterns. In the thermogravimetric (TG) curve, the calcite and dolomite show a total weight loss at up to 880 °C of 48 wt % and 44 wt %, respectively, with the limestones C_SN_1 and C_SN_2 showing equivalent weight losses. This weight loss is due to the decomposition of the calcium and magnesium carbonates in two consecutive stages. The analyses carried out on C_T (Figure 4d), which is mineralogically composed of calcite and quartz, as indicated by XRD data, show a total weight loss of 39 wt % up to 880 °C in the TG curve, which is due to the decomposition of calcium carbonate (Equation (9)).

According to the differential scanning calorimetry (DSC) analysis of limestones C_A, C_SN_1, and C_SN_2, the first endothermic reaction is observed at 803 °C, 780 °C, and 775 °C, respectively. This is due to the decomposition of magnesium carbonate associated with the decomposition of dolomite (Equation (10)) [36–38], with a weight loss, according to the TG curve, of 20 wt % for C_A, 9 wt % for C_SN_1, and 7 wt % for C_SN_2. Subsequently, for C_A two other endothermic reactions occur at 845 °C and 860 °C, respectively, while in C_SN_1 and C_SN_2 a single endothermic reaction is observed at 860 °C, which is due to the decomposition of calcium carbonate, in this case (Equation (9)) [39–44], with a weight loss of 28 wt % for C_A, 35 wt % for C_SN_1 and 37 wt % for C_SN_2.

$$CaCO_3\ (s) \rightarrow CaO\ (s) + CO_2\ (g) \tag{9}$$

$$CaMg(CO_3)_2\ (s) \rightarrow MgO\ (s) + Ca\ (s) + 2CO_2\ (g) \tag{10}$$

Comparing the DSC test curves of the C_A limestone with those of the C_SN_1 and C_SN_2 limestones, the first endothermic peak produced by the decomposition of the carbonate ions associated with magnesium appears in the range of 775–803 °C; that is, the C_A has a higher thermal reaction enthalpy. This reflects an agreement with the previous results and also the higher content of magnesium oxides in its composition. According to the differential scanning calorimetry (DSC) analysis of C_T, the endothermic reaction takes place at about 860 °C [41–44].

The DSC-TG analyses of clays A_A (Figure 4e), A_N (Figure 4f), and A_R (Figure 4g) show an endothermic peak at temperatures between 20 °C and 200 °C, demonstrating the removal of weakly bound water (dehydration), with a weight loss of 1.7 wt % for A_A, 1.3 wt % for A_N, and 1.7 wt % for A_R. The intensity is slightly higher in the A_A curve, which is possibly due to the presence of montmorillonite [45], as corroborated by XRD analysis. In the temperature range between 200 °C and 600 °C, a weight loss attributed to the removal of organic matter and the dehydroxylation of silicates is observed, as indicated by the exothermic band centered at approximately 380 °C and the endothermic band centered at approximately 550 °C, respectively [46,47]. These weight losses correspond to 1.9 wt % for A_A, 2.4 wt % for A_N, and 3.0 wt % for A_R. The largest weight loss occurs between 600 °C and 800 °C, this being due to the decomposition of carbonates, as indicated by the endothermic peak centered at approximately 730–780 °C, with losses of 9.6 wt % for A_A, 8.7 wt % for A_N, and 5.0 wt % for A_R. At temperatures between 800 °C and 1100 °C, an exothermic band associated with the crystallization of the high-temperature phases is observed, with weight losses of 0.3 wt % for A_A, 2.2 wt % for A_N, and 0.3 wt % for A_R. In this temperature zone, specifically at 900 °C and in A_A, there is a small endothermic effect associated with the conversion of low-temperature albite to its high-temperature phases, although these effects may overlap with those of calcium carbonate decomposition [48]. At temperatures above 1100 °C, no weight loss is observed, with a more pronounced endotherm in the DSC curve in A_A, which could possibly indicate the formation of a liquid phase [6].

In the DSC curve of the sample of gypsum (Y) (Figure 4h), different endothermic peaks and an exothermic peak are observed. Firstly, up to about 110 °C, an endothermic effect occurs, which is due to the loss of moisture in the material, resulting in a weight loss of 1 wt %. Secondly, there are two overlapping endothermic effects corresponding to the dehydration of the gypsum into hemihydrate and anhydrite III. The first of these, produced at 150 °C, corresponds to the loss of $1\frac{1}{2}$ H_2O to form the calcium sulfate hemihydrate phase (Equation (11)) [49,50], while the second, produced at 190 °C, corresponds to the loss of $\frac{1}{2}$ H_2O to form the β-anhydrite III or soluble anhydrite phases (Equation (12)) [49–51]. These two reactions result in a weight loss of 17.5 wt %.

$$CaSO_4{\cdot}2H_2O \rightarrow CaSO_4{\cdot}1/2H_2O + 3/2H_2O \tag{11}$$

$$CaSO_4{\cdot}1/2H_2O \rightarrow CaSO_4\ (III) + 1/2H_2O \tag{12}$$

An exothermic reaction associated with the transformation of anhydrite III (soluble) to anhydrite II (insoluble) occurs at 380 °C (Equation (13)) [49,50]. The weight loss between the temperatures of 190 °C and 380 °C is 2 wt %.

$$CaSO_4\ (soluble) \rightarrow CaSO_4\ (insoluble) \tag{13}$$

The next endothermic reaction takes place at 1220 °C and corresponds to the change from anhydrite II to anhydrite I. From this point on, the decomposition of calcium sulfate takes place (Equation (14)) [50], with a weight loss of 3 wt %.

$$CaSO_4\ (I) \rightarrow CaO + SO_2 + 1/2O_2 \tag{14}$$

In the rather complex thermogravimetric curve of the CHM waste (Figure 4i), two important inflections can be observed, corresponding first to weight loss due to the decomposition of the carbonates (4 wt %) up to 750 °C and a second weight loss of 1.5 wt % between 1100 °C and 1200 °C, due to the elimination of the last traces of structural water and the formation of the amorphous or glassy phase, and also due to the consumption produced by the continuous reactions, such as the formation of anorthite and mullite [$3Al_2O_3{\cdot}2SiO_2$]. Other compounds, such as illite and gehlenite [$Ca_2Al_2SiO_7$], decompose at temperatures above 1150 °C. According to the DSC analysis, a first thermal dehydration process produced by heating can be observed, as indicated by the endothermic reaction at a temperature below 100 °C that is associated with the release of the absorbed water. The second endothermic reaction at 575 °C is due to the polymorphic phase-change of the silica, which generates strong dimensional variations in its structure (Equation (15)) [48].

$$quartz\ (\alpha) \rightarrow (\beta)\ quartz \tag{15}$$

The third endothermic reaction at 731 °C is due to the decomposition of calcite (Equation (9)). Around 1100 °C, an exothermic reaction can be observed, indicating the formation of crystalline phases. At temperatures above 1150 °C, no weight loss is observed, but the DSC curve shows three endothermic peaks, which may indicate the formation of the liquid phase [6].

The TG curve of the MRM waste (Figure 4j) shows a main inflection due to the decomposition of carbonates (Equation (9)) with a weight loss of 44 wt %, in agreement with the results for the loss on ignition (Table 3). According to the DSC analysis, the presence of calcite is deduced through the endothermic reaction that takes place around 866 °C [6].

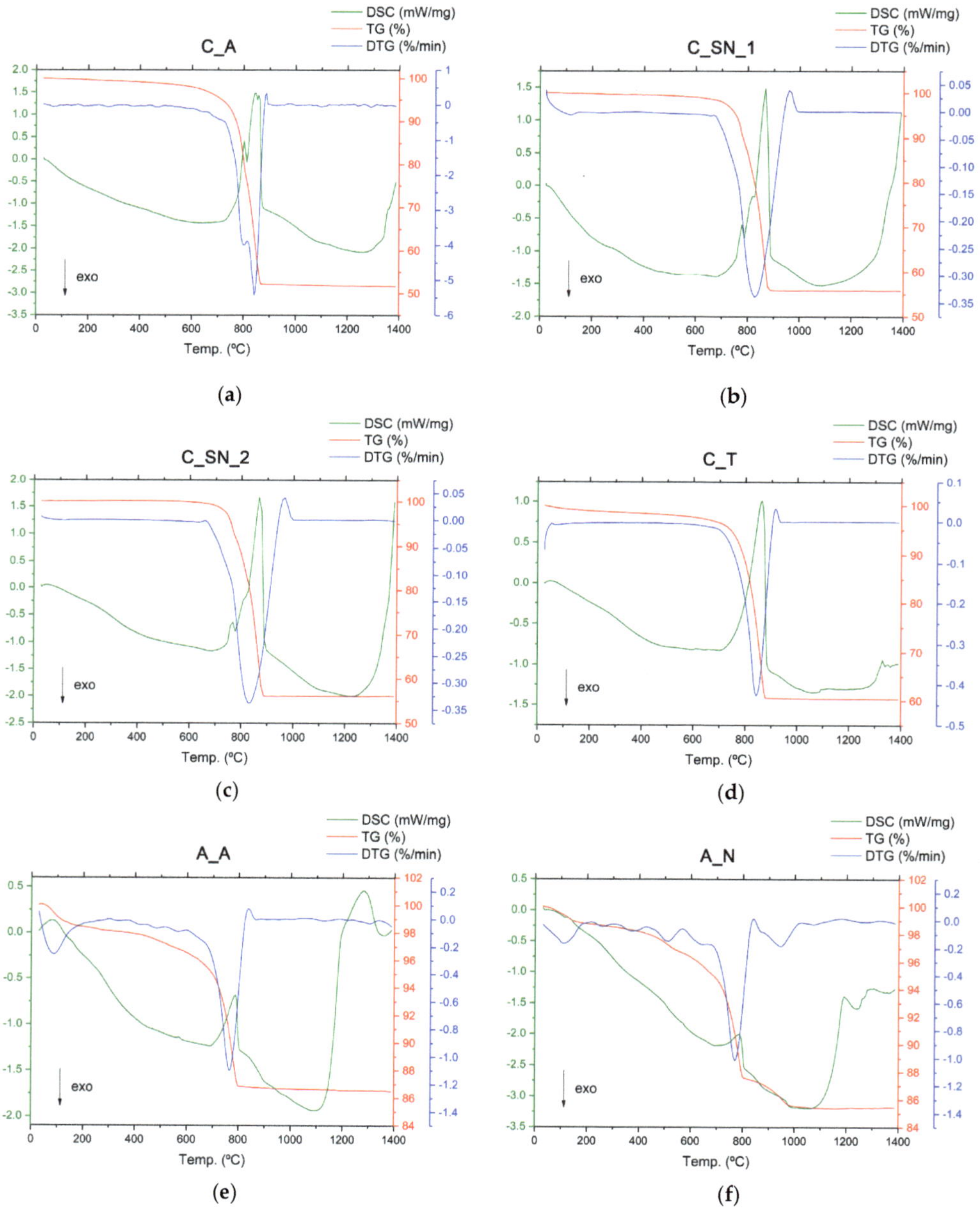

Figure 4. *Cont.*

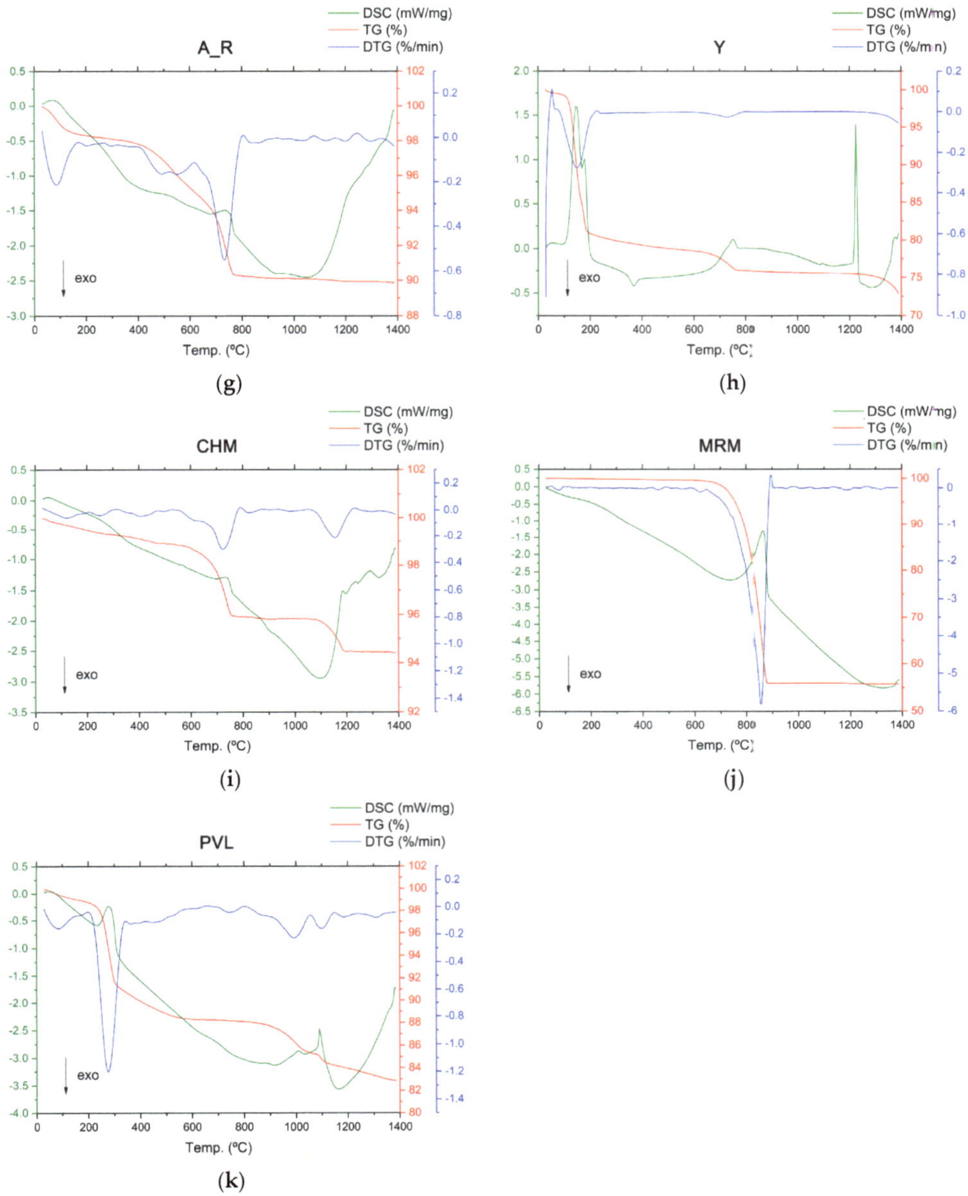

Figure 4. TG (red line), DTG (blue line), and DSC (green line) analysis of: (**a**) C_A; (**b**) C_SN_1; (**c**) C_SN_2; (**d**) C_T; (**e**) A_A; (**f**) A_N; (**g**) A_R; (**h**) Y; (i) CHM; (j) MRM; (**k**) PVL.

In the TG curve of the PVL waste (Figure 4k), a total weight loss of 17.2 wt % can be observed, up to a temperature of 1390 °C. According to the DSC curve, an endothermic reaction occurs at 290 °C that is associated with the dehydroxylation of $Al(OH)_3$ (Equation (16)) [49], which generates a weight loss of 8.5 wt % [52].

$$2Al(OH)_3 \rightarrow Al_2O_3 + 3H_2O \quad (16)$$

The sequence of the thermal transformation of alumina can be defined as obtaining γ-Al_2O_3 at 500 °C, remaining as a single phase up to 600 °C. Subsequently, at 850 °C, the δ-Al_2O_3 phase is identified, which coexists with γ-Al_2O_3. At 1000 °C, θ-Al_2O_3 is obtained, with the coexistence of the previous phases γ-Al_2O_3 and δ-Al_2O_3. Subsequently, an exothermic reaction can be seen at about 1150 °C, which is due to the transformation into α-Al_2O_3 (corundum). The transformation of metastable alumina into corundum takes place in a temperature range of 1062–1204 °C when the treatment is carried out in an air atmosphere, and between 1090 and 1181 °C when an inert atmosphere (N_2) is used [53,54].

3.1.4. Scanning Electron Microscopy and Energy-Dispersive X-ray Spectroscopy Analysis (SEM/EDS)

In general, and due to the grinding process to which the materials have previously been subjected, it can be observed via SEM that the particle size obtained in all the materials is less than 100 µm, with grains of angular appearance appearing in many cases due to the grinding process.

According to the EDS analysis carried out on C_A, the composition is dominated by Ca, Mg, and O. There are also other minority elements, such as Al, Si, and Fe. For the limestone C_SN_1, there are zones of different compositions, with Ca, Mg, and O predominating. In general, the presence of Al, Si, and Fe can also be observed, although in a minimum proportion in all the zones. The EDS analysis of sample C_SN_2 indicated that the composition is very similar to that of sample C_SN_1; it should be noted that both samples come from the same raw material. The only difference in both samples is the granulometry prior to the grinding process. The main difference is the appearance of K in one of the areas studied, although in a very small proportion (0.17 wt %). In the sample C_T that is analyzed, the particle size ratio is different from that studied for C_A, C_SN_1, and C_SN_2. A greater number of large particles are observed, although always smaller than 100 µm. This may be due to the greater hardness of the grains in C_T compared to the other three samples since the grinding time in all these samples is the same. In addition to this, it is observed that the larger grains do not have more defined edges or surfaces, which may be due to difficulties regarding the fragmentation of the grains in the grinding process and the need to induce a greater number of impacts to reduce the size, which causes the faces to collide a greater number of times and become more rounded. The smaller particles have better-defined smooth faces. In terms of the EDS analysis, the studied zones can be grouped into three differentiated zones and present a higher Ca and O composition, with a higher wt % of Ca in C_A compared to the other three studied limestones. Different percentages of Mg are also observed, although in a much lower proportion than in the other limestones, coinciding with the XRF results (Table 2), Al, Si, K, and Fe.

In the morphological study of A_A and A_N, very similar particle sizes are observed, with grains of heterogeneous size and different shapes (both rounded and angular). In A_R, the overall particle shape is smaller, being more homogeneous and rounded in most of the grains. According to the EDS analysis, in general, in A_A, A_N, and A_R, the elements with a higher wt % are Si, Al, and O, as expected in the case of clays and coinciding with the XRF analysis (Table 2). In the case of A_A, zones of differentiated composition can be identified, with different percentages of K, Ca, Fe, and, to a lesser extent, Na, Mg, Ti, and Cu. In the case of A_N, zones of differentiated composition can be identified, with different percentages of mainly K and Fe, and Na, Mg, S, Ca, and Ti in relative proportions. In the case of S, for the three clays studied, this element is only identified in the A_N in values lower than 0.5 wt %. For the analysis of A_R, it is possible to identify areas with different compositions; Fe is the element with the highest percentage, in addition to Si, Al, and O, with a higher composition than in the case of A_A and A_N. In addition, the presence of Na, Mg, K, Ca, and Ti can be observed in smaller proportions.

In the case of Y, the particles have a very heterogeneous morphology, with grains smaller than 10 µm, agglomerates, and larger particles that are still smaller than 100 µm. Some particles have an elongated shape, although most of them are rounded. The results of the EDS microanalysis show two zones of differentiated composition. In both zones, the predominant elements are S, Ca, and O, which is in agreement with the XRD and XRF (Table 3) analyses mentioned earlier.

In terms of industrial waste, Figure 5a shows a selected SEM micrograph at a 500× magnification of the CHM sample, in which particles of different sizes can be observed, the smaller ones being more numerous and all being smaller than 100 µm. The morphology of the particles is varied, some being particularly rounded and others more angular. The EDS microanalysis shows 5 zones with different compositions (Table 4), although the major elements are Si, Al, and O, followed by K, Ca, Fe, and, to a lesser extent, Na, Mg, P, S, Ti, and Cu. The distribution of the elements in the microanalysis of this sample resembles the composition in the EDS analysis that was conducted on the clays.

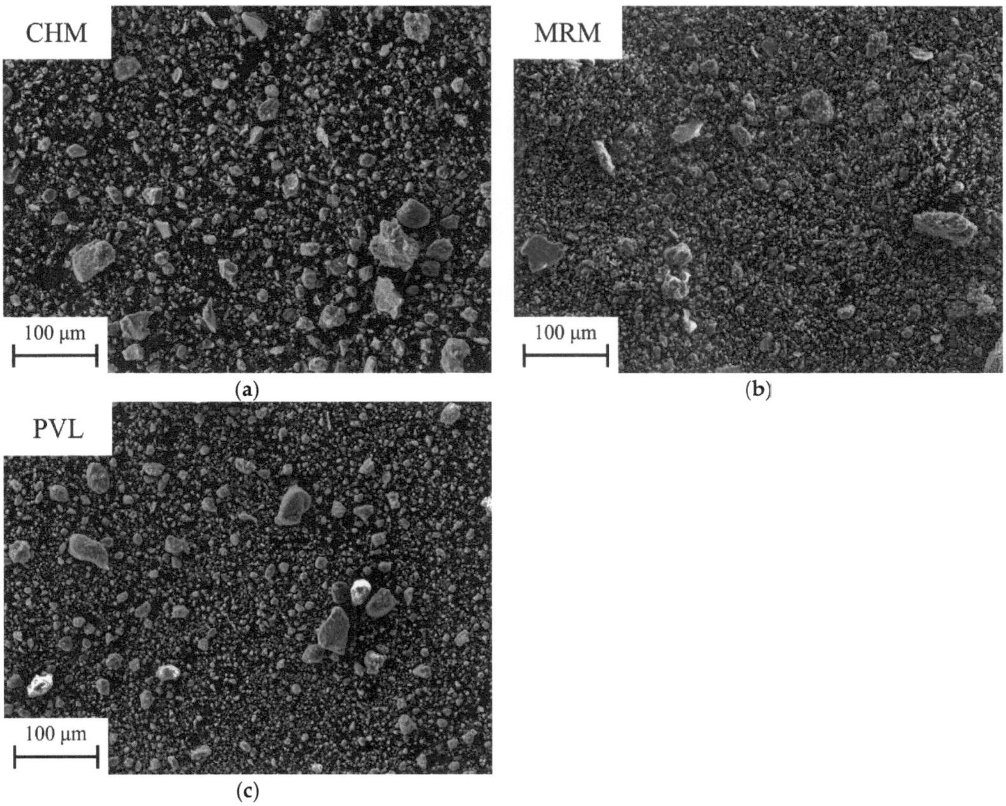

Figure 5. Selected SEM micrographs (×500) of: (**a**) CHM powder; (**b**) MRM powder; (**c**) PVL powder.

In the case of the MRM waste (Figure 5b), the particles are much smaller in size, with a more powdery appearance. In this case, most of the particles are smaller than 50 µm. Agglomerates of the particles and some larger fragments can be observed. The EDS analysis (Table 5) shows three zones with different compositions, in which the elements with the highest wt % are Ca and O, although the differences lie in the wt % of minor elements, such as Mg, S, and Cu, which are always in quantities lower than 0.50 wt %. This composition agrees with the XRD (Figure 3) and XRF (Table 3) results that correspond to calcium carbonate.

Figure 5c shows a selected SEM micrograph of the PVL. Heterogeneous particle sizes are observed, although most of them do not exceed 50 µm. A larger amount of carbon tape (black background) without adhered particles can be observed, which may be due to the shape of the particles, with defined edges, which may hinder adhesion and permanence on this tape. The EDS microanalysis (Table 6) shows 5 point zones with variable composition, although the major elements in all of them are Al and O, which finding is in agreement with the XRF analysis (Table 3). Different compositions can also be observed, depending on the zone, of Mg, Si, and K, as well as F, Na, S, Cl, Ca, Ti, Fe, Cu, and Ba, to a lesser extent,. This composition agrees with the phases identified in the XRD test (Figure 3).

Table 4. EDS microanalysis (wt %) of the areas marked in Figure 5a (CHM powder).

Element (wt %)	Zone				
	1	2	3	4	5
O	50.69 ± 0.00	39.84 ± 0.00	51.36 ± 0.00	59.38 ± 0.00	61.77 ± 0.00
Na	0.35 ± 0.00	0.18 ± 0.00	0.29 ± 0.00	0.00 ± 0.00	0.00 ± 0.00
Mg	0.99 ± 0.00	1.54 ± 0.00	1.68 ± 0.00	2.21 ± 0.00	0.14 ± 0.00
Al	13.86 ± 0.00	9.00 ± 0.00	9.82 ± 0.00	7.06 ± 0.00	0.51 ± 0.00
Si	24.58 ± 0.00	29.43 ± 0.00	21.95 ± 0.00	13.75 ± 0.00	36.71 ± 0.00
P	0.00 ± 0.00	0.00 ± 0.00	0.00 ± 0.00	0.42 ± 0.00	0.00 ± 0.00
S	0.27 ± 0.00	1.46 ± 0.00	1.49 ± 0.00	0.75 ± 0.00	0.00 ± 0.00
K	6.46 ± 0.00	4.88 ± 0.00	3.66 ± 0.00	1.95 ± 0.00	0.13 ± 0.00
Ca	1.12 ± 0.00	7.03 ± 0.00	6.63 ± 0.00	10.48 ± 0.00	0.35 ± 0.00
Ti	0.00 ± 0.00	0.53 ± 0.00	0.13 ± 0.00	0.00 ± 0.00	0.00 ± 0.00
Fe	1.68 ± 0.00	6.11 ± 0.00	2.69 ± 0.00	4.00 ± 0.00	0.20 ± 0.00
Cu	0.00 ± 0.00	0.00 ± 0.00	0.30 ± 0.00	0.00 ± 0.00	0.19 ± 0.00
Total	100.00 ± 0.00	100.00 ± 0.00	100.00 ± 0.00	100.00 ± 0.00	100.00 ± 0.00

Table 5. EDS microanalysis (wt %) of the areas marked in Figure 5b (MRM powder).

Element (wt %)	Zone		
	1	2	3
O	62.06 ± 2.67	61.94 ± 2.67	65.39 ± 0.00
Mg	0.28 ± 0.02	0.33 ± 0.02	0.00 ± 0.00
S	0.00 ± 0.00	0.16 ± 0.00	0.14 ± 0.00
Ca	37.67 ± 2.69	37.58 ± 2.69	34.17 ± 0.00
Cu	0.00 ± 0.00	0.00 ± 0.00	0.30 ± 0.00
Total	100.00 ± 0.00	100.00 ± 0.00	100.00 ± 0.00

Table 6. EDS microanalysis (wt %) of the areas marked in Figure 5c (PVL powder).

Element (wt %)	Zone				
	1	2	3	4	5
O	57.76 ± 0.00	29.56 ± 0.00	55.04 ± 0.00	50.41 ± 0.00	48.20 ± 0.00
F	0.00 ± 0.00	0.00 ± 0.00	1.83 ± 0.00	1.20 ± 0.00	1.24 ± 0.00
Na	0.72 ± 0.00	0.79 ± 0.00	1.20 ± 0.00	1.03 ± 0.00	0.55 ± 0.00
Mg	2.46 ± 0.00	10.89 ± 0.00	3.72 ± 0.00	4.64 ± 0.00	2.92 ± 0.00
Al	17.05 ± 0.00	42.68 ± 0.00	29.61 ± 0.00	35.94 ± 0.00	43.35 ± 0.00
Si	12.97 ± 0.00	7.21 ± 0.00	5.89 ± 0.00	2.07 ± 0.00	1.63 ± 0.00
S	0.00 ± 0.00	0.00 ± 0.00	0.21 ± 0.00	0.19 ± 0.00	0.00 ± 0.00
Cl	0.55 ± 0.00	0.98 ± 0.00	0.56 ± 0.00	0.63 ± 0.00	0.41 ± 0.00
K	6.60 ± 0.00	1.78 ± 0.00	0.27 ± 0.00	0.74 ± 0.00	0.39 ± 0.00
Ca	1.89 ± 0.00	1.01 ± 0.00	0.56 ± 0.00	1.37 ± 0.00	0.47 ± 0.00
Ti	0.00 ± 0.00	0.00 ± 0.00	0.16 ± 0.00	0.00 ± 0.00	0.00 ± 0.00
Fe	0.00 ± 0.00	2.63 ± 0.00	0.31 ± 0.00	0.52 ± 0.00	0.45 ± 0.00
Cu	0.00 ± 0.00	1.05 ± 0.00	0.64 ± 0.00	0.49 ± 0.00	0.39 ± 0.00
Ba	0.00 ± 0.00	1.42 ± 0.00	0.00 ± 0.00	0.77 ± 0.00	0.00 ± 0.00
Total	100.00 ± 0.00	100.00 ± 0.00	100.00 ± 0.00	100.00 ± 0.00	100.00 ± 0.00

3.1.5. Attenuated Total Reflectance Fourier Transform Infrared Spectroscopy (ATR-FTIR)

According to the XRD patterns, C_A, C_SN_1, and C_SN_2 are mineralogically composed of calcite and dolomite. In the case of the C_T limestone, the mineralogical composition is calcite and quartz. Furthermore, according to the results of the XRF analysis, it can be seen that the C_A has a high MgO content, being lower in the case of the C_SN_1 and C_SN_2 limestones, and much lower in the C_T limestone. According to the absorption spectra, C_A (Figure 6a) shows absorption bands corresponding to low-Mg calcite, specifically those centered at 1796 cm^{-1} and 878 cm^{-1}, as well as a band centered at 1429 cm^{-1} corresponding to high-Mg calcite. In the absorption spectrum of dolomite [$CaMg(CO_3)_2$], where the structure of the flat $CO_3{}^{2-}$ ions is intercalated with the Ca^{2+} and Mg^{2+} cations, the absorption band is close to 729 cm^{-1}, and a band centered at 728 cm^{-1} can be observed in the C_A band [55–57]. For C_SN_1 (Figure 6b) and C_SN_2 (Figure 6c), absorption bands are shown corresponding to low-Mg calcite, namely, those centered at 1796 cm^{-1}, 1408 cm^{-1} and 1409 cm^{-1}, 873 cm^{-1} and 874 cm^{-1} and 712 cm^{-1}, a band centered at 728 cm^{-1} corresponding to dolomite, two bands centered at 1035 cm^{-1} and 540 cm^{-1} matching the feldspars, and a band centered at 470 cm^{-1} corresponding to the angular deformation of the Si-O-Si bond [55–60]. It can be seen that the band centered at 728 cm^{-1} has a higher intensity in the spectrum of C_A compared to the spectra of C_SN_1 and C_SN_2, which is consistent with the higher proportion of dolomite in its composition. This is due to the higher concentration of MgO since, according to the results of the XRF analysis (Table 2), these concentrations vary from 19.12 wt % to 4.59 and 4.63 wt %, respectively. For the C_T sample (Figure 6d), absorption bands corresponding to low-Mg calcite are shown, namely, those centered at 1796 cm^{-1}, 1412 cm^{-1}, 874 cm^{-1}, and 712 cm^{-1}, a band centered at 1028 cm^{-1} corresponding to feldspar, and three bands centered at 800 cm^{-1}, 519 cm^{-1}, and 470 cm^{-1}, matching the angular deformation of Si-O-Si. In the case of the band centered at 800 cm^{-1}, it can be seen that it is accompanied by another band of lower intensity at approximately 780 cm^{-1}. This double band is characteristic of the symmetric Si-O stretching of quartz [55–62].

The study of clays via ATR-FTIR indicates two absorption frequency zones: the OH vibration zone (3200–3700 cm^{-1}) and the lattice vibration zone (1100–400 cm^{-1}). In general, for the spectra of the three types of clays studied, A_A (Figure 6e), A_N (Figure 6f), and A_R (Figure 6g), the characteristic bands of the silicates are observed, corroborating the presence of O-bonded Si and O-H groups, as evidenced by the Si-O stretching bands at 1200–950 cm^{-1}. The doublet bands at approximately 700–800 cm^{-1} reveal the characteristic signatures of the Si-O bond stretching, while the bands at 600–800 cm^{-1} are characteristic of the Si-O bond-bending mode [62]. In the case of the three clays, an absorption band centered at 3622–3621 cm^{-1}, corresponding to the stretching band of the structural O-H groups, can be observed [61]. The absorption band centered at 2349 cm^{-1} can be attributed to atmospheric CO_2 and can also be observed in all three clays [63]. The absorption band centered between 1008 and 987 cm^{-1} represents the in-plane stretching of Si-O in the Si-O-Si groups of the tetrahedral sheet [61]. On the other hand, three absorption bands can be distinguished that correspond to the three different modes of vibration of the C=O bond contained in the carbonate ion [64]. These absorption bands, corresponding to low-Mg calcite, are centered between 1432 and 1426 cm^{-1}, due to the asymmetric stretching of the $CO_3{}^{2-}$ band, the asymmetric bending mode of the $CO_3{}^{2-}$ band at 874 cm^{-1}, and the symmetric mode of $CO_3{}^{2-}$ centered at 712 cm^{-1}, although these last two absorption bands are only found in the A_A and A_N spectra. The presence of quartz is associated with the doublet at 800 cm^{-1} and 778 cm^{-1}, respectively, due to the Si-O symmetric stretching mode, as well as the band at 694 cm^{-1} corresponding to Si-O-Al vibrations, and also the bands centered at 522–512 cm^{-1} and 466–459 cm^{-1}, which are due to Si-O-Al (octahedral) bending and Si-O-Si angular deformation vibrations, respectively. Finally, the band located at 424–414 cm^{-1} is due to the bending modes of O-Ca-O [55–63,65].

In the case of A_N, an absorption band centered at 3621 cm^{-1} is observed, corresponding to the stretching band of the structural O-H groups [61]. Regarding the absorption band centered at 2349 cm^{-1}, this can be attributed to atmospheric CO_2 [63]. The absorption band centered at 1008 cm^{-1} represents the in-plane stretching of Si-O in the Si-O-Si groups of the tetrahedral sheet [61]. The absorption bands corresponding to low-Mg calcite are centered at 1428 cm^{-1} due to the asymmetric stretching of the $CO_3{}^{2-}$ band, the asymmetric bending mode of the $CO_3{}^{2-}$ band at 874 cm^{-1} and the symmetric mode of $CO_3{}^{2-}$ situated at 712 cm^{-1}. For quartz, the doublet at 800 cm^{-1} and 778 cm^{-1} can be observed, due to the symmetric Si-O stretching, as well as the band at 694 cm^{-1} that corresponds to the Si-O-Al vibrations and the bands centered at 520 cm^{-1} and 463 cm^{-1}, which are due to Si-O-Al (octahedral) bending and the Si-O-Si angular deformation vibrations modes, respectively. The band identified at 421 cm^{-1} is due to the O-Ca-O bending mode. In the case of A_R, an absorption band at 828 cm^{-1} is observed, which can be attributed to the OH deformation of the octahedral layers (Fe^{3+}-Fe^{3+}-OH) [55–63,65].

In the ATR-FTIR spectrum of the gypsum (Y) (Figure 6h), the presence of water can be detected and characterized by bands in the spectral region near 3500 cm^{-1} and 1600 cm^{-1}. In this case, a doublet of absorption bands is observed at 3530 cm^{-1} and 3401 cm^{-1}, the first one being due to the stretching vibrations of the hydroxyl groups, and the second one being due to the ν_1 vibrational modes of the symmetric stretching of water [61,66]. In the frequency ranging between 2000 and 1000 cm^{-1} appears in two bands at 1683 cm^{-1} and 1620 cm^{-1}, which are assigned to the bending vibrational modes of the O-H bond [66,67]. The presence of two bands indicates that there are two different types of crystallographic water. One type of water is bound to the sulfate ions by hydrogen bonding; the corresponding band has a lower frequency (also due to hydrogen bonding) and high intensity. The other is directly related to calcium ions [66]. There is also a strong band at 1100 cm^{-1}, due to the antisymmetric stretching vibration (ν_3) of the SO_4 tetrahedra. In addition, there is a very weak band at about 1003 cm^{-1}, which represents the symmetric stretching vibration mode (ν_1) of the SO_4 tetrahedra. Finally, the spectrum exhibits two antisymmetric bending vibration bands at 669 cm^{-1} and 598 cm^{-1} [66,68].

For the CHM sample (Figure 6i), an absorption band centered at 984 cm^{-1} can be observed, which represents the asymmetric Si-O stretching of the main tetrahedral silicate vibrations [61,69,70]. The absorption band centered at 1427 cm^{-1} is due to the asymmetric stretching (ν_3) of $CO_3{}^{2-}$, while the band centered at 876 cm^{-1} is due to the asymmetric bending mode (ν_2) of $CO_3{}^{2-}$. A doublet at 800 cm^{-1} and 777 cm^{-1} can be observed, due to the symmetric stretching of the Si-O bond, as well as the band at 680 cm^{-1} corresponding to the Si-O-Al vibrations. The absorption band centered at 442 cm^{-1} corresponds to the vibration produced by the angular deformation of Si-O-Si [55–63,65].

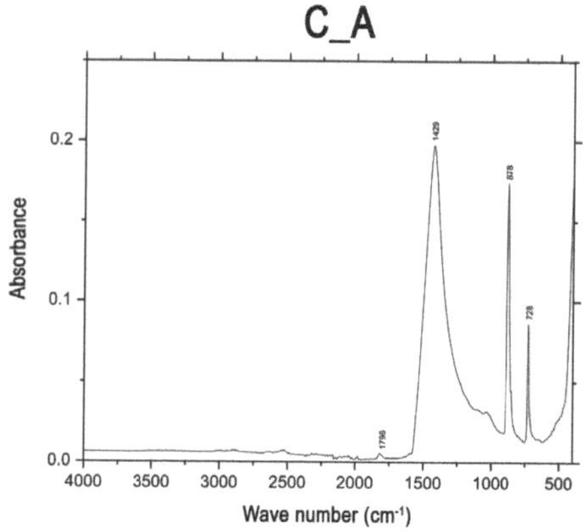

(**a**)

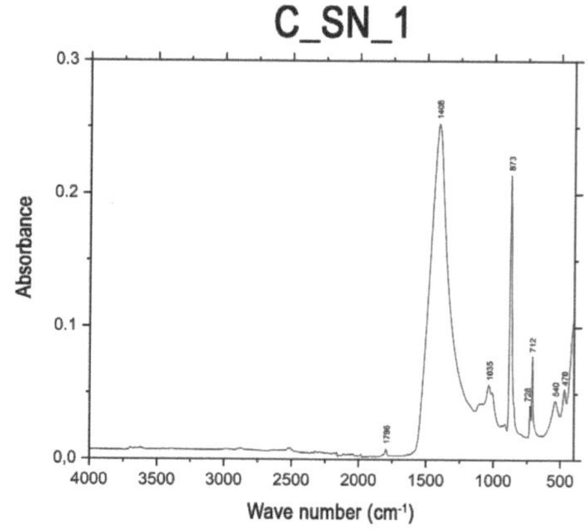

(**b**)

Figure 6. *Cont.*

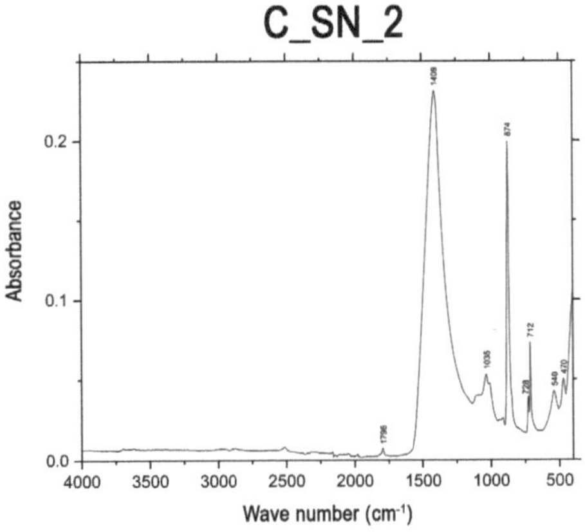

(**c**)

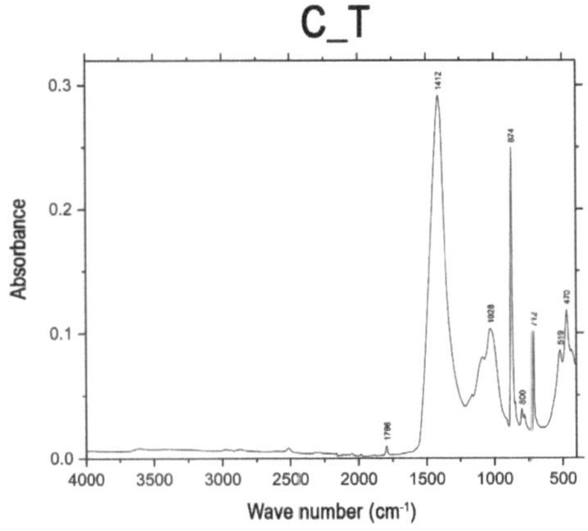

(**d**)

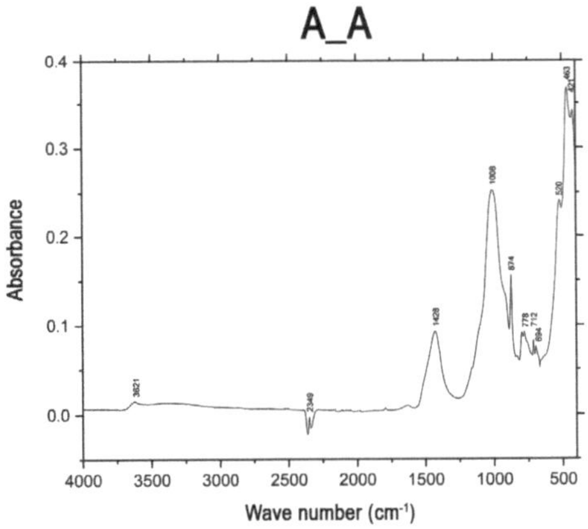

(**e**)

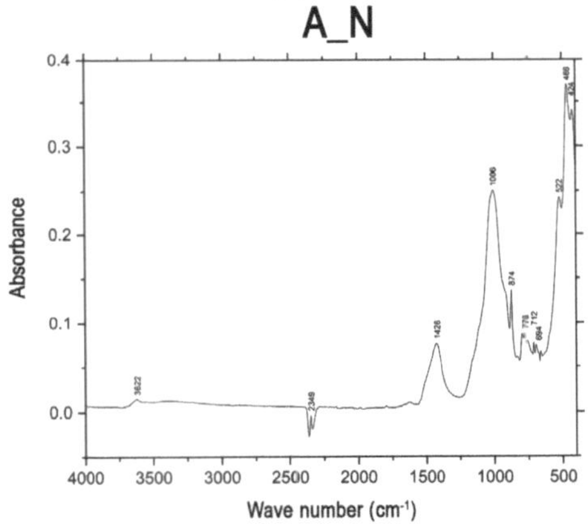

(**f**)

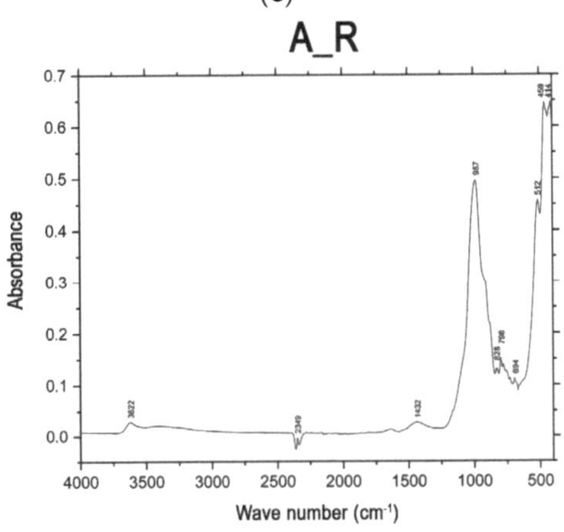

(**g**)

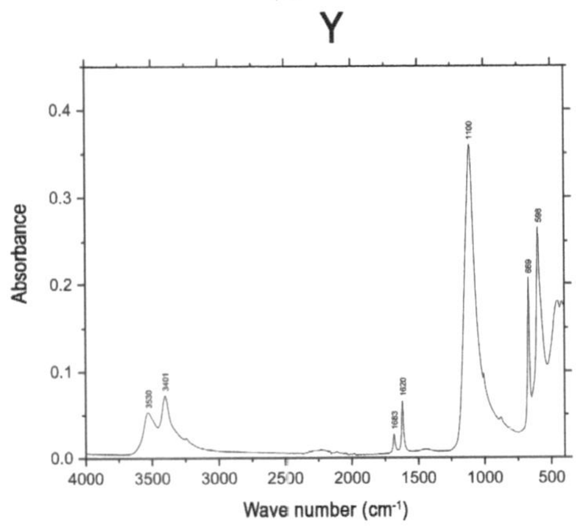

(**h**)

Figure 6. *Cont.*

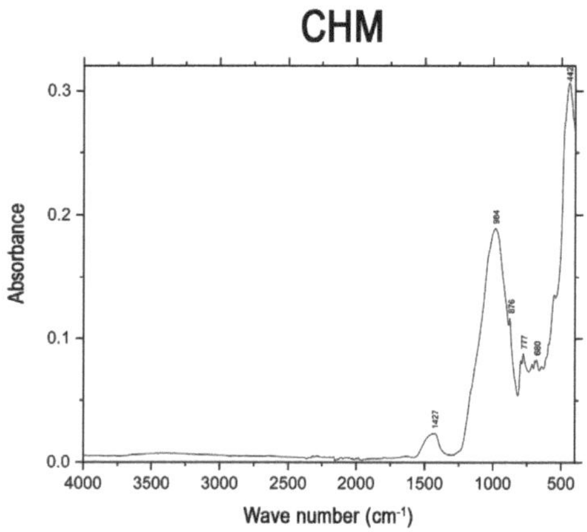

(i)

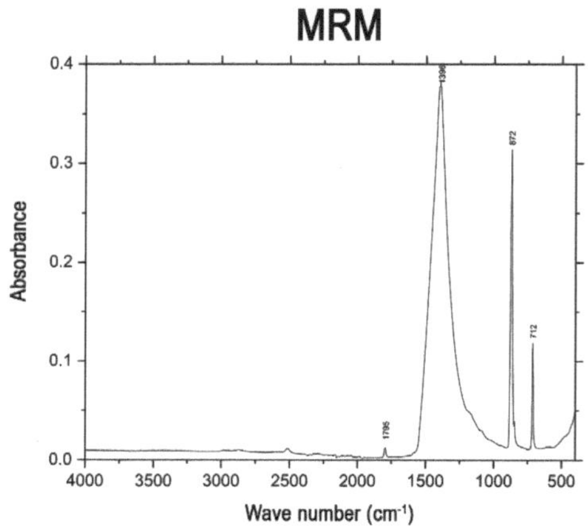

(j)

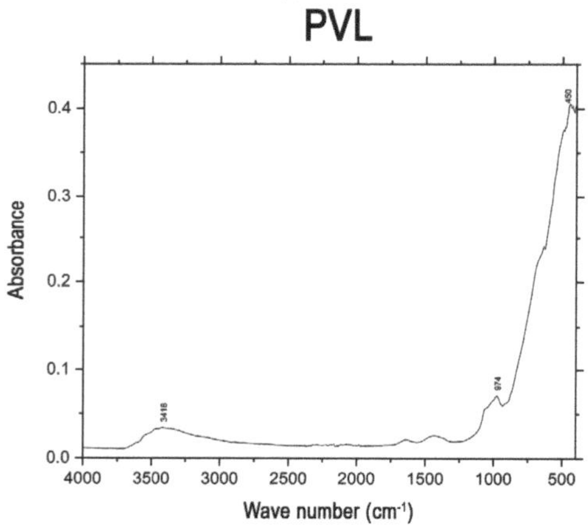

(k)

Figure 6. Infrared spectra (ATR-FTIR) of: (**a**) C_A; (**b**) C_SN_1; (**c**) C_SN_2; (**d**) C_T; (**e**) A_A; (**f**) A_N; (**g**) A_R; (**h**) Y; (**i**) CHM; (**j**) MRM; (**k**) PVL.

The analysis of the MRM (Figure 6j) shows absorption bands corresponding to calcium carbonate, as indicated by the XRD test, specifically located at 1795 cm^{-1} ($\nu_1 + \nu_4$), 1396 cm^{-1} (ν_3), 872 cm^{-1} (ν_2) and 712 cm^{-1} (ν_4) [67,71–73]. As indicated above in the analysis of the limestones, the free $CO_3{}^{2-}$ ion is flat and causes four intramolecular vibrations, namely, ν_1 (symmetrical stretching), ν_2 (bending beyond the plane), ν_3 (symmetrical stretching), and ν_4 (bending in the plane) [55].

For the PVL sample (Figure 6k), a spectrum with an absorption band centered at 3418 cm^{-1} is observed, due to the stress mode of the O-H bonds. This band indicates the presence of moisture in the analyzed sample. The band centered at 974 cm^{-1} is due to the α-Al_2O_3 phase, while the absorption band centered at 450 cm^{-1} is associated with the stretching mode of Al-O in the octahedral structure [74].

3.2. Crude Blend Design

Taking into account all the previous results and the compatibility in the results, which were obtained using raw materials and industrial wastes, the design of the raw blend for low-energy cement using industrial wastes was studied. As indicated above, the main difference between ordinary Portland cement (OPC) and low-energy cement with a high dicalcium silicate content was the preponderance of the type of silicate seen in the final clinker, with the C_3S/C_2S weight ratio in the case of OPC being approximately 60%/20% [75], while in the newly designed clinker crudes, the value of the percentage of C_2S varies up to 60–65% by weight [6]. According to the percentage of C_2S proposed, the percentage by weight of the rest of the major components of the clinker (C_3S, C_3A, and C_4AF) will vary, taking into account the limitations set by the current regulations on the need for cement clinker to be made up of at least two-thirds of calcium silicates [76]. Figure 7 classifies the different types of cement, according to the majority of phases calculated in the design, with the target zone shown in blue [77].

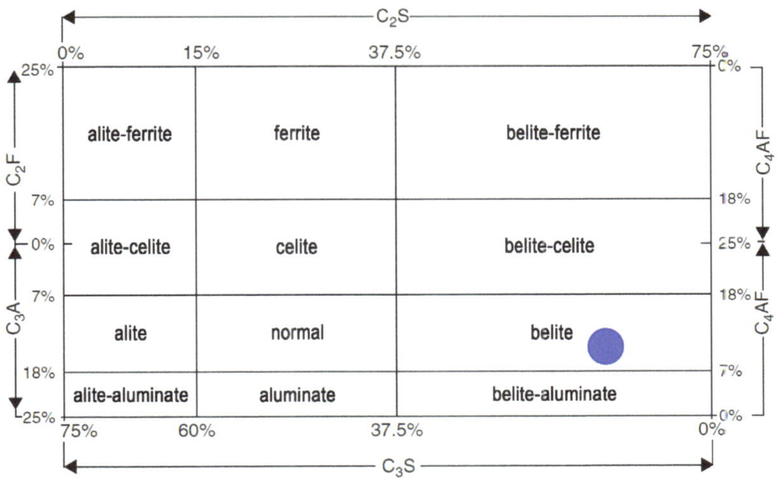

Figure 7. Classification of cement types according to the composition of their majority phases. Blue area: target cement type. (Adapted from reference [77]).

In the present study, to achieve desirable C_2S and C_3S percentages, the LSF should be between 78% and 83% [31,32,78] and, more specifically, within the range of 78–80%, this being appropriate for belite cement [30], in order to obtain high-quality hydraulic cements. Furthermore, it should be taken into account that the mass ratio (CaO)/(SiO_2) should not be less than 2.0 [76].

Table 7 presents the crude blend formulations designed using the calculation and with a Microsoft Excel® spreadsheet, as described in Section 2.3. The design of three crude blends using only raw materials (WST_0_A, WST_0_SN, and WST_0_T) can be seen, but each of them uses a different limestone (C_A, C_SN, and C_T). Due to the common characteristics of the C_SN_1 and C_SN_2 limestones, analyzed above, which come from the same mining site, the designed blend, WST_0_SN, contains only the limestone C_SN_1. In addition, a crude blend designed using 50 wt % of raw materials (C_T, A_A, A_N, and A_R) and 50 wt % of the industrial wastes previously studied (CHM, MRM, and PVL), known as WST_50_T, as well as a crude blend designed exclusively from industrial wastes, called WST_100, can be observed. These two blends are intended to demonstrate the possibility of manufacturing cement from industrial waste, with either a high percentage of 50% or a total substitution of the raw materials.

Table 8 shows the theoretical compositions of the designed raw blends, in terms of the main cement clinker phases (C_2S, C_3S, C_3A, and C_4AF), the quality indexes (LSF, AM, and SM). and the CaO/SiO_2 ratio. It is worth taking note of the design limitations for a C_2S composition of 60–65 wt %, with an LSF of 78–80 wt % and a CaO/SiO_2 ratio that is higher than 2.0.

Table 7. Composition (wt %) of the raw materials and wastes of the designed crude blends.

	C_A	C_SN	C_T	A_A	A_N	A_R	CHM	MRM	PVL	Total
WST_0_A	75.85	0.00	0.00	4.17	9.32	10.66	0.00	0.00	0.00	100.00
WST_0_SN	0.00	78.75	0.00	11.50	2.25	7.50	0.00	0.00	0.00	100.00
WST_0_T	0.00	0.00	84.00	5.00	0.75	10.25	0.00	0.00	0.00	100.00
WST_50_T	0.00	0.00	37.50	7.00	0.50	5.00	12.00	37.50	0.50	100.00
WST_100	0.00	0.00	0.00	0.00	0.00	0.00	30.75	68.25	1.00	100.00

Table 8. Composition of target phases (wt %), quality indexes (LSF, AM, and SM), and weight ratio (CaO/SiO_2) of the crude blends designed in this study.

	C_3S	C_2S	C_3A	C_4AF	Total	LSF	AM	SM	CaO/SiO_2
WST_0_A	11.32	62.73	14.22	11.73	100.00	78.00	2.03	2.13	2.55
WST_0_SN	8.76	64.14	17.93	9.17	100.00	78.24	2.88	2.11	2.58
WST_0_T	19.00	61.71	10.43	8.86	100.00	78.03	1.99	3.05	2.44
WST_50_T	16.04	61.94	12.44	9.58	100.00	78.18	2.13	2.62	2.49
WST_100	11.72	63.06	14.66	10.57	100.00	78.04	2.23	2.24	2.54

Taking into account the results of the characterization of the C_A limestone, it is necessary to highlight a fundamental fact about this raw material that will be decisive in the calculation of the crude blends. In this case, the chemical composition (Table 2) of C_A shows a percentage of 19.12 wt % of MgO. It is important to keep in mind that, to avoid future cement expansivity problems, its MgO percentage cannot exceed 5 wt % [76]. This characteristic of C_A is also corroborated in the thermal analyses that were performed, with an endothermic peak centered at 803 °C due to the decomposition of the carbonates associated with magnesium. Therefore, for the design of the crude blends with C_A, the values associated with MgO and how these influence the calculations in the designed Microsoft Excel® spreadsheet must be taken into account. Another problem associated with the high MgO content of the blends is the theoretical composition of the clinker target phases (Table 8), since the calculation of these phases is performed by means of the modified Bogue Equations taking into account the four main oxides of the crude, namely Al_2O_3, CaO, SiO_2, and Fe_2O_3. Thus, a high MgO content can imply a notorious difference between the theoretically calculated phases and the real manufactured phases. As in the limestone C_A, the chemical composition (Table 2) of C_SN_1 indicates a high percentage of MgO, specifically 4.59 wt %. This percentage is lower than the 5 wt % previously indicated as a limit in the composition of cement clinker, although the 4.59 wt % is referred to the chemical composition without calcination. Therefore, it will increase based on the calcined composition. Consequently, as with the crudes designed with C_A, the values associated with MgO, and its influence on the calculations in the Microsoft Excel® spreadsheet that has been designed, must be taken into account for the crudes designed with C_SN_1.

The results of the chemical composition of the manufactured crudes and the loss on ignition (LOI), are shown in Table 9. These results are reasonable considering the composition of the main phases of the cement clinker to be obtained. It should be noted that in both the WST_0_A and WST_0_SN blends, the percentage of MgO is 13.45 wt % and 4.60 wt %, respectively. This indicates that in both cases the percentage of MgO, according to the calcined base, will be higher than 5.0 wt %. Therefore, they will not be valid clinkers. In the case of the WST_0_T, WST_50_T and WST_100 crude samples, the results of the MgO limiting factor are ~1 wt %.

According to the DSC analysis performed on the three blends that were manufactured using only raw materials in their composition (Figure 8), it can be seen that in the WST_0_A and WST_0_SN blends, the double decomposition reactions of magnesium and calcium carbonates occur, while in the WST_0_T crude, only an endothermic reaction occurs in the same temperature range for the $CaCO_3$ decomposition, due to the low MgO composition of the mixture. This corroborates the data obtained in the XRF analysis for these three blends (Table 7), which allows us to definitively discard the WST_0_A and WST_0_SN crude samples.

Table 9. Chemical composition (XRF, wt %) of the designed crude blends.

	WST_0_A	WST_0_SN	WST_0_T	WST_50_T	WST_100
Al_2O_3	4.64	5.14	4.67	4.22	4.46
CaO	26.03	37.35	41.17	44.16	46.49
SiO_2	13.55	12.81	15.89	14.27	12.61
Fe_2O_3	1.65	1.69	1.82	1.80	1.96
Na_2O	0.14	0.18	0.09	0.13	0.20
K_2O	0.95	0.90	1.11	1.00	0.97
MgO	13.45	4.60	1.10	0.95	0.94
SO_3	0.26	0.13	0.06	0.40	0.72
TiO_2	0.19	0.25	0.23	0.22	0.29
P_2O_5	0.05	0.04	0.08	0.05	0.03
MnO	0.04	0.02	0.06	0.04	0.02
LOI	38.80	36.85	33.62	32.67	31.24

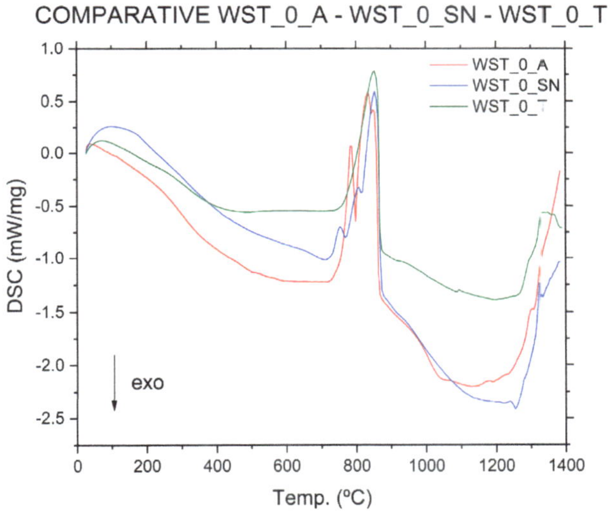

Figure 8. Comparative DSC analysis WST_0_A (red line), WST_0_SN (blue line), and WST_0_T (green line).

The mineralogical phases identified by XRD (Figure 9) for the WST_0_T crude blend, the one produced with 50 wt % waste substitution (WST_50_T) and the one produced with 100 wt % waste in its composition (WST_100), are quartz, calcite, and muscovite. It should be noted that in the case of the WST_0_T blend, part of the raw materials used are clays. In the case of the WST_50_T blend and WST_100 sample, clays are replaced by 50 wt % or 100 wt % by CHM, respectively. Thus, in the case of WST_50_T and WST_100 crudes, part of the identified muscovite will be dehydroxylated as it comes from CHM.

It can be observed that the three blends have a similar XRD pattern (Figure 9), regardless of the use or non-use of wastes in their composition and according to the chemical composition results (Table 7). It can be seen how, in these three blends, the presence of dolomite is not identified, due to the low MgO content of the C_T and the CHM and MRM wastes (Tables 2 and 3). In addition, the phases provided by PVL are not identified in the XRD diffractograms of the WST_50_T and WST_100 blends, since they only represent 0.50 wt % and 1 wt % of the total blend, respectively (Table 7).

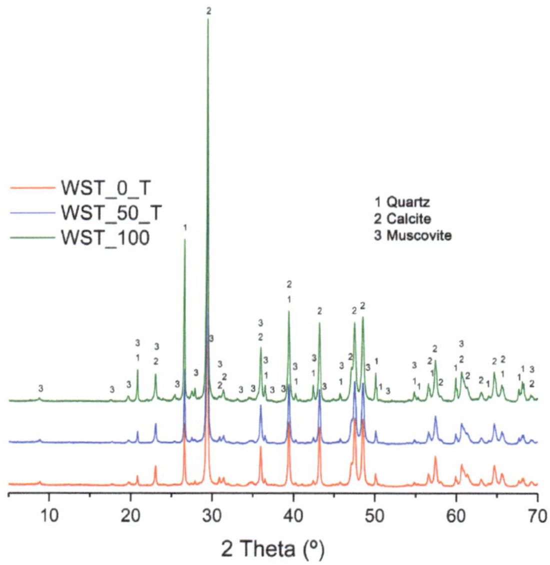

Figure 9. Comparative XRD analysis: WST_0_T (red line), WST_50_T (blue line), and WST_100 (green line).

Thus, a peak of maximum intensity corresponding to calcite in the 28–30° 2θ zone can be observed for the XRD patterns (Figure 9) of WST_0_T and WST_50_T. In the case of the sample WST_100, two peaks of maximum intensity can be observed in the 25–30° 2θ zone. The first one corresponded to quartz and the second one corresponded to a higher-intensity peak for calcite. It is consistent with the current findings that the peak of higher intensity corresponds to calcite, since it is the main mineralogical phase in the MRM, accounting for 37.50 wt % (WST_50_T) and 68.25 wt % (WST_100) of the total crude blend (Table 7). In general, as can be seen in Figure 9, the identified peaks of quartz and muscovite present a higher intensity in the XRD of the WST_100 crude blend, compared to the crude blends made with C_T. This may be caused by the higher composition of CHM in the WST_100 crude sample versus the sum of the compositions of A_A, A_N, and A_R, and CHM in the crude sample manufactured with C_T. Therefore, a lower composition of MRM in this crude sample is reached, compared to the sum of C_T and MRM in the other two crude samples.

Thermal analyses performed on the C_T blends (Figures 10 and 11), composed of quartz, calcite, muscovite, and dehydroxylated muscovite, according to the XRD patterns, show similar results for the two compositions. According to the TG test, there is a total weight loss of up to about 860 °C of 36.5 wt % for the WST_0_T crude blend and 34 wt % for the WST_50_T crude blend. This lower weight loss in the crude blend in which the wastes have been incorporated may be due to the lower ignition loss of CHM compared to clays, which is not compensated for by the higher ignition loss of MRM, compared to C_T (Tables 2 and 3). Conversely, the reactions that occur in the DSC analysis curves have similarities with the analysis curves of the other crude blends that were studied previously. The initial weight loss that occurs at temperatures between 30 °C and 200 °C is due to the elimination of weakly bound water (dehydration) [6]. The weight losses produced at temperatures ranging from 200 °C to 600 °C, associated with the endothermic band observed in the DSC analysis curves, are explained by taking into account the elimination of organic matter and the dehydroxylation of the hydroxylated silicates that come mainly from the clay components [6]. The weight loss of approximately 29 wt % for the WST_0_T and WST_50_T blends between 600 °C and 860 °C is due to the decomposition of calcium carbonate (calcite) [6,44]. As in the DSC curve of the WST_0_T crude blend, only one endothermic reaction occurs in the same temperature range for the decomposition of calcium carbonate in the WST_50_T crude blend, due to the low MgO composition of the mixture (Figure 11). Between 1050 °C and 1200 °C, different reactions occur, which are due, firstly, to the combination of CaO and SiO_2, forming the C_2S phase, along with the combination of CaO with Al_2O_3 and Fe_2O_3, to form C_4AF. At temperatures above 1200 °C, reactions occur from the combination of CaO with SiO_2, to form C_3S, and from CaO with Al_2O_3, to form C_3A aluminate [6].

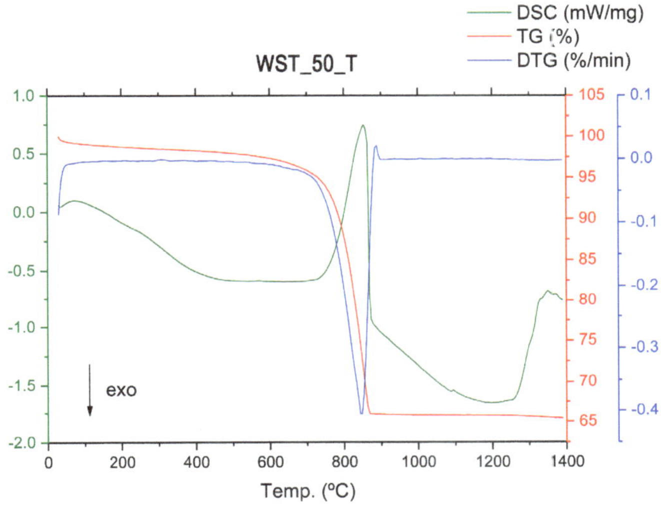

Figure 10. The TG (red line), DTG (blue line), and DSC (green line) analyses of crude-blend WST_50_T.

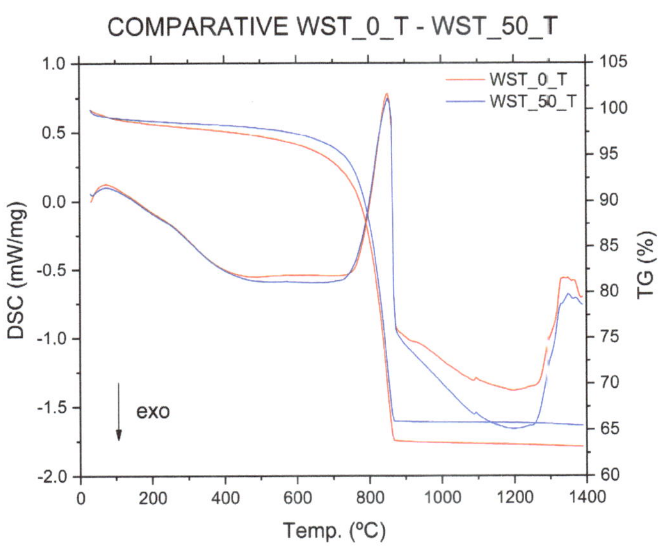

Figure 11. Analyses of WST_0_T (TG (red line), DSC (red line)) vs. WST_50_T (TG (blue line), DSC (blue line)).

According to the thermogravimetric TG curve (Figure 12), in the case of the WST_100 crude blend, there is a total weight loss of up to ~860 °C of 32 wt %. This total weight loss value for the sample is lower than the weight losses indicated in the TG curves for the WST_0_T and WST_50_T blends, which were recorded as 36.5 wt % and 34 wt %, respectively. The lower weight loss, in the WST_100 crude sample, is due to the composition of MRM being lower than that of the sum of C_T and MRM in the other two blends, as well as being due to CHM having a lower LOI compared to clays. Thus, we have a first value that indicates that the CO_2 emissions of this type of crude blend in the clinkering process, at the same process temperature, will be lower than in the rest of the blends studied.

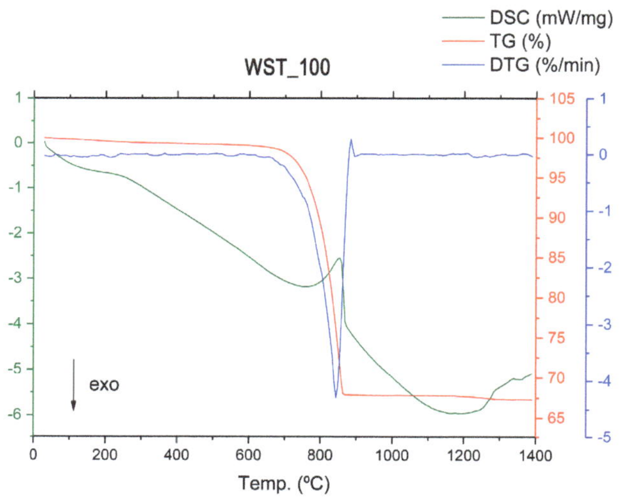

Figure 12. The TG (red line), DTG (blue line), and DSC (green line) analyses of crude-blend WST_100.

As in the previous crude blends studied, the initial weight loss that occurs at temperatures between 30 °C and 200 °C is due to the elimination of weakly bound water (dehydration). At temperatures ranging from 200 °C to 600 °C, a weight loss of approximately 5 wt % occurs, due to the elimination of organic matter and the dehydroxylation of hydroxylated silicates, which is associated with the endothermic band shown in the DSC analysis curve (Figure 12). This is followed by a weight loss of approximately 27 wt %, up to 860 °C, which is due to the decomposition of calcium carbonate (calcite). It can be seen that the endothermic effect at 860 °C has a lower intensity in the WST_100 crude sample, compared to the previously analyzed blends. This single effect reinforces the analysis of the low MgO content of the crude sample since a second effect associated with the decomposition of magnesium-associated carbonates does not appear. Between 1050 °C and 1200 °C, as well as above 1200 °C, the same reactions occur as are indicated for the WST_0_T and WST_50_T crude samples.

4. Conclusions

The results obtained from the experimental work carried out in this research, as outlined above, focused on the study of crude blends for creating low-energy cement from industrial waste, allowing us to draw the following conclusions.

The main morphological features of the raw materials and industrial wastes have been revealed via SEM studies. It has been deduced that a grinding treatment has been performed in the majority of these materials, resulting in the formation of grains of angular appearance, with a variation in average particle size lower than 100 µm. However, it has been found that polishing-marble sludges are constituted by particles smaller than 50 µm. In the case of PAVAL, heterogeneous particle sizes have been identified, although the 50 µm-diameter measurement is not exceeded. EDS analyses allowed the identification of the main chemical elements that constitute the different powders.

The mineralogical composition of all the samples has been determined by XRD. Calcite and quartz constitute the main mineralogical phases of the limestones, although dolomite has been identified in some samples. Thus, if MgO is present in the chemical composition, it is associated with dolomite. In contrast, the marble residual sludges are constituted by calcite as a single mineralogical phase. The mineralogical composition of the clays, as deduced by XRD, demonstrated the presence of muscovite (illite mica), montmorillonite, and chlorite as clay minerals, and additional silicates, such as feldspars (orthoclase and albite), apart from quartz. The main mineralogical phases identified in the chamotte are feldspars, quartz, calcite, and hematite, with dehydroxylated muscovite and anorthite. The PAVAL sample is constituted of several aluminum hydroxides and oxides, periclase, and quartz.

The application of the ATR-FTIR technique confirmed the mineralogical composition of all the samples studied in this work.

The results of the bulk chemical analysis by XRF have been found to be in good agreement with the EDS analyses of the powders. The chemical composition results have been considered to be of great interest and application in this investigation of the new types of dosages for the manufacture of more sustainable and low-energy cements. It was determined by XRF that the CaO in the limestones ranges between 34.54 and 46.54 wt %, with differences in the MgO content. This is a determining factor in the calculation of the raw material dosages, with implications and the effect of MgO in the expansivity of the produced cements.

The effect of the addition of different industrial wastes for the substitution of raw materials in the production of low-energy cements with high dicalcium silicate content has been studied. Thus, the dosage design has been performed in this investigation, using the Bogue equations and quality indexes (LSF, AM, and SM). The calculations of both the Bogue equations and quality indexes imply the knowledge of the weight percentages of CaO, SiO_2, Al_2O_3, and Fe_2O_3, as determined by XRF. In this theoretical design of the different crude samples, a dicalcium silicate ratio of 60–65 wt % and an LSF of 78–83% has been established as a limit value common to all the samples. The calculation basis for the crude blends has been based on calcined materials. Therefore, this has been established as the chemical composition, following this premise.

Thermal analysis by simultaneous TG-DTG-DSC has allowed for checking the mineralogical composition that was deduced by XRD, along with the decomposition and weight loss of calcite and dolomite in the case of the limestone samples and the dehydroxylation of clay minerals in the clay samples, by progressive heating.

By means of the thermal analysis carried out on the crude blends of cement, we have been able to verify the reactions produced in the range of ~700–850 °C as the main difference between blends, due to the decomposition of the carbonates associated with magnesium and calcium. These reactions have a greater or lesser intensity, depending on the composition of the raw materials and the waste used, which allows us to identify the critical compositions for the future characteristics of the cement. Thus, compositions using certain types of limestone with a high MgO content have been discarded. In the case of the thermal analyses, carried out on the crude-blend WST_0_T, using only the raw materials for WST_50_T, which is composed of 50 wt % of raw materials and wastes, and WST_100, which is composed entirely (100 wt %) of industrial wastes, the decomposition reaction of the carbonate associated with the magnesium is not identified. This means that there is less risk of expansivity problems in the manufactured cement.

5. Significance and Practical Application of this Investigation

Several recent research efforts have been performed in the utilization of certain industrial wastes for eco-friendly cement production [79–81]. For instance, El-Gamal and Selim [79] used ground and granulated blast-furnace slag to produce geopolymer cement at ambient temperature and at about 100% relative humidity. Azad et al. [80] studied the utilization of industrial by-products/waste to manufacture geopolymer cement/concrete.

These authors, on the basis of the sustainability indicators, suggested that most of the geopolymers developed using industrial waste leave a positive impact on the environment, economy, and society. More recently, Pinto Ferreira Brasileiro et al. [81] studied self-healing concrete as a way to establish planning for the synthesis and/or characterization of self-healing concretes, reductions in the consumption of cement, and the production of CO_2. However, in connection with the present investigation, it is interesting that Singh and Subramanian [82] previously explored the production of cement and clinker on a large scale, using post-industrial waste materials from different sources. This research provided the fundamental basis for an environmentally sustainable utilization of these kinds of industrial waste in the production of clinker, thereby making it suitable for use in construction.

In the present work, different types of Spanish and Greek raw materials and industrial wastes (8 raw materials and 3 industrial waste samples) have been identified with the necessary physicochemical characteristics, as well as high generation capacity, to study their possible use as alternative raw materials in the manufacture of low-energy cement. The wastes utilized in this study can be sourced all over the world, as the ceramic industry, the marble extraction and processing industry, and the secondary aluminum industry can be found in all five continents. It should be noted that selected industrial wastes are currently being deposited in landfill sites.

The clinkerization process, as described in the present investigation and in previous precedents [82], is one of the main options for their valorization and their consideration as "technological nutrients", with researchers working to include them as raw materials within the circularity of the cement industry's production process. Thus, in this way, the amount of traditional raw materials to be extracted from nature will be reduced, in turn, reducing the volumes to be deposited in landfill sites, with the consequent environmental improvement of the area and a reduction in all the polluting processes associated with mining extraction works. In this sense, wastes are used as raw materials for the production of a more useful material (in the present study, low-energy cement), taking into account the main principle of the circular economy. In fact, a new, 100% waste-based cement clinker blend has been found, using industrial waste as a practical application. This is in line with research concerning news on 100% waste-based materials, such as waste-based mortars and pastes [83].

The next step in this investigation will be the clinkerization process and the production of innovative cements using the waste materials at the laboratory on a pilot scale, as this previous characterization work has revealed a set of promising results.

Author Contributions: Conceptualization, S.M.-M.; methodology, S.M.-M., D.E.-Q. and P.J.S.-S.; investigation, S.M.-M.; data analysis, S.M.-M., L.P.-V. and D.E.-Q.; writing—original draft preparation, S.M.-M.; writing—review and editing, S.M.-M. and P.J.S.-S.; supervision, L.P.-V. and D.E.-Q. All authors have read and agreed to the published version of the manuscript.

Funding: The support of the European Union and the Spanish Government (Ministry of Universities): this research was supported, in part, through the European Recovery Instrument "Next-Generation EU" (UJAR05MS).

Institutional Review Board Statement: Not applicable.

Informed Consent Statement: Not applicable.

Data Availability Statement: Not applicable.

Acknowledgments: Sergio Martínez-Martínez acknowledges the support of the European Union and the Spanish Government (Ministry of Universities): this research was supported, in part, through the European Recovery Instrument "Next-Generation EU". Furthermore, it is grateful for the support shown by George N. Angelopoulos and the staff of the METLAB laboratory (Angeliki Christogerou, Dimitra G. Kanellopoulou, and Maria D. Kamitsou) of the University of Patras (Patras, Greece), during their research stay for the achievement of the objectives set out in this research. Research Group TEP 204 acknowledges the support of Junta de Andalucía.

Conflicts of Interest: The authors declare no conflict of interest.

Appendix A

Table A1. List of notations/abbreviations ordered according to their occurrence in the text.

	Notation/Abbreviation
Greenhouse gas emissions	GHG
Carbon dioxide	CO_2
Conference of the Parties to the United Nations Framework Convention on Climate Change	COP
Conference of the Parties serving as the meeting of the Parties to the Kyoto Protocol	COP-MOP
European Union	EU
Megatonne	Mt
Percentage	%
Calcium carbonate, calcite	$CaCO_3$
Calcium oxide	CaO
Kilogram	kg
Kilowatt hours	kWh
Ordinary Portland cement	OPC
Silicon dioxide, quartz	SiO_2
Aluminum oxide	Al_2O_3
Ferric Oxide, hematite	Fe_2O_3
Alite or tricalcium silicate	C_3S
Belite or dicalcium silicate	C_2S
Tricalcium aluminate	C_3A
Tetracalcium alumino-ferrite	C_4AF
Belitic Portland cement	BPC
Celsius	°C
Kilojoule	kJ
Belite polymorphs	α, α'H, α'L, β, and γ
Hour	h
Percentage by weight	Wt %
Magnesium oxide, periclase	MgO
Limestone ANAYA	C_A
Association Limited	S.L.
Limestone SANTO NICASIO 1	C_SN_1
Limestone SANTO NICASIO 2	C_SN_2
Limestone TITAN	C_T
Yellow clay	A_A
Black clay	A_N
Red clay	A_R
Chamotte	CHM
Société Anonyme	SA
Cutting and polishing marble sludges	MRM
Millimeter	mm
Secondary aluminium production of salt slag	PVL
Gypsum	Y
Initial material sample weight	m_i
Final material sample weight	m_f
Micrometer	µm
X-ray fluorescence	XRF
United Kingdom	UK
X-ray diffraction	XRD
Thermogravimetric analysis	TG
Derivative thermogravimetric analysis	DTG
Differential scanning calorimetry	DSC
Scanning electron microscopy	SEM
Energy dispersive X-ray spectroscopy analysis	EDS
Attenuated total reflectance Fourier transform infrared spectroscopy	ATR-FTIR
United States of America	USA

Table A1. *Cont.*

	Notation/Abbreviation
Lime Saturation Factor	LSF
Aluminum Modulus	AM
Silica Modulus	SM
Loss on ignition	LOI
Not detected	N.D.
Sodium Oxide	Na_2O
Potassium oxide	K_2O
Sulfur trioxide	SO_3
Titanium dioxide, rutile	TiO_2
Phosphorus pentoxide	P_2O_5
Manganese oxide	MnO
Zirconium	Zr
Parts per million	ppm
Chlorine	Cl
Chromium	Cr
Strontium	Sr
Cerium	Ce
Lead	Pb
Nickel	Ni
Copper	Cu
Zinc	Zn
Barium	Ba
Rubidium	Rb
Bismuth	Bi
Fluorine	F
Tin	Sn
Dolomite	$CaMg(CO_3)_2$
Muscovite (mica, illite)	$KAl_2(AlSi3O_{10})(OH)_2$
Orthoclase	$KAlSi_3O_8$
Albite	$NaAlSi_3O_8$
Montmorillonite	$MgOAl_2O_{35}SiO_2xH_2O$
Bassanite or gypsum hemihydrate	$CaSO_4.1/2H_2O$
Chlorite	$(Al,Fe,Mg)_{4–6}(Al,Si,Fe)_4O_{10}(OH,O)_8$
Calcium sulphate dihydrate	$CaSO_4$-$2H_2O$
Anhydrite	$CaSO_4$
Anorthite	$CaAl_2Si_2O_8$
Corundum	α-Al_2O_3
Bayerite	γ-$Al(OH)_3$
Nordstrandite	$Al(OH)_3$
Diaoyudaoite	$NaAl_{11}O_{17}$
Fluorite	CaF_2
Halite	NaCl
Arbitrary units	a.u.
Solid	s
Gas	g
Mullite	$3Al_2O_3{\cdot}2SiO_2$
Gehlenite	$Ca_2Al_2SiO_7$
Molecular nitrogen	N_2
Reciprocal centimeter or wave number	cm^{-1}
Symmetrical stretching	ν_1
Bending beyond the plane	ν_2
Symmetrical stretching	ν_3
Bending in the plane	ν_4
Blend containing only raw materials and C_A	WST_0_A
Blend containing only raw materials and C_SN	WST_0_SN
Blend containing only raw materials and C_T	WST_0_T
Blend containing 50 wt % of raw materials, 50 wt % of wastes and C_T	WST_50_T
Blend containing only wastes	WST_100

References

1. United Nations. Climate Change. Available online: www.un.org/es/sections/issues-depth/climate-change/index (accessed on 21 December 2020).
2. Government of Spain. Third Vice-Presidency of the Government of Spain. Ministry of Ecological Transition and Demographic Challenge. Main Elements of the Paris Agreement. Available online: www.miteco.gob.es/es/cambio-climatico/temas/el-proceso-internacional-de-lucha-contra-el-cambio-climatico/naciones-unidas/elmentos-acuerdo-paris.aspx (accessed on 30 November 2022).
3. European Commission. A European Green Deal. Striving to Be the First Climate-Neutral Continent. Available online: www.ec.europa.eu/info/strategy/priorities-2019-2024/european-green-deal_en (accessed on 30 November 2022).
4. European Commission. Eurostat. Circular Economy—Overview. Available online: www.ec.europa.eu/eurostat/web/circular-economy (accessed on 30 November 2022).
5. IEA (2022), Cement, IEA, Paris, License: CC BY 4.0. Available online: www.iea.org/reports/cement (accessed on 30 November 2022).
6. Martínez-Martínez, S.; Pérez-Villarejo, L.; Eliche-Quesada, D.; Sánchez-Soto, P.J.; Christogerou, A.; Kanellopoulou, D.G.; Angelopoulos, G.N. New waste-based clinkers for the preparation of low-energy cements. A step forward toward circular economy. *Int. J. Appl. Ceram. Technol.* **2020**, *17*, 12–21. [CrossRef]
7. Andrew, R.M. Global CO_2 emissions from cement production. *Earth Syst. Sci. Data* **2018**, *10*, 195–217. [CrossRef]
8. Kamitsou, M.D.; Kanellopoulou, D.G.; Christogerou, A.; Angelopoulos, G.N. A Contribution towards a More Sustainable Cement: Synergy of Mill Scales, Greek Wet Fly Ash, Conventional Raw Materials and Clinkering Temperature. *Minerals* **2022**, *12*, 324. [CrossRef]
9. Del Strother, P. Manufacture of Portland Cement. In *Lea's Chemistry of Cement and Concrete*, 5th ed.; Hewlett, P., Liska, M., Eds.; Elsevier: Amsterdam, The Netherlands, 2019; Volume 1, pp. 31–56.
10. Locher, F.W. *Cement: Principles of Production and Use*, 1st ed.; Vbt Verlag Bau U. Technik: Dusseldorf, Germany, 2005.
11. TITAN Cement Group. Accelerating Our Efforts to Tackle Climate Change. Available online: www.titan-cement.com/sustainability/environment/climate-change/ (accessed on 10 October 2021).
12. The European Cement Association. About Our Industry. Key Facts & Figures. Available online: https://cembureau.eu/about-our-industry/key-facts-figures/ (accessed on 30 November 2022).
13. Martínez-Martínez, S.; Pérez-Villarejo, L.; Eliche-Quesada, D.; Bonet-Martínez, E.; Sánchez-Soto, P.J. Eco-cemento sostenible con el medio ambiente como visión de futuro. In *Cuadernos de Didáctica. Medio Ambiente y Sociedad*, 1st ed.; Jordán-Vidal, M.M., Pardo-Fabregat, F., Eds.; Editorial Club Universitario: San Vicente, Spain, 2017; Volume 1, pp. 243–253.
14. Hewlett, P.C. *Lea's Chemistry of Cement and Concrete*, 4th ed.; Elsevier: Amsterdam, The Netherlands, 2003.
15. Taylor, H.F.W. *Cement Chemistry*, 2nd ed.; Thomas Telford Publishing: London, UK, 1997.
16. Popescu, C.D.; Muntean, M.; Sharp, J.H. Industrial trial production of low energy belite cement. *Cem. Concr. Compos.* **2003**, *25*, 689–693. [CrossRef]
17. Stark, J.; Müller, A.; Schrader, R.; Rümpler, K. Existence conditions of hydraulically active belite cement. *Zem.-Kalk-Gips.* **1981**, *34*, 476–481.
18. Martínez-Martínez, S.; Pérez-Villarejo, L.; Eliche-Quesada, D.; Sánchez-Soto, P.J.; Christogerou, A.; Kanellopoulou, D.; Kamitsou, M.; Angelopoulos, G.N. Desarrollo preliminar y estudio térmico de diferentes tipos de clínker de ecocemento sostenible con el medio ambiente utilizando residuos industriales y piedra caliza con alto porcentaje de óxido de magnesio en su composición. In Proceedings of the Congreso Nacional del Medio Ambiente CONAMA 2018, Madrid, Spain, 26–29 November 2018.
19. Odler, I. Hydration, Setting and Hardening of Portland Cement. In *Lea's Chemistry of Cement and Concrete*, 4th ed.; Hewlett, P.C., Ed.; Elsevier: Amsterdam, The Netherlands, 2003; Volume 1, pp. 241–297.
20. Government of Spain. Ministry of Transport, Mobility and Urban Agenda. National Geographic Institute. National Center for Geographic Information (CNIG). National Atlas of Spain. Thematic Structure. Natural Environment. Land Structure and Landforms. Geology. Map of the Major Geological Regions of the Iberian Peninsula and the Balearic Islands. Available online: www.centrodedescargas.cnig.es/CentroDescargas/busquedaRedirigida.do?ruta=PUBLICACION_CNIG_DATOS_VARIOS/aneTematico/Espana_Grandes-regiones-geologicas-de-la-peninsula-iberica-y-baleares_2004_mapa_13966_spa.pdf (accessed on 30 November 2022).
21. Government of Spain. Ministry of Science and Innovation. Geological and Mining Institute of Spain. Geological Cartography. MAGNA 50k (2ª Serie). Hoja 886 (BEAS DE SEGURA). Memoria Asociada al Mapa de la Hoja 886. Available online: www.info.igme.es/cartografiadigital/datos/magna50/memorias/MMagna0886.pdf (accessed on 30 November 2022).
22. Government of Spain. Ministry of Science and Innovation. Geological and Mining Institute of Spain. Geological Cartography. MAGNA 50k (2ª Serie). Hoja 946 (MARTOS). Memoria Asociada al Mapa de la Hoja 946. Available online: www.info.igme.es/cartografiadigital/datos/magna50/memorias/MMagna0946.pdf (accessed on 30 November 2022).
23. Government of Spain. Ministry of Science and Innovation. Geological and Mining Institute of Spain. Geological Cartography. MAGNA 50k (2ª Serie). Hoja 905 (LINARES). Memoria Asociada al Mapa de la Hoja 905. Available online: www.info.igme.es/cartografiadigital/datos/magna50/memorias/MMagna0905.pdf (accessed on 30 November 2022).
24. Government of Spain. Ministry of Science and Innovation. Geological and Mining Institute of Spain. Geological Cartography. MAGNA 50k (2ª Serie). Hoja 911 (CEHEGÍN). Memoria Asociada al Mapa de la Hoja 911. Available online: www.info.igme.es/cartografiadigital/datos/magna50/memorias/MMagna0911.pdf (accessed on 30 November 2022).

25. Sigüenza-Amichis, F. La extracción del mármol en las canteras de Macael. In *El Sector Industrial en la Economía de Almería: III Jornadas Sobre Economía en Almería, Almería, Spain, November 1997*; Instituto de Estudios Almerienses: Almería, Spain, 2000; pp. 23–32.
26. Morales-Güeto, J. *Tecnología de los Materiales Cerámicos*, 1st ed.; Diaz de Santos: Madrid, Spain, 2005.
27. Fernández-Paris, J.M. Dosificación de crudos de cemento portland mediante el módulo de cal. *Mater. Constr.* **1967**, *125*, 611–670. [CrossRef]
28. Bogue, R.H. Calculation of the compounds in Portland cement. *Ind. Eng. Chem. Anal.* **1929**, *1*, 192–197. [CrossRef]
29. Tobón, J.I.; López, F. Replanteamiento de las ecuaciones de Bogue en el cálculo mineralógico del clínker para una cementera colombiana. *Dyna* **2007**, *74*, 53–60.
30. Iacobescu, R.I.; Koumpouri, D.; Pontikes, Y.; Saban, R.; Angelopoulos, G.N. Utilization of EAF metallurgical slag in "green" belite cement. *UPB Sci. Bull. Series B Chem. Mater. Sci.* **2011**, *73*, 187–194.
31. Lawrence, C.D. The production of low-energy cements. In *Lea's Chemistry of Cement and Concrete*, 4th ed.; Hewlett, P.C., Ed.; Elsevier: Amsterdam, The Netherlands, 2003; Volume 1, pp. 421–470.
32. Iacobescu, R.I.; Koumpouri, D.; Pontikes, Y.; Saban, R.; Angelopoulos, G.N. Valorisation of electric arc furnace steel slag as raw material for low energy belite cements. *J. Hazard. Mater.* **2011**, *196*, 287–294. [CrossRef]
33. Haque, M.A.; Chen, B.; Javed, M.F.; Jalal, F.E. Evaluating the mechanical strength prediction performances of fly ash-based MPC mortar with artificial intelligence approaches. *J. Clean. Prod.* **2022**, *355*, 131815. [CrossRef]
34. Sinka, M.; Van den Heede, P.; De Belie, N.; Bajare, D.; Sahmenko, G.; Korjakins, A. Comparative life cycle assessment of magnesium binders as an alternative for hemp concrete. *Resour. Conserv. Recycl.* **2018**, *133*, 288–299. [CrossRef]
35. *Beneficios del Uso del PAVAL para la Fabricación de Cementos. Guías técnicas*, 1st ed.; Instituto Español del Cemento y sus Aplicaciones (IECA): Madrid, Spain, 2014.
36. Criado, J.M.; Trillo, J.M. Influencia de la dilución de la muestra en los diagramas de ATD de reacciones reversibles e irreversibles de descomposición térmica de sólidos. *Bol. Soc. Esp. Ceram. Vidr.* **1975**, *14*, 513–516.
37. Formosa, J.; Chimenos, J.M.; Lacasta, A.M.; Haurie, L.; Rosell, J.R. Novel fire-protecting mortars formulated with magnesium by-products. *Cem. Concr. Res.* **2011**, *41*, 191–196. [CrossRef]
38. Formosa, J.; Chimenos, J.M.; Lacasta, A.M.; Haurie, L. Thermal study of low-grade magnesium hydroxide used as fire retardant and in passive fire protection. *Thermochim. Acta.* **2011**, *515*, 43–50. [CrossRef]
39. Valverde, J.M.; Sánchez-Jiménez, P.E.; Pérez-Maqueda, L.A. Limestone Calcination Nearby Equilibrium: Kinetics, CaO Crystal Structure, Sintering and Reactivity. *J. Phys. Chem. C* **2015**, *119*, 1623–1641. [CrossRef]
40. Durán-Martín, J.D.; Sánchez-Jimenez, P.E.; Valverde, J.M.; Perejón, A.; Arcenegui-Troya, J.; García-Triñanes, P.; Pérez-Maqueda, L.A. Role of particle size on the multicycle calcium looping activity of limestone for thermochemical energy storage. *J. Adv. Res.* **2020**, *22*, 67–76. [CrossRef]
41. Popescu, M.A.; Isopescu, R.; Matei, C.; Fagarasan, G.; Plesu, V. Thermal decomposition of calcium carbonate polymorphs precipitated in the presence of ammonia and alkylamines. *Adv. Powder Technol.* **2014**, *25*, 500–507. [CrossRef]
42. Halikia, I.; Zoumpoulakis, L.; Christodoulou, E.; Prattis, D. Kinetic study of the thermal decomposition of calcium carbonate by isothermal methods of analysis. *Eur. J. Min. Process. Env. Protect.* **2001**, *1*, 89–102.
43. Yue, L.; Shui, M.; Xu, Z. The decomposition kinetics of nanocrystalline calcite. *Thermochim. Acta* **1999**, *335*, 121–126. [CrossRef]
44. Allibert, M.; Verein Deutscher Eisenhüttenleute (VDEh). *Slag Atlas*, 2nd ed.; Verlag Stahleisen GmbH: Dusseldorf, Germany, 1995.
45. González-García, S. Análisis mineralógico de arcillas de suelos andaluces. I. Tierras rojas. *An. De Edafol. Y Agrobiología* **1960**, *1*, 191–212.
46. Dweck, J. Qualitative and quantitative characterization of Brazilian natural and organophilic clays by thermal analysis. *J. Therm. Anal. Calorim.* **2008**, *92*, 129–135. [CrossRef]
47. Arsenović, M.; Pezo, L.; Mančić, L.; Radojević, Z. Thermal and mineralogical characterization of loess heavy clays for potential use in brick industry. *Thermochim. Acta* **2014**, *580*, 38–45. [CrossRef]
48. Jordán, R.; Faria, T.; Rodríguez, G.; Díaz, G.C.; Zayas, M.E. Study of the thermal transformations in the elaboration of a soda-lime glass for containers with cuban natural zeolite. *Ingeniería y ciencia.* **2007**, *3*, 91–105.
49. López-Delgado, A.; Medina, J.; Alonso, P.; Tayibi, H.; Pérez, C.; López, F.A. Estudio del comportamiento térmico del polvo de aluminio estabilizado con el yeso. *Rev. Metal.* **2005**, *41*, 330–334. [CrossRef]
50. Ould Chamekh, M.L.; Taibi, M.; Aride, J.; Boukhari, A. Caracterización fisicoquímica del yeso natural de Mauritania. *Mater. Constr.* **1998**, *48*, 61–65. [CrossRef]
51. Ruiz de Cauka, A. La determinación del contenido en yeso hemihidratado y dihidratado en los cementos por análisis térmico. *Mater. Constr.* **1973**, *23*, 614–633.
52. Collier, N.C. Transition and decomposition temperatures of cement phases—A collection of thermal analysis data. *Ceram. Silik.* **2016**, *60*, 338–343. [CrossRef]
53. López-Delgado, A.; Fillali, L.; Jiménez, J.A.; López-Andrés, S. Synthesis of α-alumina from a less common raw material. *J. Sol-Gel Sci. Technol.* **2012**, *64*, 162–169. [CrossRef]
54. Jiménez, J.A.; Padilla, I.; López-Delgado, A.; Fillali, L.; López-Andrés, S. Characterization of the aluminas formed during the thermal decomposition of boehmite by the Rietveld refinement method. *Int. J. Appl. Ceram. Technol.* **2015**, *12*, E178–E186. [CrossRef]
55. Stanienda-Pilecki, K.J. The Importance of Fourier-Transform Infrared Spectroscopy in the Identification of Carbonate Phases Differentiated in Magnesium Content. *Spectroscopy.* **2019**, *34*, 32–42.

56. Ji, J.; Ge, Y.; Balsam, W.; Damuth, J.E.; Chen, J. Rapid identification of dolomite using a Fourier Transform Infrared Spectrophotometer (FTIR): A fast method for identifying Heinrich events in IODP Site U1308. *Mar. Geol.* **2009**, *258*, 60–68. [CrossRef]
57. Farmer, V.C. *The Infrared Spectra of Minerals*, 1st ed.; Mineralogical Society of Great Britain and Ireland: London, UK, 1974.
58. Santos, F.D.; Da Conceição, L.R.V.; Ceron, A.; De Castro, H.F. Chamotte clay as potential low cost adsorbent to be used in the palm kernel biodiesel purification. *Appl. Clay Sci.* **2017**, *149*, 41–50. [CrossRef]
59. Aroke, U.O.; El-Nafaty, U.A. XRF, XRD and FTIR properties and characterization of HDTMA-Br surface modified organo-kaolinite clay. *Int. J. Emerg. Technol. Adv. Eng.* **2014**, *4*, 817–825.
60. Nayak, P.S.; Singh, B.K. Instrumental characterization of clay by XRF, XRD and FTIR. *Bull. Mat. Sci.* **2007**, *30*, 235–238. [CrossRef]
61. Ritz, M.; Vaculíková, L.; Plevová, E. Application of infrared spectroscopy and chemometric methods to identification of selected minerals. *Acta Geodyn. Geomater.* **2011**, *8*, 47–58.
62. Yeşilbaş, M. Thin Water and Ice Films on Minerals: A Molecular Level Study. Ph.D. Thesis, Umea University, Umea, Sweden, 13 April 2018.
63. Rezende, J.C.T.; Ramos, V.H.S.; Oliveira, H.A.; Oliveira, R.M.P.B.; Jesus, E. Removal of Cr(VI) from Aqueous Solutions Using Clay from Calumbi Geological Formation, N. Sra. Socorro, SE State, Brazil. *Mater. Sci. Forum.* **2018**, *912*, 1–6. [CrossRef]
64. Galván-Ruiz, M.; Velázquez-Castillo, R. Lime, an Ancient Material as a Renewed Option for Construction. *Ing. Invest. Tecnol.* **2011**, *12*, 93–102.
65. Dikmen, S.; Gunay, A.; Ersoy, B.; Erol, I. Determination of equilibrium, kinetic and thermodynamic parameters of acid red 88 adsorption onto montmorillonitic clay. *Environ. Eng. Manag. J.* **2015**, *14*, 1097–1110. [CrossRef]
66. Liu, Y.; Wang, A.; Freeman, J.J. Raman, MIR, and NIR spectroscopic study of calcium sulfates: Gypsum, bassanite, and anhydrite. In Proceedings of the 40th Lunar and Planetary Science Conference, The Woodlands, TX, USA, 23–27 March 2009.
67. La Russa, M.F.; Ruffolo, S.A.; Barone, G.; Crisci, G.M.; Mazzoleni, P.; Pezzino, A. The Use of FTIR and Micro-FTIR Spectroscopy: An Example of Application to Cultural Heritage. *Int. J. Spectrosc.* **2009**, *2009*, 893528. [CrossRef]
68. Kamel, N.A.; Abou-Aiaad, T.H.; Iskander, B.A.; Khalil, S.K.H.; Mansour, S.H.; Abd-El-Messieh, S.L.; Abd-El-Nour, K.N. Biophysical studies on bone cement composites based on polyester fumarate. *J. Appl. Polym. Sci.* **2010**, *116*, 876–885. [CrossRef]
69. Adazabra, A.N.; Viruthagiri, G.; Shanmugam, N. Infrared analysis of clay bricks incorporated with spent shea waste from the shea butter industry. *J. Environ. Manag.* **2017**, *191*, 66–74. [CrossRef]
70. Sedmale, G.; Randers, M.; Rundans, M.; Seglinš, V. Application of differently treated illite and illite clay samples for the development of ceramics. *Appl. Clay Sci.* **2017**, *146*, 397–403. [CrossRef]
71. Rodriguez-Blanco, J.D.; Shaw, S.; Benning, L.G. The kinetics and mechanisms of amorphous calcium carbonate (ACC) crystallization to calcite, *via*vaterite. *Nanoscale* **2011**, *3*, 265–271. [CrossRef] [PubMed]
72. Al Dabbas, M.; Eisa, M.Y.; Kadhim, W.H. Estimation of Gypsum-Calcite Percentages Using a Fourier Transform Infrared Spectrophotometer (FTIR), in Alexandria Gypsiferous Soil -Iraq. *Iraqi J. Sci.* **2014**, *55*, 1916–1926.
73. Pérez-Villarejo, L.; Takabait, F.; Mahtout, L.; Carrasco-Hurtado, B.; Eliche-Quesada, D.; Sánchez-Soto, P.J. Synthesis of vaterite $CaCO_3$ as submicron and nanosized particles using inorganic precursors and sucrose in aqueous medium. *Ceram. Int.* **2018**, *44*, 5291–5296. [CrossRef]
74. Djebaili, K.; Mekhalif, Z.; Boumaza, A.; Djelloul, A. XPS, FTIR, EDX, and XRD analysis of Al_2O_3 scales grown on PM2000 alloy. *J. Spectrosc.* **2015**, *2015*, 868109.
75. Castañón, A.M.; García-Granda, S.; Guerrero-Bustos, A.M.; Gómez-Fernández, F. Estudio de las fases mineralógicas del clínker en una cementera española, utilizando el método de Rietveld. *Dyna* **2012**, *173*, 41–47.
76. AENOR. *Cemento. Parte 1: Composición, Especificaciones y Criterios de Conformidad de los Cementos Comunes. UNE-EN 197-1*; AENOR: Madrid, Spain, 2011.
77. Kurdowski, W. *Cement and Concrete Chemistry*, 1st ed.; Springer Science & Business: Berlin, Germany, 2014.
78. Iacobescu, R.I.; Koumpouri, D.; Pontikes, Y.; Angelopoulos, G.N. Hydraulic and leaching behaviour of belite cements produced with electric arc furnace steel slag as raw material. *Ceram. Silik.* **2013**, *57*, 126–132.
79. El-Gamal, S.M.A.; Selim, F.A. Utilization of some industrial wastes for eco-friendly cement production. *Sustain. Mater. Technol.* **2017**, *12*, 9–17. [CrossRef]
80. Azad, N.M.; Samarakoon, S.M.S.M.K. Utilization of Industrial By-Products/Waste to Manufacture Geopolymer Cement/Concrete. *Sustainability* **2021**, *13*, 873. [CrossRef]
81. Pinto Ferreira Brasileiro, P.; Brandao, Y.B.; Sarubbo, L.A.; Benachour, M. Self-Healing Concrete: Background, Development, and Market Prospects. *Biointerface Res. Appl. Chem.* **2021**, *11*, 14709–14725.
82. Singh, G.V.P.B.; Subramaniam, K.V.L. Production and characterization of low-energy Portland composite cement from post-industrial waste. *J. Clean. Prod.* **2019**, *239*, 118024. [CrossRef]
83. Soriano, L.; Font, A.; Tashima, M.M.; Monzó, J.; Borrachero, M.V.; Payá, J. One-part blast furnace slag mortars activated with almond-shell biomass ash: A new 100% waste-based material. *Mater. Lett.* **2020**, *272*, 127882. [CrossRef]

materials

Article

Combining an Intensive Green Roof with Seismic Retrofitting of Typical Reinforced Concrete Buildings in Israel

Svetlana Pushkar [1,2], Ido Halperin [1,*] and Yuri Ribakov [1]

1 Department of Civil Engineering, Ariel University, Ariel 40700, Israel; svetlanap@ariel.ac.il (S.P.); ribakov@ariel.ac.il (Y.R.)
2 Faculty of Architecture and Town Planning, Technion Israel Institute of Technology, Haifa 3200003, Israel
* Correspondence: idoh@ariel.ac.il; Tel.: +972-74-723-3092

Abstract: This study suggests an intensive green roof as part of a sustainable and hazard-resistant conceptual design for the retrofitting of old buildings in Israel. The roof is suggested to be built with waste-based materials. A five-story reinforced concrete residential building was retrofitted with: Case 1: concrete wall strengthening (CWS)-conventional concrete + conventional green roof; Case 2: CWS-waste-included concrete + waste-based green roof; Case 3: seismic isolation columns (SIC)-conventional concrete + conventional green roof; and Case 4: SIC-waste-included concrete + waste-based green roof. Palekastro, Nuweiba, Tabas, and Erzincan ground motions were used for a structural dynamic time-history analysis of the retrofitted buildings. Life cycle assessments of cases 1–4 were performed using ReCiPe 2016 midpoint and endpoint evaluations. A two-stage analysis of variance (ANOVA) was used to analyze the ReCiPe endpoint results. According to the seismic results, Case 3 and Case 4 were much more preferable to Case 1 and Case 2, whereas according to the environmental evaluations, Case 4 was the most preferable to the other cases.

Keywords: retrofitted buildings; intensive green roofs; waste-based materials; seismic design; life-cycle analysis; ANOVA

Citation: Pushkar, S.; Halperin, I.; Ribakov, Y. Combining an Intensive Green Roof with Seismic Retrofitting of Typical Reinforced Concrete Buildings in Israel. *Materials* **2022**, *15*, 889. https://doi.org/10.3390/ma15030889

Academic Editor: Raffaele Landolfo

Received: 1 December 2021
Accepted: 19 January 2022
Published: 24 January 2022

1. Introduction

Nowadays, the seismic retrofitting of old buildings toward their improved hazard-resistant design is an important issue in many countries. Israel uses a nationwide statutory plan known as Tama 38 for retrofitting buildings built during the 1970s [1]. According to Tama 38, building practitioners should add reinforced concrete (RC) stiffening elements to the retrofitted structure [2]. Except for this conventional method, there are certain modern retrofitting methods which are expected to supersede the older ones. Installing dampers that are based on the energy dissipation principle, such as high-damping rubber bearings that isolate the structure from ground motion [2–4], are examples of such contemporary approaches.

Recently, in addition to the improved seismic capacity they provide, the environmental performance of retrofitting methods has become a very pressing issue. Plainly, this is because they are based on concrete and steel, whose production might significantly harm the environment. For example, the production of 1 kg of Portland cement and 1 kg of steel can release 0.833 and 1.106 kg CO_2-equivalent, respectively, thereby increasing the global warming potential (GWP) [5].

In this respect, Wei et al. [6] conducted a life cycle assessment (LCA) of CO_2 emissions related to the conventional concrete retrofitting jacketing method. Vitiello et al. [7] studied LCAs of fiber-reinforced polymer (FRP)-based strengthening sheets, FRP-reinforced concrete (RC) jacketing, RC shear wall, rubber bearings, and friction isolators, applied to a reinforced building in Italy. They concluded that the rubber bearings and friction isolators were the most sustainable alternatives. Salgado et al. [8] studied RC column jacketing, RC shear walls and beam weakening, revealing that the shear wall strengthening was the unsustainable alternative. Ribakov et al. [2,9] analyzed concrete wall strengthening (CWS) and viscous dampers connected to steel chevron braces (VD&SB) and to concrete chevron braces (VD&CB), high-damping rubber bearings (HDRBs), and seismic isolation columns (SIC). It was reported that VD&SB, VD&CB, HDRBs, and SIC were more sustainable alternatives than CWS.

Welsh-Huggins and Liel [10] suggested incorporating green roofs as an additional element to hazard-resistant design toward further development in a sustainable seismic design. A typical office building in Los Angeles was designed with intensive and extensive green roofs (a new building) and extensive green roof (a retrofitted building). The average roof dead load for the intensive green roof was 1000 kg/m^2 due to the deeper soil layer (up to 200 cm), whereas for the extensive green roof, it was 250 kg/m^2 due to the shallower soil layer (up to 20 cm). Also, the average live load for the intensive green roof was higher than for the extensive green roof (500 and 100 kg/m^2, respectively). Thus, Welsh-Huggins and Liel [10] concluded that in contrast to the extensive green roof, the building with intensive green roof manifested an improved seismic performance. However, these results were relevant only for a typical new building located in Los Angeles, which had a configuration specific to the city.

The aforementioned sustainable seismic design studies are country-specific because of the country-dependent nature of local building technologies, building configurations, and probable seismic events. Therefore, their findings cannot be generalized across all building configurations and countries [10]. In this respect, the present paper suggests a new concept of seismic as well as environmental solution–incorporation of an intensive green roof into the seismic design of retrofitted buildings in Israel. Such seismic design incorporating a green roof has not been environmentally evaluated yet for a retrofitted building. The substantial of the analysis, conducted here, is conceptual. It discusses a class of buildings, rather than a specific structure and intends to provide incipient results, which will underpin future studies.

2. Methods

2.1. Research Aims and Framework

This paper presents results of a conceptual study inspecting the inclusion of an intensive green roof into the retrofitting of typical, earthquake susceptible, existing old buildings in Israel. This class of buildings is represented here by a symmetric model of a five-story RC residential building, which accords construction approaches accepted in Israel during the 1970s and 1980s. As mentioned above, the main concern of this study is a new concept of seismic and environmental retrofitting. It includes a seismic resistance and environmental impact evaluation of the applied retrofitting.

The following two seismic retrofitting alternatives were selected—CWS and seismic isolation columns (SIC)—and bring into account an addition of an intensive green roof to the structure. Replacing natural aggregates with recycled aggregates in the considered hazard-resistant design is also under consideration in this evaluation.

To sum up, the retrofitted design alternatives considered here are: (1) Case 1: CWS-conventional concrete + conventional intensive green roof; (2) Case 2: CWS-waste-included concrete + waste-based intensive green roof; (3) Case 3: SIC-conventional concrete + conventional intensive green roof; and (4) Case 4: SIC-waste-included concrete + waste-based intensive green roof. Table 1 provides an overview of these retrofitting-material combinations. CWS-conventional concrete, SIC-conventional concrete, and conventional intensive green roofs contains natural aggregates such as gravel, perlite, and sand, whereas CWS-waste-included concrete, SIC-waste-included concrete, and waste-based intensive green roofs contain recycled aggregates and waste materials such as concrete waste-based aggregates (CWAs), fly ash-based aggregates (FAAs), coal bottom ash (CBA), and fly ash (FA). The original, non-retrofitted structure, serves as the control case.

Table 1. Case studies.

Materials \ Retrofitting Method	CWS	SIC
Conventional concrete + conventional intensive green roof	Case 1	Case 3
Waste-included concrete + waste-based intensive green roof	Case 2	Case 4

Palekastro (October 2021, Greece), Nuweiba (1995, Egypt), Tabas (1978, Iran) and Erzincan (1992, Turkey) ground motions were used for structural dynamic time-history analysis of the retrofitted buildings. Life cycle analysis (LCA) was conducted to evaluate the environmental damage related to the four retrofitted design alternatives. The LCA results of the retrofitted design alternatives were compared through a two-stage nested mixed ANOVA model.

2.2. CWS and SIC

During the 1970s and 1980s, many residential mid-rise RC buildings in Israel were constructed with open ground floor and slender columns. These buildings usually suffer from seismic vulnerability, do not comply with the modern seismic-design codes, and therefore require suitable engineering intervention.

The model, analyzed in this work, represents such a building. Its ground floor is open and, besides stairway walls, comprises only square-sectioned, 30/30 cm columns, made of C20 concrete. Each floor contains 60 columns. Their concrete Young's Modulus is 23.8 GPa [11]. The floors' heights are 2.6 m and the spans are 6 m in both directions. The floor's area is 1548 m^2. The overall mass of a typical story is 1904 ton and of the roof, 1333 ton. It is assumed that internal infill walls do not contribute to the structural system's rigidity as a long time had passed since the structure's construction, as well, low-quality construction and supervision place a question on their ability to operate as part of the lateral load bearing system. These assumptions lead to a structure whose mechanical properties can be described by a symmetric model. A schematic of this model is depicted in the left side of Figure 1a–c where the dashed-dotted lines represent the planes of symmetry. The right half of the non-retrofitted model is the left's reflection. A planar dynamic model was chosen by virtue of the above symmetry. In accordance with common structural practice [12], the ceilings are assumed to be rigid and the mass is concentrated at the ceilings' levels. All the above boils down to a dynamic shear model whose details are provided below, in Section 2.4.

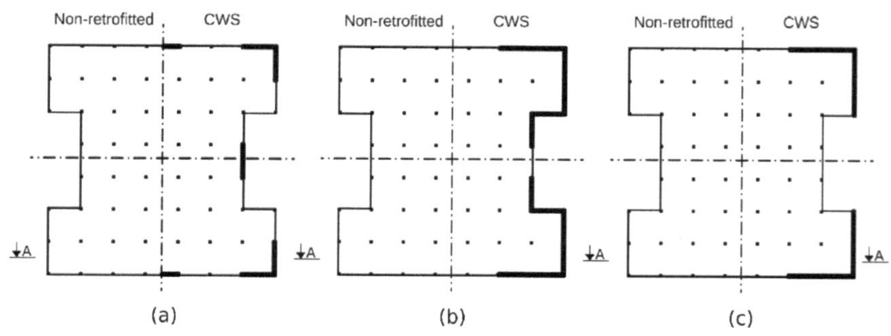

Figure 1. Floor plans of the analyzed typical building: on the right, concrete wall strengthening (Cases 1 and 2); on the left, the non-retrofitted structure (control case). (**a**) Ground floor (20 cm thick walls), (**b**) first floor (10 cm thick walls), and (**c**) floors two to four (10 cm thick walls).

The CWS method, which is rather classical and common, is based on strengthening the structure through improved strength elements, mainly concrete diaphragms. This is the retrofitting used in Cases 1 and 2. A vertical section of this configuration is depicted in Figure 2a. While the same live load, 490 kg_f/m^2, is applied to both cases, the dead load is different, due to the different compound of materials used in each one. In Case 1, the dead load is 2340 kg_f/m^2, whereas in Case 2, it is 1945 kg_f/m^2. However, the distribution of the strengthening RC diaphragms and the other retrofitting elements remains identical. It is illustrated by the right side of Figure 1. The left half of the CWS-retrofitted model is the right's reflection. The retrofitting retains the structure's symmetricity and accounts for two main practical issues, commonly evoked in such retrofitting. First, many structures slated for retrofitting are already occupied, thereby reducing the number of structural elements that are accessible for retrofitting operations. That is why in many retrofitting solutions, braces are implemented at the perimeter of the structure and in its internal staircase, as they are usually more accessible than other places. Another problem is that in many buildings, the ground floor is used for car parking. Hence, even though the elements in such a floor are more accessible, the need for parking places conflicts with the need to add stiffening walls, as it would block parking spots. In the suggested retrofitting, RC diaphragms are placed only at the outer walls, and no retrofitting elements are placed in the internal staircase. Such retrofitting is found sufficient for this case from a performance point of view. The number of RC diaphragms on the ground floor is reduced by half, compared to the first floor, for minimizing their impact on the parking places. However, in order to compensate for the reduced stiffness and strength, the thickness of the added diaphragms in this floor is doubled, and the perimeter beams in its ceiling level are modified accordingly. Shotcreting [13] of C30 concrete on existing walls is assumed as a method for forming these RC diaphragms. In total, 12 diaphragms of 20 cm thickness are embedded at the ground floor, 24 of 10 cm at the first floor, and 16 of 10 cm at each of the other floors. Additionally, in order to bond the reinforcing elements into a structural system, 32 beams are added at the perimeter of each ceiling. The beams' cross-section is square, 40/40 cm in dimension, and are made of C30 concrete. Typical reinforcing elements are shown in Figure 3.

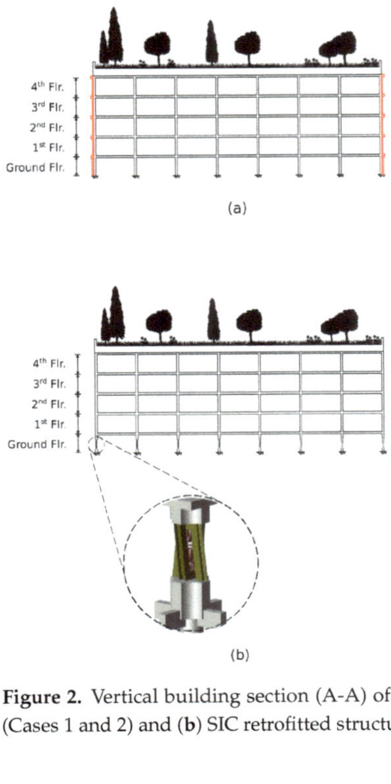

Figure 2. Vertical building section (A-A) of (**a**) concrete wall strengthening with deep green roof (Cases 1 and 2) and (**b**) SIC retrofitted structure with a deep green roof (Cases 3 and 4).

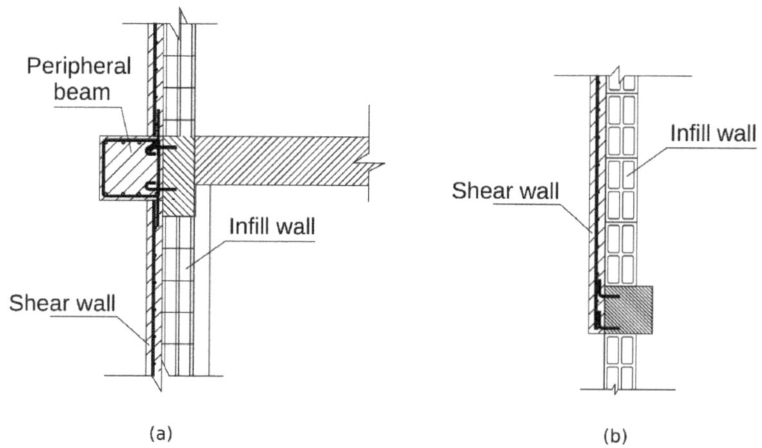

Figure 3. Concrete wall strengthening (Cases 1 and 2); vertical section of typical shear wall and peripheral beams (**a**); and horizontal section of typical shear wall-existing column interface (**b**).

Seismic isolation by seismic isolation columns (SIC) [14] is used for increasing the seismic capacity in Cases 3 and 4. Their schematic is depicted in Figure 2b. These isolators were especially developed for retrofitting buildings with open ground floor, as in the control model. They are founded on the well-known concept of friction-pendulum [15]. Essentially, the idea is to transform each existing column into a seismic-isolation device by replacing its middle portion by two steel V-shaped frames, connected by a high-load bearing chain. The remaining parts of the column are jacketed by concrete to increase their load capacity. As in Cases 1 and 2, the retrofitting of Cases 3 and 4 is identical, except for a difference in the roofs' dead loads. The roof dead load for Case 3 is 2340 kg_f/m^2 and for Case 4, it is 1945 kg_f/m^2.

A total of 288.2 m^3 of concrete and 22.5 tons of steel are required by the CWS retrofitting, and 19.2 m^3 of concrete and 22.3 tons of steel by the SIC retrofitting. The mix proportions for conventional concrete with a density of 2421 kg/m^3 and compressive strength of 43.2 MPa as well as for recycled concrete with density of 2334 kg/m^3 and compressive strength of 35.5 MPa were adopted from Turk [16]. Thus, CWS and SIC-related mix proportions for 288.2 m^3 concrete and 19.2 m^3 concrete, respectively, were evaluated and are presented in Table 2.

2.3. *Intensive Green Roof*

Tables 3 and 4 detail the layers and materials required by the conventional and waste-based intensive green roofs, respectively. As was mentioned above, the roof area in both cases is 1548 m^2. The weight of the roofs, 2340 kg_f/m^2 (the conventional green roof) and 1945 kg_f/m^2 (waste-based green roof), serves as a dead load in the seismic analysis. In addition, for both types of roofs, 490 kg_f/m^2 is used for live load.

Table 2. Materials required for retrofitting with CWS and SIC-based conventional and waste-included concretes.

Retrofitting Method	Aggregates (kg)	Recycled Aggregates (kg)	Fly Ash (kg)	Cement (kg)	Water (kg)	Plasticizer (kg)
CWS-conventional concrete	555,073	–	–	92,224	50,435	576
CWS-waste-included concrete	365,726	156,781	23,056	69,168	57,352	576
SIC-conventional concrete	36,979	–	–	6144	3360	38
SIC-waste-included concrete	24,365	10,445	1536	4608	3821	38

Table 3. Materials required for the conventional green roof.

Layer	Thickness (m)	Material	Weight (kg/m^2)	Weight (kg/Roof Area)
Substrate	1	*Gravel*	1325	*2,051,100*
		Sand	625	*967,500*
		Perlite	275	*425,700*
Filter layer	–	Polypropylene sheet	0.60	–
Drainage	0.1	*Perlite*	110	*170,280*
Protection layer	–	Polypropylene sheet	0.60	–
Roof barrier	0.0004	HDPE	0.385	–
Insulation	0.05	Polystyrene	1.5	–
Total	–	–	2340	–

Note: HDPE–high-density polyethylene. Italic font—layers with their respective weights that were included in the LCA evaluations.

Table 4. Materials required for the waste-based green roof.

Layer	Thickness (m)	Material	Weight (kg/m^2)	Weight (kg/Roof Area)
Substrate	1	*CWAs*	1193	*1,846,764*
		CBA	335	*518,580*
		FAAs	296	*458,208*
Filter layer	–	Polypropylene sheet	0.60	–
Drainage	0.1	*FAAs*	118	*182,664*
Protection layer	–	Polypropylene sheet	0.60	–
Roof barrier	0.0004	HDPE	0.385	–
Insulation	0.05	Polystyrene	1.5	–
Total	–	–	1945	–

Note: CWAs–concrete waste-based aggregates, CBA–coal bottom ash, FAAs–fly ash-based aggregates, HDPE–high-density polyethylene. Italic font—layers with their respective weights that were included in the LCA evaluations.

In the environmental evaluations, only two different layers, substrate and drainage, were considered, omitting the other layers, as those are the same for both roof types. According to Table 3, the substrate of the conventional green roof includes gravel (2650 kg/m^3), sand (2500 kg/m^3), and perlite (1100 kg/m^3). Perlite is included in the drainage layer as well. The densities of these natural materials accord those noted by SI 1045-0 [17]. According to Table 4, the substrate of the waste-based green roof includes CWAs (2385 kg/m^3 [18]), CBA (1340 kg/m^3 [19]), and FAAs (1182 kg/m^3 [20]). FAAs are included in the drainage layer as well.

2.4. Seismic Response Evaluation

Five dynamic models were formulated, one for the control case and four for Cases 1–4, taking into account each case's special traits. Figure 4 presents dynamic schemes used for each one of the cases [12]. In this figure, z represents a dynamic-degree-of-freedom, m for the floor's mass, k for a floor's lateral stiffness, and F for the friction force in a friction damper. Different parameters were assigned for reflecting the dynamic attributes of each case. In the control case, the lateral stiffness of each floor is 60 times the lateral flexural stiffness of an elastic beam element, having two fixed ends and a square section made of concrete. The floor masses represent the dead load of the ceilings, flooring and infill walls, as well as 30% of the live load for residential buildings. A shear building model, given in Figure 4a, is used and its parameters are detailed in Tables 5 and 6. Case 1's lateral stiffness and masses are the same as in the control case, with the addition of the contribution of the shear walls and the conventional deep green roof. Assuming small displacements, the shear walls remain elastic. In the wall's plane, their stiffness was modeled as a deep beam, connected through hinges to the original structure. The wall's stiffness was neglected perpendicular to this plane. The stiffness of the shear walls yields a full stiffness matrix, rather than a tri-diagonal one. Hence, the model in Figure 4b is used for this case and the stiffness in its DOFs is given by the matrix:

$$K^{CWS} = \begin{bmatrix} 6013.4 & -2221.5 & -6.88 & 1.967 & -3.491 \\ -2221.5 & 9384 & -7232.5 & -392.11 & 696 \\ -6.88 & -7232.5 & 9444.3 & -1840.8 & -609.73 \\ 1.967 & -392.11 & -1840.8 & 6187 & -3886 \\ -3.491 & 696 & -609.73 & -3886 & 3678.5 \end{bmatrix}, \text{[kN/mm]}$$

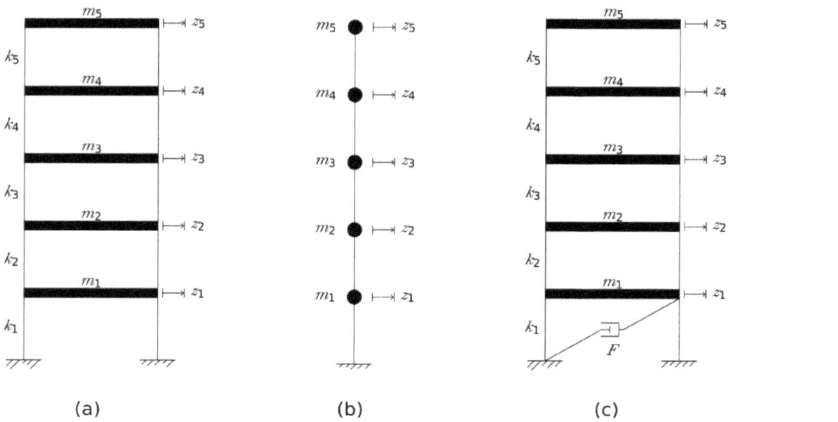

Figure 4. Dynamic schemes used for seismic response simulation of (**a**) the control case, (**b**) Cases 1 and 2, and (**c**) Cases 3 and 4.

Table 5. Mass values used in the dynamic models of each of the cases.

	Control Case	Case 1	Case 2	Case 3	Case 4
m_1, ton	1904	2066	2066	1904	1904
m_2, ton	1904	2066	2066	1904	1904
m_3, ton	1904	2036	2036	1904	1904
m_4, ton	1904	2036	2036	1904	1904
m_5, ton	1333	5846	5234	5714	5102

Table 6. Stiffness and friction values used in the dynamic models of the control case, Case 3 and Case 4.

	Control Case	Case 3	Case 4
k_1, kN/mm	658.1	46.245	46.245
k_2, kN/mm	658.1	658.1	658.1
k_3, kN/mm	658.1	658.1	658.1
k_4, kN/mm	658.1	658.1	658.1
k_5, kN/mm	658.1	658.1	658.1
F, kN	–	1255.2	1198

The mass of the retrofitting system and the conventional green roof was added to the masses of the control case. The total floor masses are detailed in Table 5. Case 2's parameters are as in Case 1, except for the mass of the waste-based green roof that has a different density. Therefore, the same model from Case 1 is used in Case 2 with the values detailed in Table 5. In Case 3 each SIC was modeled following the approach suggested by Briman and Ribakov [14], which yielded an equivalent elastic spring coefficient and friction damper force reflecting the column properties. The total stiffness and damping of the ground floor are represented by this floor's lateral stiffness and a friction damper with damping force F. The rest of the floors have stiffness similar to the control case. The same is for the model's masses, except the fifth mass that takes into account the conventional green roof's mass. Therefore, the model described in Figure 4c is used with the values given in Tables 5 and 6. Case 4's parameters are similar to those of Case 3 except for a waste-based green roof instead of the conventional one. Thus, same model from Case 3 is used for Case 4, modifying the parameters accordingly. The difference between Cases 3 and 4 in the ground floor's horizontal stiffness emerges from the different axial forces in the SICs' chains, leading to different lateral re-centering force. Additionally, each model was assumed to have inherent damping, typical to concrete structures. Rayleigh damping matrix was formulated for each case, assuming a damping ratio of 5% in the first two modes of vibration. The spectral traits of each model are given in Table 7. The change in the natural frequencies expresses the effect of each retrofitting. Cases 1 and 2 manifest higher frequencies compared to the control case, whereas a clear decrease is observed in Cases 3 and 4, as expected.

Table 7. Natural frequencies and mode shapes for each case.

	f_n, **Hz**			φ^T		
	0.89	0.17	0.33	0.46	0.55	0.59
	2.6	0.46	0.57	0.24	−0.27	−0.58
Control case	4	0.58	0.083	−0.57	−0.16	0.55
	5.1	0.51	−0.49	−0.026	0.52	−0.48
	5.7	0.28	−0.49	0.57	−0.51	0.31
	1.2	0.12	0.31	0.39	0.56	0.65
	4.3	0.33	0.67	0.6	0.063	−0.28
Case 1	8.7	0.89	−0.041	−0.26	−0.36	0.11
	9.7	0.34	−0.26	−0.23	0.86	−0.19
	14	0.14	−0.69	0.7	−0.091	−0.012
	1.2	0.12	0.31	0.39	0.56	0.65
	4.3	0.33	0.67	0.59	0.039	−0.3
Case 2	8.7	0.89	−0.051	−0.27	−0.34	0.12
	9.8	0.31	−0.25	−0.22	0.86	−0.21
	14	0.14	−0.69	0.7	−0.09	−0.014
	0.28	0.4	0.43	0.45	0.47	0.48
	1.5	0.68	0.54	0.27	−0.082	−0.41
Case 3	3.2	0.58	−0.042	−0.62	−0.49	0.2
	4.6	0.43	−0.58	−0.18	0.66	−0.1
	5.6	0.23	−0.57	0.65	−0.45	0.046
	0.28	0.41	0.43	0.45	0.47	0.48
	1.6	0.67	0.53	0.24	−0.11	−0.44
Case 4	3.2	0.58	−0.052	−0.62	−0.47	0.22
	4.6	0.43	−0.58	−0.17	0.66	−0.12
	5.6	0.23	−0.56	0.65	−0.45	0.053

Note: The mode shapes are written as row vectors, next to the pertaining natural frequency.

These models were coded into suitable state-space models in MATLAB [21]. Whilst dynamic linear models accord the control case, Case 1, and Case 2, dynamic nonlinear models are required for Cases 3 and 4. This is due to the nonlinear behavior entailed by the friction in the SIC columns. The model's response was computed by integrating the state equations with common LTI methods (the control case and Cases 1 and 2) or fourth order Runge-Kutta method [22] (Cases 3 and 4), for each of the given accelerograms.

As in many other works, an ensemble of past ground accelerations records was used to test the performance of each alternative [23]. It allows to examine how the inspected alternative would perform, had it been constructed and subjected to those earthquakes. This ensemble comprises Palekastro (October 2021, Greece), Nuweiba (1995, Egypt), Tabas (1978, Iran), and Erzincan (1992, Turkey) ground motions, which are accelerograms recorded in earthquakes occurred at the vicinity of discussed region. Different PGAs were recorded in these accelerograms. Hence, they were scaled to 0.29 g, for obtaining a similar comparison basis. This PGA value was set in accordance with the Israeli design code defining it as the highest design PGA in Israel [23]. The data was analyzed and transformed into common structural performance indices.

2.5. Environmental Evaluation

2.5.1. LCA: Stages and Methodology

According to [24], a cradle-to-grave LCA of concrete components consists of: (i) design, (ii) production/execution, (iii) usage, and (iv) end of life. However, this study evaluated a cradle-to-gate LCA in the production stage only. This means that the usage and end-of-life stages were excluded from the scope of this study. The usage stage of both retrofitted alternatives was supposed to have similar lifetimes: CWS has a design lifetime of 50–70 years and SIC has a design lifetime of 50–100 years [25]. The usage stage of both roofs was also supposed to have similar maintenances during their similar lifetimes as these are related to the same extensive type of green roof. In addition, the end-of-life stage was also excluded from the evaluation because of its nonsignificant environmental damage in concrete-related cradle-to-grave LCAs [26].

Thus, goal and scope definition includes the definition of the functional unit (FU) and a description of system boundaries. An inventory analysis (life cycle inventory (LCI)) collects input and output of raw materials and emissions related to the FU, an impact assessment converts LCI results into life cycle impact assessment (LCIA) results, and an interpretation step decides on the best and worst alternatives.

2.5.2. LCA: Goal and Scope

In this study, the FUs were the four retrofitted design alternatives (i.e., the materials required by Cases 1–4) that could improve the retrofitted building-related seismic performance in comparison to the seismic performance of the control case. The LCA methodology recommends that only different products/processes should be compared [27]. All the composite materials (aggregates, recycled aggregates, fly ash, cement, water, and plasticizer (Table 2)) were evaluated for the LCA analyses of Cases 1–4 of the retrofitted design alternatives. However, only the substrate and drainage layers (Tables 3 and 4) were evaluated for the LCA analyses of Cases 1–4 of the conventional and waste-based green roofs.

In summary, the system boundary included the production of materials for Cases 1–4 (Tables 2–4) and transportation of these materials to the building site. The transportation distance from the producer/supplier to the retrofitted building was posited as 50 km for gravel, sand, and perlite, 100 km for cement and CWAs, and 200 km for steel, plasticizer, fly ash, CBA, and FAAs.

2.5.3. LCA: Life Cycle Inventory

LCIs of these retrofitting materials were based on the Ecoinvent v3.2 database and the literature sources (Table 8), and were modeled on the SimaPro platform [28]. Such secondary data were used due to the absence of local Israeli data. Nevertheless, the use of the secondary data is sufficiently appropriate for this kind of comparative study because Cases 1–4 included mostly the same building materials, such as concrete, steel, and aggregates. Note that in this study, CBA and fly ash were considered as waste from coal-fired power plants. Therefore, in concrete plant, the LCI of production processes of these wastes were omitted for Cases 1–4. LCI of FAAs production was modeled as a cold bonding process as suggested by Frankovič et al. [20]. In this process, a mixture comprising 90% FA, 10% Portland cement, and water is prepared and cured for 28 days. After that, the mixture is crushed into FAAs [20].

Table 8. Production stage: data input (Ecoinvent v3.2, SimaPro v9.1, 2019) for modeling life-cycle inventory (LCI) for Cases 1–4.

Material/Process	Process
Gravel	Gravel, crushed, at mine/CH U
Sand	Sand, at mine/CH U
Perlite	Perlite, at mine/DE U
Cement	Portland cement, strength class Z 42.5, at plant/CH U
CWAs	Disposal, building, concrete, not reinforced, to recycling/CH U
Water	Top water, at user/CH U
FAAs	FAAs, production (modeled according to Frankovič et al. [17]).
Plasticizer	Polycarboxylates, 40% active substance (RER)
Steel	Steel rebar/EU
Transportation	Lorry transport, Euro 0, 1, 2, 3, 4 mix, 22 t total weight, 17.3 t

Note: CWAs, concrete waste-based aggregates; FAAs, fly ash-based aggregates. FAAs were modeled on SimaPro platform according to Frankovič et al. [20].

Table 9 shows the resulting LCI for the production of 1 kg of the materials used in Cases 1–4 and for moving these materials 1 tkm toward the building's retrofitting site.

Table 9. Life-cycle inventory (LCI) for assessment of Cases 1–4 (Ecoinvent v3.2 database).

Material/Process	GWP (kg CO_2)	TE (kg 1,4-DCB)	WC (m^3)	LU (m^2a crop eq)
Gravel	0.00445	0.0167	0.081	0.000362
Sand	0.00242	0.0067	0.026	0.000337
Perlite	0.00169	0.00706	0.00409	0.00312
Cement	0.828	0.297	1.4	0.00218
Water	0.000171	0.000409	0.00448	0.0000166
CWAs	0.00403	0.00343	0.00136	0.00000676
FAAs	0.025	0.00661	0.0113	0.000031
Plasticizer	1.15	3.75	0.0137	0.0204
Steel	2.31	0.381	0.00246	0.0211
Transportation	0.0663	0.00587	0.00000529	0

Note: GWP, global warming potential; FRS, fossil resource scarcity; WC, water consumption; TE, terrestrial ecotoxicity; 1,4-DCB, 1,4-dichloro-benzine equivalent.

2.5.4. LCA: Life Cycle Impact Assessment

The LCI results of Cases 1–4 were converted to LCIA results using ReCiPe 2016 [28]. ReCiPe 2016 can perform midpoint and endpoint single-score evaluations.

On the midpoint, 22 environmental impacts, including global warming (human health), global warming (terrestrial ecosystems), global warming (freshwater ecosystems), stratospheric ozone depletion, ionizing radiation, ozone formation (human health), fine particulate matter formation, ozone formation (terrestrial ecosystems), terrestrial acidification, freshwater eutrophication, marine eutrophication, terrestrial ecotoxicity, freshwater ecotoxicity, marine ecotoxicity, human carcinogenic toxicity, human non-carcinogenic toxicity, land use, mineral resource scarcity, fossil resource scarcity, water consumption (human health), water consumption (terrestrial ecosystem), and water consumption (aquatic ecosystems) can be evaluated. Among those, global warming potential, terrestrial ecotoxicity, fossil resource scarcity, and water consumption are the most significant environmental impacts for the environmental evaluation of the concretes studied here [28]. On the endpoint, the 22-midpoint evaluated environmental impacts can be grouped into damage to human health, ecosystem quality, and resources, and then converted to a single-score evaluation. The converting to single-score evaluation is performed by applying individualist (I), egalitarian (E), and hierarchist (H) perspectives on environmental problems. The I perspective evaluates only short-term damage (a 20-year time horizon), the E perspective considers all possible long-term damages (an 1000-year time horizon) and the H perspective balances between short- and long-term damage (an 100-year time horizon). In addition, the I perspective considers only substances with complete proof of their effects, the E perspective considers all reported substances, and the H perspective accounts for substances that are recognized by international health bodies. Moreover, the ReCiPe 2016 uses two types of weighting sets: a perspective-relevant (e.g., the E, H, and I weightings) and an average (A) weighting set. In the I weighting, human health is the most important damage; in the H and E weightings, ecosystem quality is the most significant damage. Taking into account all these factors, the ReCiPe 2016 converts damage to human health, ecosystem quality, and resources in individualist/average (I/A), hierarchist/average (H/A), egalitarian/average (E/A), individualist/individualist (I/I), hierarchist/hierarchist (H/H), and egalitarian/egalitarian (E/E) single-score evaluations [28]. Both midpoint and single-score evaluations have advantages and disadvantages. The midpoint evaluation has lower uncertainty, whereas the results of the endpoint single-score evaluation are much easier to interpret [29].

Consequently, the four retrofitting alternatives were evaluated using both of these methods. The midpoint H evaluation was performed considering global warming potential, terrestrial ecotoxicity, fossil resource scarcity, and water consumption. The endpoint single-score evaluation was performed applying all six evaluations: I/A, H/A, E/A, I/I, H/H, and E/E.

2.5.5. Statistical Evaluation

Figure 5 presents the application of the ReCiPe 2016 endpoint single-score design structure for comparing the LCIAs of Case 1 and Case 2. This design structure permits a pairwise comparison of the ReCiPe 2016 endpoint single-score evaluations using two-stage nested mixed ANOVA [30]. The primary sampling unit included the average and particular weighting set subunits and the two subunits utilized the six individual subunit evaluations (I/A, H/A, E/A, I/I, H/H, and E/E). This design structure was recently applied to environmental comparison between strengthened pre-stressed normal-strength concrete beams with different steel-fibered concrete layers [31].

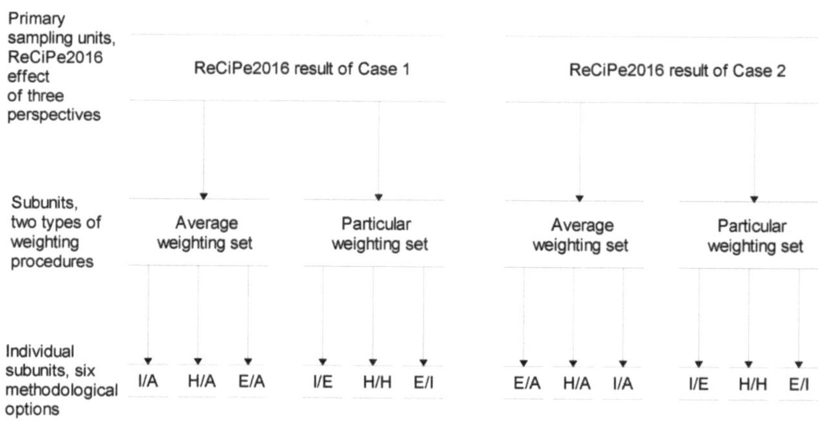

Figure 5. Two-stage analysis of variance (ANOVA) hierarchical design structure for environmental evaluation of Case 1 and Case 2 (ReCiPe 2016 single-score evaluation).

Prior to statistical analysis, the ReCiPe 2016 endpoint single-score results were multiplied by 10^3 and log10 transformed. Then, using the neo-Fisherian paradigm, the statistical differences between the ReCiPe 2016 endpoint single-score results of Cases 1 and 2 were compared. The neo-Fisherian paradigm (1) does not fix α, (2) does not describe p-values as significant or nonsignificant, (3) does not accept null hypotheses based on high p-values but only suspends judgment, (4) interprets significance tests according to three-valued logic, and (5) presents effect-size information if necessary [32]. As a result, the p-values were evaluated according to three-valued logic: either it seems to be positive (i.e., there appears to be an environmental difference between Case 1 and Case 2), it seems to be negative (i.e., there does not appear to be an environmental difference between Case 1 and Case 2), or judgment is suspended regarding the environmental difference between the two cases.

3. Results and Discussion

3.1. Seismic Response Evaluation

In slender buildings, roof displacements of a seismically excited structure provide a general sense of the response's intensity. Smaller roof displacements reflect a milder response. Time histories of the roof displacements are depicted in Figure 6. As Cases 3 and 4 are base isolated, their roof displacement is increased due to the reduced rigidity of the first floor, leading to an increased displacement at the isolated floor. Hence, the roof displacements for these cases are given after offsetting the base isolation displacement, to have a better picture of the vibration intensity of the superstructure. It can be seen that Case 3 and Case 4 perform better than the others. Comparing to the control case, Cases 1 and 2 obtained an improvement of 18–31% for two earthquakes but performed worse for two others (an increase of 120–173%). Cases 3 and 4, however, performed well for all earthquakes, showing a reduction of 43–75%.

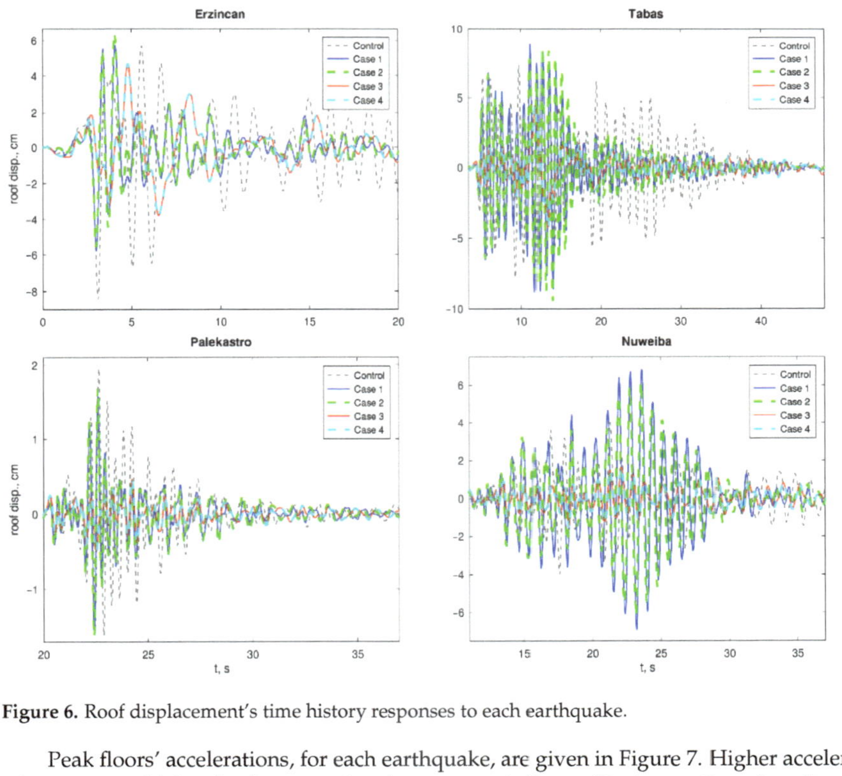

Figure 6. Roof displacement's time history responses to each earthquake.

Peak floors' accelerations, for each earthquake, are given in Figure 7. Higher accelerations express higher loads, exerted to the structure's floors. During earthquakes, these loads are a combination of the seismic load and the structural response, which is trying to alleviate the induced vibration. Case 3 and Case 4 experienced lower accelerations, meaning that the seismic loads are lower with such retrofittings. Thereby, Case 3 and Case 4 have responded better, in this context.

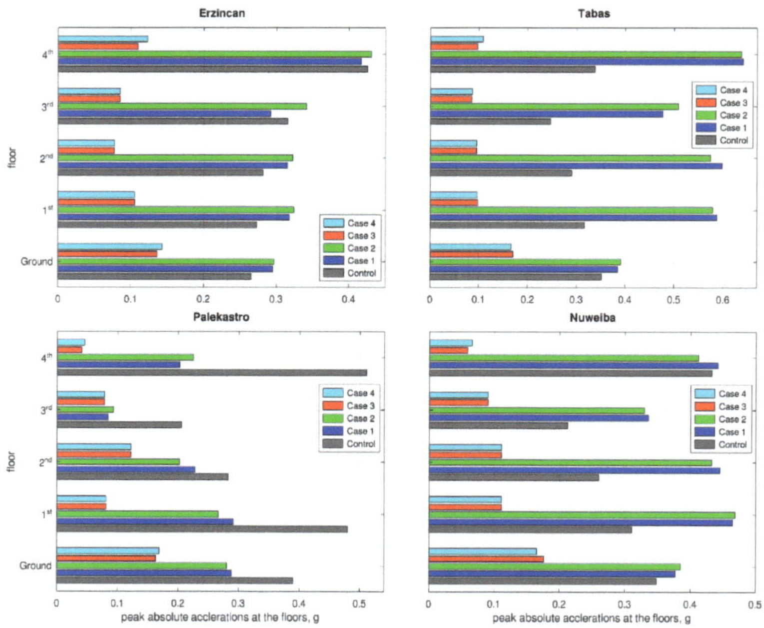

Figure 7. Peak floors' acceleration responses to each earthquake.

Mechanical failure of structural elements is a major hazard during earthquakes. Intensive vibrations impose deformations on the structure's skeleton, generate large stresses, and when these exceed the material's limits, the structure's stability is endangered. A prevalent approach for assessing these response components during earthquakes is to inspect the building's inter-story drifts. Large inter-story drifts in regular structure are highly related with large bending stresses in the relevant columns. The same is true for the converse case. The peak inter-story drifts in the structure's floors are described in Figure 8. The inter-story drifts for Case 3 and Case 4 outperformed the other cases, except for the first floor. However, this is not surprising because Case 3 and Case 4 are base isolated, allowing significant displacement to occur in the isolated floor. This phenomenon is part of the retrofitting strategy and, unlike standard floors, is not related with large stresses. Thus, the large displacements in the first floor in Case 3 and Case 4 are acceptable. Another important issue in the seismic response of structures is the loads applied to the foundations' system. A common way to generally evaluate this is to inspect the horizontal load generated at the building's ground floor, known as base-shear. A large base-shear means that the foundations must carry higher loads during the vibration, and if they are unable to do so, a foundation failure might follow. The peak base-shear forces are provided by Figure 9. Smaller base-shear values were observed in Cases 3 and 4, compared to the others, meaning that the foundations system is under lower risk in these cases.

Totaling the above, it can be seen that a general improvement was obtained by all the retrofitting methods, comparing to the control case. Case 3 and Case 4, however, performed much better than the others. Additionally, the responses of Case 3 and Case 4 are essentially all but identical. The same can be said for Case 1 and Case 2. This is explained by the small difference in the mechanical properties of each pair of cases, i.e., Cases 3 and 4 and Cases 1 and 2. However, there is a significant difference between the pairs due to the different retrofitting approaches.

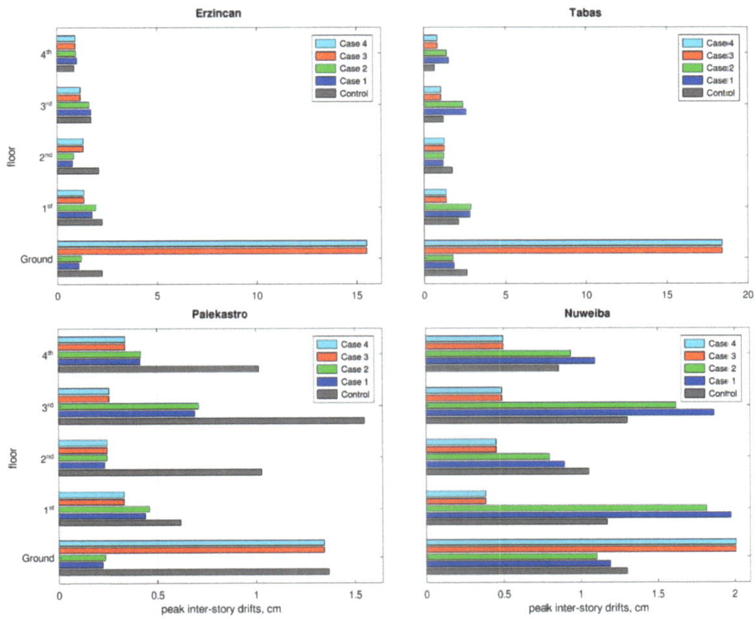

Figure 8. Peak inter-story drift responses to each earthquake.

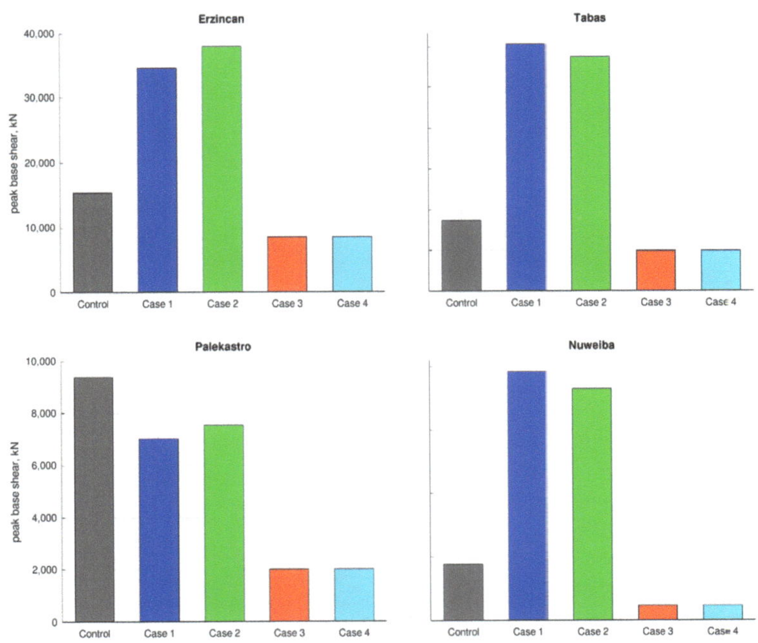

Figure 9. Peak base-shear responses to each earthquake.

3.2. Environmental Evaluation

3.2.1. ReCiPe 2016 Midpoint Results

Figure 10 shows that, comparing to Cases 1 and 3 (CWS and SIC conventional concrete-based retrofitting methods, respectively, with conventional green roofs), Case 2 and Case 4

(CWS and SIC waste-based retrofitting methods, respectively, with waste-based roofs) have lower terrestrial ecotoxicity, water consumption, and land use. On the other hand, Case 1 and Case 3 have lower global warming potential. In particular, compared to Case 1, the impact of ecotoxicity, water consumption, and land use of Case 2 are lower by 44, 63, and 79%, respectively, and compared to Case 2, the impact of the global warming potential of Case 1 is lower by 5%. Similarly, compared to Case 3, the impact of ecotoxicity, water consumption, and land use of Case 4 is lower by 58, 91, and 85%, respectively, and compared to Case 4, the impact of the global warming potential of Case 3 is lower by 21%.

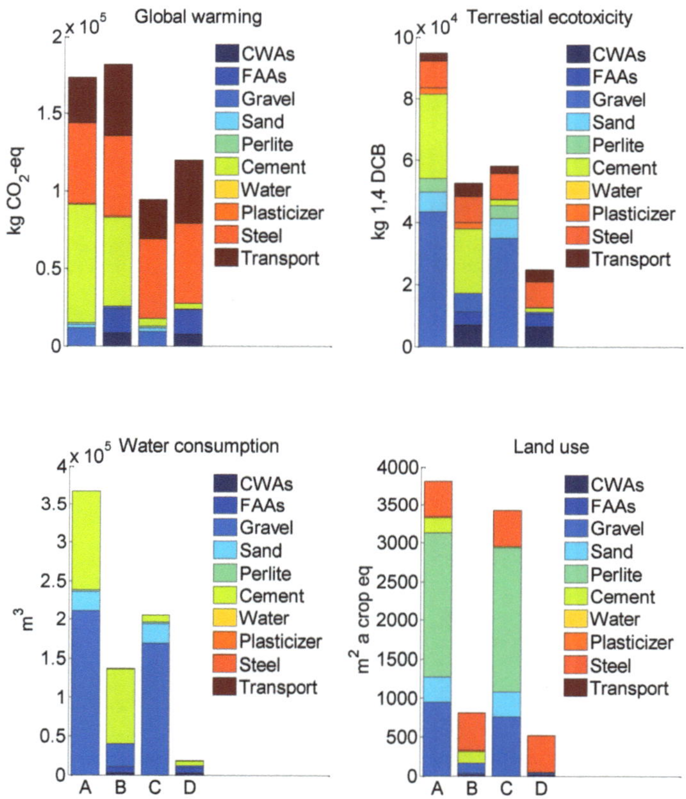

Figure 10. (**A**) Case 1: concrete wall strengthening (CWS)-conventional concrete + conventional green roof, (**B**) Case 2: CWS-waste-included concrete + waste-based green roof, (**C**) Case 3: seismic isolation columns (SIC)-conventional concrete + conventional green roof, and (**D**) Case 4: SIC-waste-included concrete + waste-based green roof evaluated with ReCiPe 2016 midpoint method.

These differences in impacts are due to different materials, natural and waste-based, used for the CWS and SIC retrofitting methods as well as for the green roofs (Tables 2–4). The natural material-based retrofitting methods with natural material-based green roofs (Case 1 and Case 3) mostly rely on the use of conventional concrete, steel, and natural aggregates (gravel, sand, and perlite), whereas the waste-based retrofitting methods with waste-based green roofs (Case 2 and Case 4) mostly rely on the use of steel and involve additional transport to deliver the wastes to the building site.

Among three basic concrete components, such as aggregates, water, and cement, the cement production process is the largest consumer of energy and producer of the highest amount of CO_2 emissions [33]. To produce clinker, calcareous and argillaceous materials need to be heated to 1450 °C. If the heating energy is produced with burning fossil fuels (oil, coal, or gas), a large amount of CO_2 is released into the atmosphere. As well, the clinker-related calcination process itself is responsible for an additional release of CO_2 into the atmosphere [34]. The high CO_2 emissions of cement production was recognized by many research studies, which confirmed that 74–93% of the total CO_2 is released from concrete production [35–37]. Cement production also requires using a large amount of water during grinding and mixing raw calcareous and argillaceous materials for clinker production. In this way, cement production is a huge contributor to GWP, TE, and WC impacts (Figure 10, Case 1 and Case 2 are based on large amounts of concrete).

Steel production is another noticeable contributor to GWP, TE, and LU (Figure 10, Cases 1–4). This is because almost the same steel quantity was used in both retrofitting methods, 22.5 tons in CWS and 22.3 tons in SC, in combination with conventional green roofs and waste-based green roofs. As was noted by Liang et al. [38], the steel-making blast furnace process, which is responsible for 70% of steel produced all over the world, mostly contributes to the global warming potential (40%), abiotic depletion potential (35%), and human toxicity potential (17%). In addition, according to the LCI results, the production of 1 kg of steel from iron ore mined from Earth required 0.0211 m^2 of crop equivalent land use, thereby resulting in high LU impact for all four retrofitted cases (Table 4).

Natural aggregates (gravel, sand, and perlite) that were mostly used for Cases 1 and 3 contribute to TE, WC, and LU (Figure 10). In contrast, GWP, which is related to the production of natural aggregates, was found to be smaller by 9% and 13% of the total concrete related GWP for CWS and SIC methods, respectively. These results are in line with the results presented by other research studies, which confirmed that aggregate production is responsible for 2 to 20% [37] of the total concrete's CO_2 emissions.

Wastes used in FA, which was used in waste-based concrete of retrofitting methods, and CBA, which was used in waste-based green roofs, are considered here as wastes from coal-burning electricity production process. Therefore, only the impacts related to the transportation of these wastes to the building site were accounted. As a result, the GWP pertaining transporting in Case 2 and Case 4 is higher than the GWP related to transport of Case 1 and Case 3.

3.2.2. ReCiPe 2016 Endpoint Single-Score Results

Figure 11 shows that Case 4 is the best retrofitted alternative with the smallest environmental damage, while Case 1 is the worst retrofitted alternative with the biggest environmental damage. Moreover, the difference between these two cases holds for all six methodological options (I/A, H/A, E/A, I/I, H/H, and E/E). However, the results for Case 2 and Case 3 are varying with respect to the different methodological options. Case 2 evaluated with E/E and E/A options has lower environmental damage than Case 3, whereas evaluation with I/I and I/A options reveals that Case 3's environmental damage is lower than that of Case 2.

Table 10 demonstrates p-values of the differences in single-score evaluation (simultaneous evaluation of the six methodological options) between pairs of four retrofitted design alternatives. According to these results, a negative difference exists only when comparing Case 2 against Case 3 ($p = 0.4141$), whereas there are positive differences between the other compared pairs, such as Case 1 and Case 2, Case 1 and Case 3, Case 1 and Case 4, Case 2 and Case 4, and Case 3 and Case 4 ($0.0002 \leq p \leq 0.0011$).

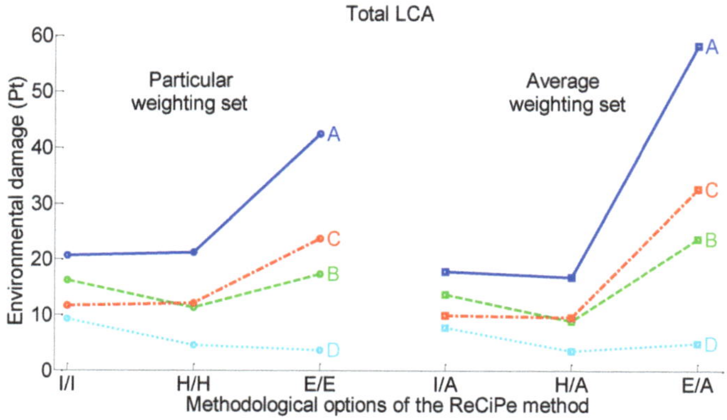

Figure 11. (**A**) Case 1: concrete wall strengthening (CWS)-conventional concrete + conventional green roof, (**B**) Case 2: CWS-waste-included concrete + waste-based green roof, (**C**) Case 3: seismic isolation columns (SIC)-conventional concrete + conventional green roof, and (**D**) Case 4: SIC-waste-included concrete + waste-based green roof evaluated with ReCiPe 2016 endpoint single-score method.

Table 10. *p*-values of the differences in single-score evaluation between pairs of four retrofitted design alternatives for the life-cycle assessment (LCA) (production stage). The LCAs were evaluated via the six ReCiPe2016 single-score methodological options.

Retrofitted Design Alternatives	Case 1	Case 2	Case 3	Case 4
Case1: CWS-conventional concrete + conventional green roof	X	**0.0011**	**0.0010**	**0.0002**
Case 2: CWS-waste-included concrete + waste-based green roof		X	0.4141	**0.0006**
Case 3: SIC-conventional concrete + conventional green roof			X	**0.0005**
Case 4: SIC-waste-included concrete + waste-based green roof				X

Note: Normal font represents a negative difference between the compared cases; bold font represents a positive difference between the compared cases.

This means that according to evaluation conducted by the ReCiPe 2016 endpoint single-score method, Case 4 is the first preferable alternative as a less damaging retrofitting method. This conclusion is independent of the applied perspective with regard to different views on the significance of the environmental problem. The next preferable alternatives are Case 2 and Case 3 as no environmental difference was revealed between this pair of the retrofitting methods. Finally, all six methodological options point out Case 1 as the most damaging retrofitting method, hence, the most inappropriate alternative.

4. Conclusions

This study evaluated seismic and environmental performances of four retrofitting methods: Case 1: concrete wall strengthening (CWS)-conventional concrete + conventional green roof, Case 2: CWS-waste-included concrete + waste-based green roof, Case 3: seismic isolation columns (SIC)-conventional concrete + conventional green roof, and Case 4: SIC-waste-included concrete + waste-based green roof.

According to the seismic analysis, Case 1 and Case 2 demonstrated certain small improvements in the seismic-bearing capacity of the retrofitted building. Thus, incorporation of an intensive green roof, as an environmental enhancement that is incorporated into the seismic design of such typical buildings in Israel, can be considered as a positive measure.

However, much better improvements in the seismic-bearing capacity of the retrofitted building was shown in Cases 3 and 4. Thus, according to the seismic analysis, using SIC is favorable for the seismic retrofitting of typical old Israeli buildings with the described attributes. Moreover, that fact is not diminished by the existence of an intensive green roof in the building.

According to the environmental performance results evaluated with the hierarchist (H) methodological option of the ReCiPe 2016 midpoint, Case 2 and Case 4 (CWS and SIC waste-based retrofitting methods, respectively, with waste-based roofs) had lower terrestrial ecotoxicity, water consumption, and land use, compared to Case 1 and Case 3 (CWS and SIC conventional concrete-based retrofitting methods, respectively, with conventional green roofs). Case 1 and Case 3, however, had lower global warming potential. Thus, replacing conventional concrete with waste-based concrete in CWS and SIC retrofitting methods, and natural aggregates with recycled aggregates in intensive green roofs, can improve three of the four environmental impacts considered in this study.

Furthermore, the ReCiPe 2016 endpoint single-score results allowed to compare Case 2 against Case 4 and pointed out Case 4 as the best retrofitting alternative. For this evaluation, the environmental preferability of Case 4 was confirmed for all six methodological options (I/A, H/A, E/A, I/I, H/H, and E/E). It was found that the environmental damage caused by production processes related to Case 4 was significantly lower than that of Case 2. This result was found to be consistent across three interpretations (egalitarian, individualist, and hierarchist) of the significance of the environmental problem. The next preferable alternatives were Case 2 and Case 3. It is interesting that comparing to Case 2 (in which CWS was built with waste-included concrete and waste-based green roof), Case 3 (in which SIC was built with conventional concrete and green roof with natural aggregates) had similar environmental damage. This is because of the lower concrete quantity used by the SIC method. However, as was already noted, the seismic performance of Case 3 was much better compared to that of Case 2.

Thus, according to both seismic and environmental performances, Case 4 retrofitting alternative, which includes SIC retrofitting, waste-included concrete, and an intensive green roof built with recycled aggregates, can be classified as the best building retrofitting method. These findings were obtained for a class of buildings located in Israel and may not be applicable to buildings with different configurations in other locations. However, it demonstrates that as the construction industry uses many other building-related retrofitting measures, seeking to improve the seismic durability and the environmental performance of the applied measures should not be ignored.

Author Contributions: Conceptualization, S.P. and Y.R.; methodology, S.P., I.H. and Y.R.; software S.P. and I.H.; validation, S.P. and I.H.; formal analysis, S.P. and I.H.; investigation, S.P. and I.H.; resources, S.P. and Y.R.; data curation, S.P.; writing—original draft preparation, S.P. and I.H.; writing—review and editing, S.P., I.H. and Y.R.; visualization, S.P. and I.H.; supervision, S.P. and Y.R.; project administration, S.P. and Y.R. All authors have read and agreed to the published version of the manuscript.

Funding: This research received no external funding.

Institutional Review Board Statement: Not applicable.

Informed Consent Statement: Not applicable.

Conflicts of Interest: The authors declare no conflict of interest.

References

1. Margalit, T.; Mualam, N. Selective rescaling, inequality and popular growth coalitions: The case of the Israeli national plan for earthquake preparedness. *Land Use Policy* **2020**, *99*, 105123. [CrossRef]
2. Ribakov, Y.; Halperin, I.; Pushkar, S. Seismic resistance and sustainable performance of retrofitted buildings by adding stiff diaphragms or seismic isolation. *J. Archit. Eng.* **2018**, *24*, 04017028. [CrossRef]
3. Naeim, F.; Kelly, J.M. *Design of Seismic Isolated Structures: From Theory to Practice*; Wiley: Hoboken, NJ, USA, 1999.
4. Cherry, S.; Filatrault, A. Seismic response control of buildings using friction dampers. *Earthq. Spectra* **1993**, *9*, 447–466. [CrossRef]
5. Stengel, T.; Schiessl, P. Life cycle assessment (LCA) of ultra high performance concrete (UHPC) structures. In *Eco-Efficient Construction and Building Materials*; Woodhead Publishing: Cambridge, UK, 2014; Volume 22, pp. 528–564, ISBN 978-0-85709-045-4.
6. Wei, H.H.; Shohet, I.M.; Skibniewski, M.J.; Shapira, S.; Yao, X. Assessing the lifecycle sustainability costs and benefits of seismic mitigation designs for buildings. *J. Archit. Eng.* **2016**, *22*, 04015011. [CrossRef]

7. Vitiello, U.; Salzano, A.; Asprone, D.; Di Ludovico, M.; Prota, A. Life-cycle assessment of seismic retrofit strategies applied to existing building structures. *Sustainability* **2016**, *8*, 1275. [CrossRef]
8. Salgado, R.A.; Apul, D.; Guner, S. Life cycle assessment of seismic retrofit alternatives for reinforced concrete frame buildings. *JOBE* **2020**, *28*, 101064. [CrossRef]
9. Ribakov, Y.; Halperin, I.; Pushkar, S. Using Eco-indicator 99 and a two-stage nested analysis of variance test to evaluate building mitigation measures under hazard risks. *Adv. Struct. Eng.* **2016**, *19*, 860–870. [CrossRef]
10. Welsh-Huggins, S.J.; Liel, A.B. A life-cycle framework for integrating green building and hazard-resistant design: Examining the seismic impacts of buildings with green roofs. *Struct. Infrastruct. Eng.* **2017**, *13*, 19–33. [CrossRef]
11. *SI 466 Part 1—Concrete Code: General Principles*; Standards Institute of Israel: Tel-Aviv, Israel, 2003.
12. Chopra, A.K. *Dynamics of Structures: Theory and Applications to Earthquake Engineering*; Prentice-Hall: Englewood Cliffs, NJ, USA, 1995.
13. ElGawady, M.; Lestuzzi, P.; Badoux, M. A review of conventional seismic retrofitting techniques for URM. In Proceedings of the 13th International Brick and Block Masonry Conference, Amsterdam, The Netherlands, 4–7 July 2004.
14. Briman, V.; Ribakov, Y. Seismic isolation columns for earthquake-resistant structures. *Struct. Des. Tall Spec. Build.* **2008**, *17*, 99–116. [CrossRef]
15. Al-Hussaini, T.M.; Zayas, V.A.; Constantinou, M.C. Seismic isolation of multi-story frame structures using spherical sliding isolation systems. In *Technical Report NCEER 94-0007*; State University of New York at Buffalo, Department of Civil Engineering: Buffalo, NY, USA, 1999.
16. Turk, J.; Cotič, Z.; Ana Mladenovič, A.; Šajna, A. Environmental evaluation of green concretes versus conventional concrete by means of LCA. *Waste Manag.* **2015**, *45*, 194–205. [CrossRef]
17. *SI 1045 Part 0—Thermal Insulation of Buildings: General*; Standards Institute of Israel: Tel-Aviv, Israel, 2011.
18. Malešev, M.; Radonjanin, V.; Broćeta, G. Properties of recycled aggregate concrete. *Contemp. Mater.* **2014**, *2*, 239–249. [CrossRef]
19. Marto, A.; Kassim, K.A.A.; Makhtar, M.; Wei, L.F.; Lim, Y.S. Engineering characteristics of Tanjung Bin coal ash Electron. *J. Geotech. Eng.* **2010**, *15*, 1117–1129.
20. Frankovič, A.; Bosiljkov, B.V.; Ducman, V. Lightweight aggregates made from fly ash using the cold-bond process and their use in lightweight concrete. *Mater. Technol.* **2017**, *51*, 267–274. [CrossRef]
21. *MATLAB*; Version 9.7 (R2019b); The MathWorks Inc.: Natick, MA, USA, 2019.
22. Butcher, J.C. A history of Runge-Kutta methods. *Appl. Numer. Math.* **1996**, *20*, 247–260. [CrossRef]
23. *SI 413 Part 1—Design Provisions for Earthquake Resistance of Structures*; Standards Institute of Israel: Tel-Aviv, Israel, 1995.
24. ISO 13315-1 (International Organization for Standardization). *Environmental Management for Concrete and Concrete Structures 2012, Part 1: General Principles*; International Organization for Standardization: Geneva, Switzerland, 2012.
25. Martin, J. *Long Term Performance of Rubber in Seismic and Non-Seismic Bearings: A Literature Review*; U.S. Department of Commerce National Institute of Standards and Technology, Building and Fire Research Laboratory: Galthenburg, MD, USA, 1999.
26. Napolano, L.; Menna, C.; Asprone, D.; Prota, A.; Manfredi, G. Life cycle environmental impact of different replacement options for a typical old flat roof. *Int. J. Life Cycle Assess.* **2015**, *20*, 694–708. [CrossRef]
27. ISO 14040 (International Organization for Standardization). *Environmental Management Life Cycle Assessment Principles and Framework*; International Organization for Standardization: Geneva, Switzerland, 2006.
28. *SimaPro*; Version 9.0; PRé Consultants: Amersfoort, The Netherlands, 2019.
29. Huijbregts, M.A.J.; Steinmann, Z.J.N.; Elshout, P.M.F.; Stam, G.; Verones, F.; Vieira, M.; Zijp, M.; Hollander, A.; van Zelm, R. ReCiPe2016: A harmonised life cycle impact assessment method at midpoint and endpoint level. *Int. J. Life Cycle Assess.* **2017**, *22*, 138–147. [CrossRef]
30. Picquelle, S.J.; Mier, K.L. A practical guide to statistical methods for comparing means from two-stage sampling. *Fish. Res.* **2011**, *107*, 1–13. [CrossRef]
31. Pushkar, S.; Ribakov, Y. Life-cycle assessment of strengthening pre-stressed normal-strength concrete beams with different steel-fibered concrete layers. *Sustainability* **2020**, *12*, 7958. [CrossRef]
32. Hurlbert, S.H.; Lombardi, C.M. Final collapse of the Neyman-Pearson decision theoretic framework and rise of the neoFisherian. *Ann. Zool. Fenn.* **2009**, *46*, 311–349. [CrossRef]
33. Chen, C.; Habert, G.; Bouzidi, Y.; Jullien, A.; Ventura, A. LCA allocation procedure used as an incitative method for waste recycling: An application to mineral additions in concrete. *Resour. Conserv. Recycl.* **2010**, *54*, 1231–1240. [CrossRef]
34. Van den Heede, P.; De Belie, N. Environmental impact and life cycle assessment (LCA) of traditional and 'green' concretes: Literature review and theoretical calculations. *Cem. Concr. Compos.* **2012**, *34*, 431–442. [CrossRef]
35. O'Brien, K.R.; Menache, J.; O'Moore, L.M. Impact of fly ash content and fly ash transportation distance on embodied greenhouse gas emissions and water consumption in concrete. *Int. J. Life Cycle Assess.* **2009**, *14*, 621–629. [CrossRef]
36. Celik, K.; Meral, C.; Gursel, A.P.; Mehta, P.K.; Horvath, A.; Monteiro, P.J.M. Mechanical properties, durability, and life-cycle assessment of self-consolidating concrete mixtures made with blended Portland cements containing fly ash and limestone powder. *Cem. Concr. Compos.* **2015**, *56*, 59–72. [CrossRef]
37. Gursel, A.P.; Ostertag, C. Impact of Singapore's importers on life-cycle assessment of concrete. *J. Clean. Prod.* **2016**, *118*, 140–150. [CrossRef]
38. Liang, T.; Wang, S.; Lu, C.; Jiang, N.; Long, W.; Zhang, M.; Zhang, R. Environmental impact evaluation of an iron and steel plant in China: Normalized data and direct/indirect contribution. *J. Clean. Prod.* **2020**, *264*, 121697. [CrossRef]

materials

MDPI

Article

Rheology of Alkali-Activated Blended Binder Mixtures

Biruk Hailu Tekle *, Ludwig Hertwig and Klaus Holschemacher

Structural Concrete Institute (IfB), Leipzig University of Applied Sciences (HTWK Leipzig), 04277 Leipzig, Germany; ludwig.hertwig@htwk-leipzig.de (L.H.); klaus.holschemacher@htwk-leipzig.de (K.H.)
* Correspondence: birukh.tekle@gmail.com

Abstract: Alkali-activated cement (AAC) is an alternative cement that has been increasingly studied over the past decades mainly because of its environmental benefits. However, most studies are on heat-cured AACs and are focused on mechanical properties. There is a lack of research on the fresh properties of ambient-cured AAC systems. This study investigates the rheological properties of ambient-temperature-cured alkali-activated blended binder mixtures activated with sodium silicate and sodium hydroxide solutions. The influence of binder amount, alkaline solid to binder ratio (AS/B), sodium silicate to sodium hydroxide solids ratio (SS/SH), and total water content to total solid (from the binding materials) ratio (TW/TS) on the rheological properties are investigated. The effect of borax as an admixture and silica fume as a replacement for fly ash is also investigated. The results showed that both the yield stress and plastic viscosity are mainly affected by the binder content and TW/TS ratio decreasing with the increase of each parameter. The yield stress increased with the increase of the SS/SH ratio. Borax significantly reduced the yield stress, while silica fume's effect was dependent on its dosage.

Keywords: rheology; alkali-activated cement; plastic viscosity; yield stress; thixotropy

Citation: Tekle, B.H.; Hertwig, L.; Holschemacher, K. Rheology of Alkali-Activated Blended Binder Mixtures. *Materials* **2021**, *14*, 5405. https://doi.org/10.3390/ma14185405

Academic Editor: Jean-Marc Tulliani

Received: 2 September 2021
Accepted: 15 September 2021
Published: 18 September 2021

1. Introduction

Concrete is the most commonly used construction material. However, with today's environmental consciousness and the high carbon dioxide (CO_2) emission associated with the production of concrete ingredients, especially ordinary portland cement (OPC), this material faces a significant challenge. Alkali-activated cement (AAC) is a relatively new type of alternative cement to replace OPC. It is mainly produced from byproduct or secondary product materials rich in silicon and aluminum such as fly ash (FA) and ground granulated blast furnace slag (GGBS). In the presence of alkaline solutions such as sodium silicate and sodium hydroxides, these materials undergo dissolution. By combining with oxygen and incorporating sodium, potassium, or calcium ions depending on the materials used, the dissolved ions form calcium aluminosilicate hydrate (C-A-S-H) or sodium aluminosilicate hydrate (N-A-S-H) [1]. The main advantage of AAC is its environmental benefits, with an 80% or greater reduction in CO_2 emission compared to OPC [2,3]. Further to this, AAC concrete has desirable mechanical and durability behaviors such as high strength, good bond performance, excellent fire resistance, and acid resistance [3–6]. These behaviors have inspired enormous research interest in this alternative binder.

AAC concrete has been reported to suit various applications ranging from conventional cast-in-place concrete to 3D printing material and more [7]. Certain applications require special attention on the deformation and flow behavior of the concrete, i.e., rheology. For instance, concrete for 3D printing requires high yield stress at rest and low plastic viscosity during flow [8]. The rheological properties of the 3D printing concrete can be used for achieving a balance between pumpability, extrudability, and buildability [9]. Early concrete processing, such as spreading, molding, and compaction, depends on its rheological properties.

Interest in the rheological behaviors of AAC systems is increasing as the mortar and concrete based on this binder are considered for various applications [10–12]. The fresh

AAC mixture is a solid-liquid suspension system and its rheology depends on various parameters such as type and amount of source materials and activators used, chemical admixtures, temperature, solid contents, water content, and so on [12–16]. Ground granulated blast furnace slag (GGBS), fly ash (FA), and silica fume (SF) are some of the commonly used source materials. Sodium silicate and sodium hydroxide, mixed or separately, are the most common activators used. Using GGBS enhances AAC mixtures' ambient curing behavior [17]. This is due to the high calcium oxide content and fineness of GGBS, improving the binder's reactivity and enhancing the reaction product formation at ambient curing conditions. However, the same reason also reduces workability and affects the rheology [18–20]. Sodium hydroxide or a combination of sodium hydroxide and sodium-silicate-activated AACs follows the Bingham model [11,21], meaning the relation between shear stress and shear rate is linear, while sodium-silicate-activated AACs follow the Herschel–Bulkley model [16], i.e., a nonlinear relation between shear stress and shear rate.

Achieving desired rheological behaviors is relatively easy in OPC concrete as various admixtures are available. However, admixture technologies are still in development for AAC concrete. Palacios et al. [22] investigated the stability of different superplasticizers (polycarboxylate, vinyl copolymer, melamine, and naphthalene derivatives) in water and different alkaline solutions. All the superplasticizers except the naphthalene-based one are unstable at high pH (greater than 13). Tong et al. [23] stated that there are almost no superplasticizers that can improve the workability of sodium-silicate-activated slag, showing that the performance of the superplasticizers is also dependent on the alkaline activators. Hence, developing the required rheological behaviors of AAC systems has a critical challenge.

In this study, the rheological behavior of AAC is studied. Different mix design factors such as water to binder ratio, sodium silicate to hydroxide ratio, activator to binder ratio, and binder content are studied for their effect on rheology. Furthermore, the effect of SF, borax (a potential set retarder), and a new type of superplasticizer is also investigated.

2. Experimental Details

2.1. Materials

FA, GGBS, and SF were used in the preparation of the alkali-activated concretes. The FA, GGBS, and SF comply with the EN 450-1 [24], EN 15167-1 [25], and EN 13263-1 [26] requirements, respectively. The chemical compositions of these source materials are summarized in Table 1. The chemical analysis was performed using energy dispersive X-Ray analysis (EDX) (Noran System SIX) and a scanning electron microscope (Jeol JSM–IT 100, JEOL Ltd., Freising, Germany). Figure 1 shows the grain size distributions for each ingredient obtained by laser diffraction method. The activator solution used is a mixture of sodium silicate and sodium hydroxide. The sodium silicate solution includes 26.82% silicate, 8.2% sodium oxide, and 64.98% water, while the sodium hydroxide is a 50% by weight solution. New superplasticizer underdevelopment (Geo-1) supplied by Sika Germany was used. Furthermore, sand with a maximum aggregate size of 2 mm (Sand 0/2 in Figure 1) was used. Additional fines (shown in Figure 1 as Fines) were also used at 10% of total sand. The chemical composition of the binders is shown in Table 1. Figure 2 shows the morphology of FA and GGBS observed by scanning electron microscope (SEM). The FA particles are spherical, and the GGBS particles have an irregular shape with high angularity.

Table 1. Chemical composition of ingredients [27].

Composition	FA (%)	GGBS (%)	SF * (%)
SiO_2	49.79	34.48	93.81
Al_2O_3	26.71	11.48	0.48
Fe_2O_3	8.57	-	1.49
MgO	2.47	7.08	0.46
CaO	4.34	42.43	0.30
K_2O	3.36	0.66	0.77
Na_2O	1.28	0.56	0.42
SO_3	1.49	2.17	0.20
TiO_2	1.23	1.14	-
Specific surface area (m^2/g)	0.45	0.46	19.40
Specific gravity (g/cm^3)	2.28	2.91	2.20

* Manufacturer's specification.

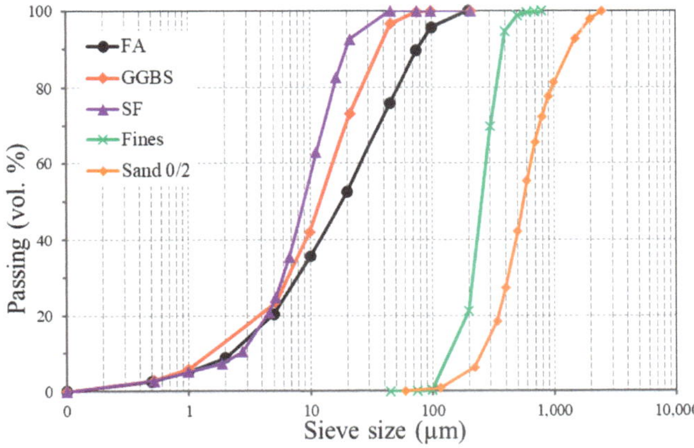

Figure 1. Particle size distribution.

Figure 2. SEM image of source materials.

2.2. Investigated Parameters and Mix Proportions

The research is part of an ongoing project for developing a concrete mixture for use in textile-reinforced concrete. Textile reinforcement requires fine-grained concretes due to their smaller openings; hence, no coarse aggregate was used. The constituents of the precursor were set at 55% FA, 40% GGBS, and 5% SF based on Tekle et al. [28]. Three mixes were designed targeting a medium- and two high-strength classes (AAC-1, AAC-2, and AAC-3). Their respective 28 days strengths were 40 MPa, 72 MPa, and 92 MPa. The mix parameters of AAC-2 were then varied to understand their effect on the rheology of the mixture. The parameters were binder (precursors) content (B), alkaline solid to binder ratio (AS/B), sodium silicate to sodium hydroxide solids ratio (SS/SH), and total water to total solid binders ratio (TW/TS). TW is the amount of free water and water from the alkaline solution, while TS is the total amount of solid binding material in the mixture, i.e., B and AS. AAC-2 was taken as a reference mix, as shown in Table 2.

Table 2. Mix proportions.

Mix No.	Binder (kg/m^3)	Sand (kg/m^3)	AS/B	SS/SH	TW/TS	SP (%)	Borax (%)
AAC-1	550	1104	0.14	2.5	0.55	-	-
AAC-2	650	1159	0.14	2.5	0.35	-	-
AAC-3	750	1195	0.18	2.5	0.25	-	-
TW/TS.40	650	1062	0.14	2.5	0.40	-	-
TW/TS.45	650	966	0.14	2.5	0.45	-	-
SS/SH-1.5	650	1157	0.14	1.5	0.35	-	-
SS/SH-2.0	650	1158	0.14	2.0	0.35	-	-
AS/B.18	650	1117	0.18	2.5	0.35	-	-
AS/B.22	650	1074	0.22	2.5	0.35	-	-
B750	750	941	0.14	2.5	0.35	-	-
B550	550	1377	0.14	2.5	0.35	-	-
SP-Borax	650	1159	0.14	2.5	0.35	4	2
SP	650	1159	0.14	2.5	0.35	4	-
Borax	650	1159	0.14	2.5	0.35	-	2
SF15	SF (15)	1159	0.14	0.14	2.5	0.35	-
SF25	SF (25)	1159	0.14	0.14	2.5	0.35	-

Binder: 55% FA, 40% GGBS, and 5% SF. AS: the solid part of the alkaline solution, TS: total content of the binder solids (binder + AS), TW: total water, SS and SH: solid parts of sodium silicate and sodium hydroxide.

2.3. Test Methods and Specimen Preparation

The sodium silicate and sodium hydroxide solution were mixed in the required proportion at least 24 h before mixing. The specimens were prepared by first mixing the dry materials (sand and binder) in a mixer for about two minutes. Afterward, the prepared alkaline solution was mixed with the additional water, added slowly to the dry mixture, and mixed for about four minutes. In the case of mixes with superplasticizer (SP), the SP was added halfway during the wet mixing.

The consistency of the fresh mixture is determined by flow tests according to EN 1015-3 [29]; however, without the jolting, as the mixes were relatively workable. Figure 3 shows AS/B.18 mix during the flow test. The fresh mixture was then removed from the flow table and added back to the mixing bowl and mixed for an additional 15 s. The rheology behavior of the mixtures was tested using a rotational viscometer (Viskomat NT) equipped with a cylindrical container of 100 mm in height and 83 mm in diameter. The paddle consists of four symmetrically arranged curved rods; a fishbone paddle. The test was carried out about 10 min after the addition of water into the dry ingredients. The rotational speed was ramped from 0 to 120 rpm in 1 min (step 1 in Figure 4). It was kept at 120 rpm for 4 min. Then, the speed was ramped down to 0. The rotational speed was then ramped up again from 0 to 80 rpm and ramped down in steps. The torque under each imposed speed was measured. The rheology parameters, yield stress, and plastic viscosity were then obtained according to the torque–rotational speed relationship.

Figure 3. AS/B.18 mix during flow test.

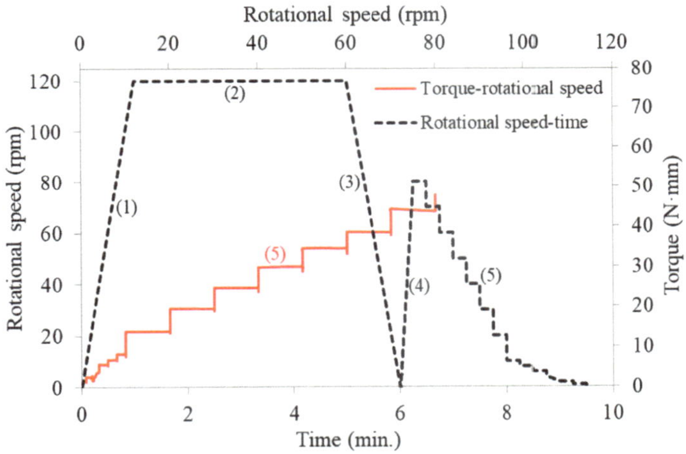

Figure 4. Typical torque–rotational speed relationship (AS/B.18).

3. Results and Discussion

3.1. Results

The rheology of AAC is affected by the type of activator and the type of precursor used [21,30]. In Na_2SiO_3-activated systems, the Herschel–Bulkley model is a better fit [21]. In NaOH-activated systems, Bingham model was reported to fit better, meaning that their rheology is defined by two physical parameters, i.e., yield stress and plastic viscosity [21]. The AAC in this study uses both these activators. Figure 4 shows the five steps used for the flow profile and the torque–rotational speed relationship for the last step (stepwise descending part) of the flow curve. The torque increased and decreased with the respective increase and decrease of the rotational speed. For the evaluation of the rheological properties, the data points of the downward stepwise ramp (step 5) were selected. The duration of each of the steps in the downward curve is 15 s (Figure 4). A linear regression of the torque and the rotational speed in the 20–80 rpm range was performed. The regression lines showed a linear relation, showing that the structure is broken down to equilibrium by the applied shear stress during the test cycle. This behavior was observed in all the mixes, showing that the relationship between the shear stress (τ) and the shear rate ($\dot{\gamma}$) for the studied AAC mixture can be described using the Bingham model, as shown in Equation (1).

$$\tau = \tau_0 + \eta\dot{\gamma} \quad (1)$$

where τ_0 is the yield stress, and η is the plastic viscosity.

The data from the Viskomat in the current study are for torque and rotational speed. Such data are only relative values as they are geometry-specific. Converting them to shear stress and shear rate is complex, especially with probes such as the one used in the current study (fishbone). Haist et al. [31] conducted an interlaboratory study for different geometries and recommended conversion factors for absolute value shear stress and shear rate determination based on the affine-translation approach. In the current study, the yield stress and plastic viscosity values were obtained by using these conversion factors. The flow values and the yield stress and viscosity parameters for each of the mixes are summarized in Table 3.

Table 3. Flow, yield stress, and plastic viscosity parameters.

Mix No.	Flow (mm)	Yield Torque (N·mm)	Viscosity (N·mm·min)	Yield Stress (Pa) [31]	Viscosity (Pa·s) [31]
AAC-1	310	1.55	0.05	16.48	0.67
AAC-2	245	7.35	0.41	78.13	5.51
AAC-3	210	6.84	1.74	72.71	23.36
TW/TS.4	280	4.18	0.18	44.43	2.42
TW/TS.45	300	1.36	0.07	14.46	0.94
SS/SH-1.5	240	4.32	0.50	45.92	6.71
SS/SH-2	245	4.87	0.49	51.77	6.58
AS/B.18	265	5.19	0.43	55.17	5.77
AS/B.22	280	5.17	0.39	54.96	5.24
B750	280	3.08	0.26	32.74	3.49
B550	210	21.64	0.94	230.03	12.62
SP-Borax	275	2.74	0.28	29.13	3.76
SP	275	3.93	0.30	41.78	4.03
Borax	265	3.73	0.32	39.65	4.30
SF15	200	4.49	0.22	47.73	2.95
SF25	145	7.74	0.20	82.28	2.69

3.2. *Flowability*

Figure 5 shows the main effect plots for the flow values of the mix proportion parameters studied. All the investigated parameters increased the flow. Binder content and the TW/TS ratio are the most significant factors. The least significant factor is the SS/SH ratio, with a minor increase as the ratio increased. Nath and Sarker [19] observed an opposite effect of sodium silicate to hydroxide solution ratio on the slump and concrete flow. This difference could be due to the difference in source materials used (no SF was used in [19], and the proportion of GGBS and FA is different) or the type of activators used.

The binder content showed a significant effect on the flow. The mixture became more workable as the binder content increased. An increase in binder content by keeping the AS/B and TW/TS ratio constant increased the paste content, ultimately raising the workability. This is consistent with previous study [32].

The AS/B is another significant factor affecting the flowability of the mixture. The flow values increased with the increase of the AS/B. Previous studies reported that the increase in alkaline liquid to binder ratio (AL/B) increased workability [18,19,33]. However, it is not easy to verify if this effect comes solely from the increase in water or the increase in the alkaline solids. It is well known that an increase in water increases workability; however, the current study also confirms that the increase in alkaline solid also increases workability.

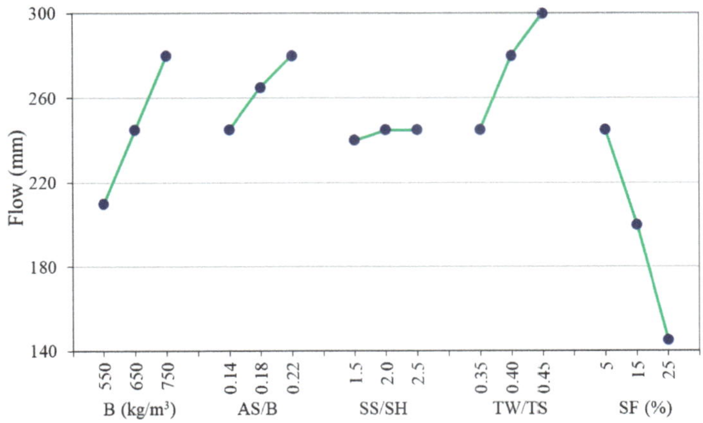

Figure 5. Effect of parameters on the flow.

The addition of the SP improved the flowability of the mixture from 245 mm flow to 275 mm. Borax also improved the flowability, and the addition of both the borax and the SP did not show additional improvement when compared to the SP-only mixture. The increase in SF from 5% of the binder (control mixture AAC-2) to 15% and 25% reduced the flow value from 245 mm to 200 mm and 145 mm, respectively. This is due to the small particle size, high specific surface area (Table 1), and high reactivity of SF, resulting in higher water requirements and accelerating the formation of flocculants [34,35].

3.3. *Rheology Results and Analysis*

Figure 6 shows the torque–rotational speed curves for the ascending and descending parts for the different mix proportion parameters. The curves show a distinct variation on parameters such as TW/TS and binder content as the parameter is varied. Figures 7 and 8 show the effect of the different parameters on the yield stress and plastic viscosity, respectively. The yield stress and the plastic viscosity mainly show a decreasing trend with the increase of the parameters except SS/SH ratio. The yield stress increased as SS/SH ratio increased. Further discussion on each parameter is presented in the following sections.

3.3.1. Effect of AS/B

The yield stress decreased with the increase of AS/B ratio from 0.14 to 0.22 (Figure 7). A similar effect has been observed by Li et al. [10] when the alkaline content to binder ratio was increased from 0.1 to 0.2 on a one-part alkali-activated slag. Zhang et al. [15] also observed a decrease in both yield stress and plastic viscosity as the alkaline content increased. The amount of microbubbles increases with the increase of alkaline solution [10]. Such microbubbles may decrease the yield stress of the mixture through their ball bearing and lubrication effects. Zhang [15] reported more agglomerations of precursor particles with a lower alkali concentration because of the larger electrostatic force between the particles. This may be one reason for why the yield stress of the mixtures decreased with the increase of AS/B. The plastic viscosity showed no significant change as the AS/B increased (Figure 8).

3.3.2. Effect of SS/SH

The yield stress increased with the SS/SH ratio while the plastic viscosity showed a minor decrease. Palacios et al. [21] reported that the initial yield stress values for Na_2SiO_3-activated AAC were higher than the values for NaOH-activated AAC. Hence, the high yield stress in the current study is caused by the higher silicate amount as the SS/SH ratio increased. Increasing the proportion of sodium silicate solution generally encourages gel formation [36]. Palacios et al. [21] and Puertas et al. [30] reported that AAC activated with sodium silicate has a lower setting time than those with sodium hydroxide. This is due to the formation of primary calcium silicate hydrate (C-S-H) gel in the early stages of the reaction due to the bonding of the Ca^{2+} ions from the slag to the silicate ions in the sodium silicate. The faster reaction and formation of reaction products in sodium silicate could be the reason for the higher yield stress observed. The plastic viscosity of the mixtures showed a minor decrease with the increase of the SS/SH ratio.

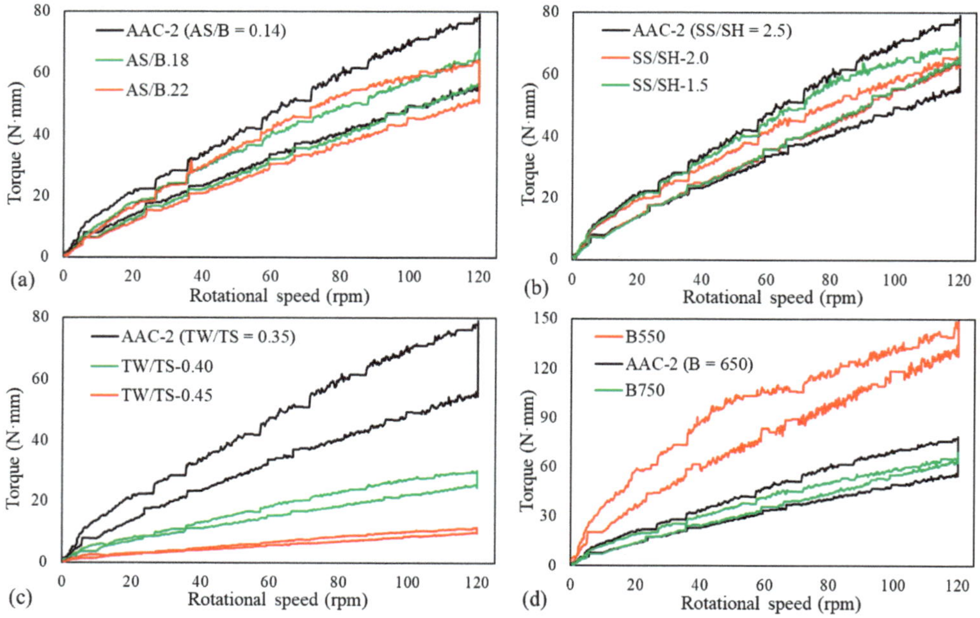

Figure 6. Torque–rotational speed curves for mix proportion parameters: (**a**) AS/B, (**b**) SS/SH, (**c**) TW/TS, (**d**) binder content.

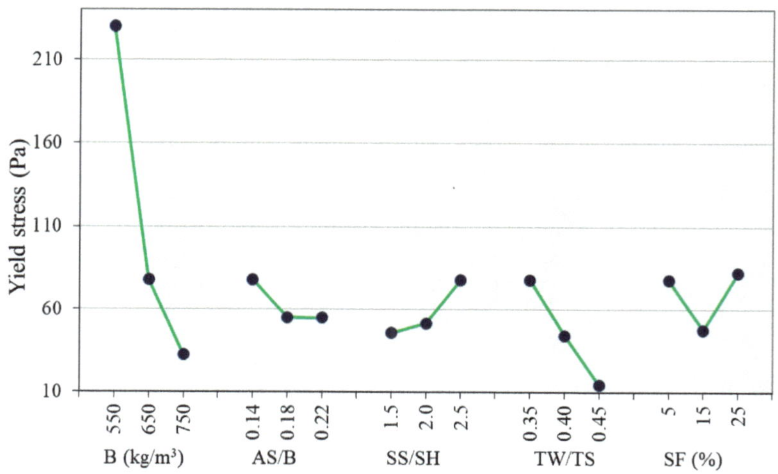

Figure 7. Effect of parameters on the yield stress.

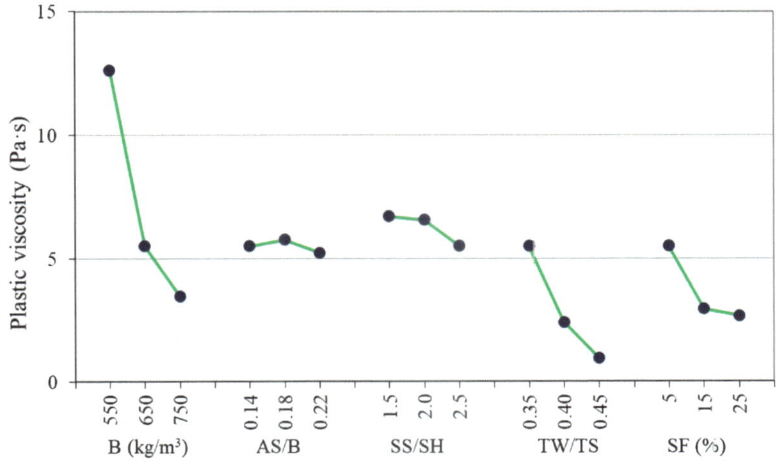

Figure 8. Effect of parameters on the plastic viscosity.

3.3.3. Effect of TW/TS

Both the mixtures' yield stress and plastic viscosity decreased with the increase of the TW/TS ratio (Figures 7 and 8). Yield stress and plastic viscosity increase with the increase of solid volume fraction [37]. As TW/TS increases, the solid volume decreases. The distance between solids increase, resulting in a lower particle interaction and hence higher fluidity. The higher fluidity also creates more air bubbles in the mixture. The air bubbles have a ball bearing effect [38]. This reduces the yield stress and the plastic viscosity of the mixtures. Flocculation and, to some extent, reaction product formation are enhanced by a lower TW/TS ratio [39]. Hence, as this ratio increases, the amount of flocs and reaction products becomes lower, resulting in lower rheological parameters. This behavior makes it possible to adjust the rheological behaviors of AAC mixture by adjusting the water content. However, this is not always possible as the water content also controls other essential behaviors such as strength and durability.

3.3.4. Effect of Binder Content

Binder content significantly influenced both yield stress and plastic viscosity (Figures 7 and 8). This is due to the associated paste content increase with the increase of the binder content. Koehler [40] confirmed that an increase in paste volume results in an increase in slump, reduction in yield stress, and plastic viscosity. The paste is an essential component of the mixture which coats the sands and fills the spaces between the sands and provides workability. With the increase of the paste content, the distance between the sand particles is increased. The paste acts as a lubricant between the sand particles. This reduces the resistance of the mixture to the shearing applied, hence reducing the yield stress and plastic viscosity. With the decrease of binder content, the sand increases, and the internal friction also increases, resulting in higher viscosity and yield stress.

3.3.5. Effect of SP and Borax

Both the SP and borax reduced the mixtures' yield stress and plastic viscosity (Table 3). The addition of 4% SP relative to the total binder content reduced the yield stress by 46%, while only 2% borax reduced it by about 49%, showing that borax is more effective than the SP in reducing the yield stress. Adding both SP and borax reduced the yield stress by about 63%. The plastic viscosity also decreased (by about 25%) with the addition of either or both of the admixtures.

Previous works of the authors [41] show that borax is an effective retarder for the AAC mixture under investigation. The results from the current study further show that borax can reduce yield stress and plastic viscosity. Furthermore, borax has no negative effect on the strength of AAC mixtures; in fact, it slightly improves the compressive strength [41–43]. This is of high importance, as the admixture technology in AAC is still in its early stages. Borax reacts with calcium ions, forming a calcium-based borate layer [42]. This additional layer on the surface of the calcium at the early stage of the reaction could be the cause of both retardation and lower yield stress and viscosity.

3.3.6. Effect of SF

SF is an extremely fine and spherical powder. The particles can fill the voids between other particles, improve the gradation, increase the packing density, and even create a lubrication effect [44]. The high specific surface area and high chemical activity, compared to FA, could also result in higher water demand and inter-particle friction [45]. As shown in Figure 7, when the SF content increased from 5% to 15%, the yield stress decreased. However, when the SF is further increased to 25%, the yield stress increased back. Up to 15%, the SF improved the fluidity of the mixture, probably due to the lubrication effect. Further increase of the SF may have increased the mixture's water demand, hence the higher yield stress. The plastic viscosity of the mixes decreased with the increase of the SF content, especially from 5% to 15%. The plastic viscosity decreased despite the decrease in the flow of the mixture as the SF content increased. This shows that the addition of SF is an effective way to reduce the viscosity of AAC mixtures. However, as can be observed in Figure 8, the reduction in viscosity was minor when the SF increased from 15% to 25%. This could be due to the two opposing effects of the SF, i.e., the lubrication and the higher water demand effects.

Previous studies have also observed a similar effect of SF. Li et al. [10] observed that the addition of SF significantly increased the yield stress and decreased the plastic viscosity of alkali-activated slag/glass powder pastes. Without the SF, Li et al. [10] observed that the mixtures followed the modified Bingham model; however, the relationship significantly changed with the SF, which fitted the Bingham model better. Correa-Yepes et al. [34], in their study on self-compacting OPC concrete with FA and SF, observed similar behavior with the increase of SF and suggested that the higher reactivity and fineness of SF accelerating the formation of flocs is a possible reason for the higher yield stress. Memon et al. [35], in their study on the behavior of FA-based AAC with SF replacement, observed the decrease in slump flow and increase in V-funnel flow time. Slump flow is inversely proportional to the yield stress, while V-funnel flow time is directly proportional to viscosity [46–49]. This shows the increase in both yield stress and viscosity with the increase of SF.

3.3.7. Thixotropy

Figure 6 shows the hysteretic response of the mixtures. The up-ramp was always higher than the down-ramp. The viscosity in the down-ramp is smaller than the up-ramp, showing the time-dependent nature of the mixture's rheology behavior. Such a response usually means that the material is thixotropic, meaning that its behavior depends on the shear history [50]. Another way to visualize this is by plotting the constant speed region of the profile (i.e., 120 rpm in Figure 4) and studying the change in torque at this constant speed. As shown in Figure 9, most of the mixes showed a decrease in the torque value despite maintaining the constant speed, thus showing the thixotropic nature of the mixture. Understanding concrete thixotropy is essential in modern construction. It influences various aspects of concrete construction, such as formwork pressure, multi-layered castings, and segregation resistance [51].

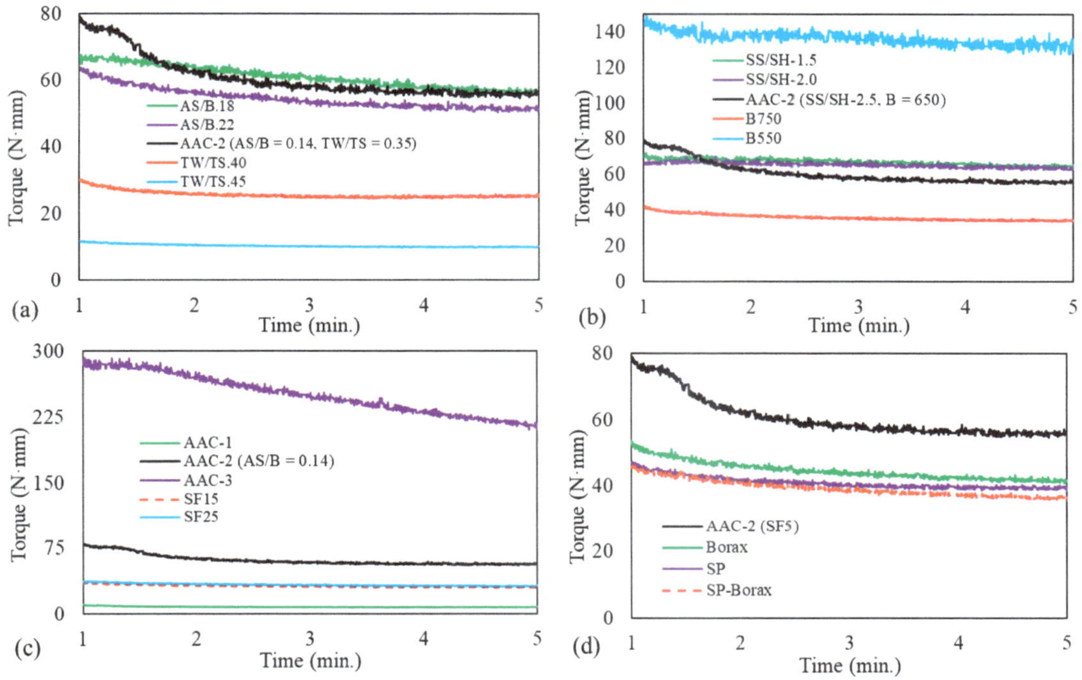

Figure 9. Thixotropic behavior for mix proportion parameters: (**a**) AS/B and TW/TS, (**b**) SS/SH and binder, (**c**) AAC and SF, (**d**) borax and SP.

There are different types of thixotropy and measurement methods [51]. Hysteresis areas may give a primary indication of the thixotropic behavior; however, quantitative determination of thixotropy is not always certain due to variations of hysteresis loops and the coupling of the shear rate and time [51,52]. For a comparative analysis of thixotropy of the mixes in the current study, the difference in area between the up- and down-ramp including the constant region (Figure 6) is used (Table 4). This area is referred as the thixotropic area; it provides a measure of the work required to break the initial linkages and internal friction of the mixtures before equilibrium is attained. As can be observed in Figure 9, after an initial high torque, the torque decayed to a minimum value at the constant speed of 120 rpm. This can be taken as the equilibrium torque that is independent of the shear history.

The maximum thixotropic area was observed for the AAC-3 mixture. AAC-3 has a low TW/TS ratio (0.24 compared to 0.35 for AAC-2), resulting in high resistance to shearing, which, in turn, resulted in the high thixotropic area observed. From the AAC-2 parametric study mixtures (mixes excluding AAC-1 and AAC-3), the maximum was observed for the B550 mixture, which means that lower binder content results in higher thixotropy. A similar result was reported for self-consolidating concrete [53]. Such behavior is attributed to the relative increase in sand volume that can lead to a greater level of internal friction. The other significant factor in the thixotropic area was the TW/TS. The highest TW/TS ratio, TW/TS.45, showed the lowest thixotropic area. The TW/TS ratio was increased by increasing the TW content and keeping the TS content constant. To keep the unit volume, the sand content decreased as TW/TS ratio increased. The decrease in sand content and the increase in TW resulted in a lower thixotropic area due to the lower degree of internal friction resulting in lower initial shear stress. It is well documented that the binder and aggregate contents and water to binder ratio are among the important factors affecting the degree of thixotropy and structural breakdown [54].

Table 4. Thixotropy.

Mix	Drop in Torque (N·mm)	Thixotropic Area (N·mm/min.)
AAC-1	2.4	677
AAC-2	22.3	1624
AAC-3	75.2	6314
TW/TS.4	5.0	354
TW/TS.45	1.5	124
SS/SH-1.5	5.7	957
SS/SH-2	2.3	571
AS/B.18	12.1	849
AS/B.22	12.2	1252
B750	7.5	632
B550	13.1	2314
SP-Borax	8.9	838
SP	7.5	740
Borax	12.0	1258
SF15	5.6	557
SF25	4.9	694

From the SS/SH ratio variations, SS/SH ratio of 2.5 showed the highest drop in torque and the highest thixotropic area (Figure 9 and Table 4). Alkali-activated mixes activated with sodium silicate form a primary C-S-H gel at an early age due to the interaction of the silicate ions from the activator and the calcium ions from the GGBS [16]. This gel is broken when the mix is subjected to shearing at a constant rate. Due to the early C-S-H formation, sodium-silicate-activated mixes are also known to have a faster setting time. Their setting time and workability can be increased by lengthening their mixing times [16]. Hence the higher initial torque of the SS/SH 2.5 (higher silicate) mix is due to the early C-S-H formation, while the lower equilibrium torque is due to the constant shearing (lengthening of mixing).

The thixotropic area showed good correlation with the viscosity (Figure 10a). The thixotropic nature of the mixtures increased with the increase of viscosity. This is due to the increase of initial torque value with the increase of plastic viscosity. Figure 10b shows the relationship between the drop in torque and thixotropic area. The two methods for thixotropy analysis showed a strong correlation with a high coefficient of correlation (R^2) value of 0.95. Hence, similar discussions made for the thixotropic area can also be made for the drop in torque.

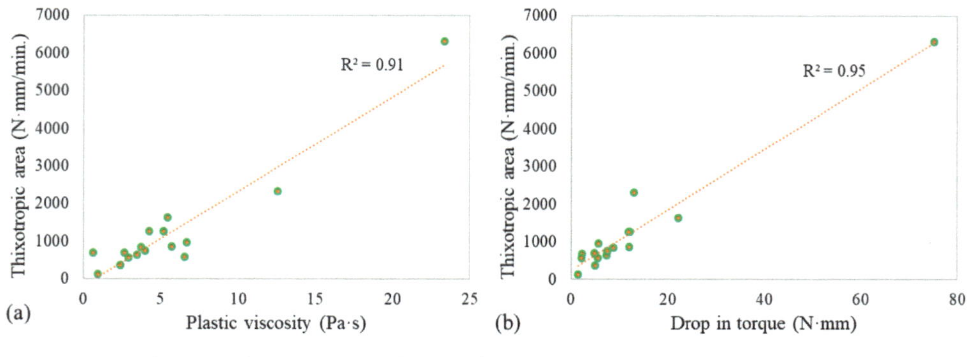

Figure 10. Thixotropic area relation with (**a**) plastic viscosity, (**b**) drop in torque.

3.3.8. Rheology and Flowability

Generally, the workability of cementitious materials shows an inverse relation with yield stress [55]. As can be observed from Figure 11, both the yield stress and plastic viscosity decreased with the increase of the flow values. The viscosity showed a stronger correlation with the flow than the yield stress. Studies on OPC systems showed that the slump is inversely (either linearly, logarithmically, or exponentially) related to the yield stress while the viscosity showed no apparent trends [56]. It is not clear why this difference is observed between AAC and OPC systems in terms of the viscosity-flow relations, and the results of the current study are not enough to give a conclusive remark on the matter. Hence, further research on the topic is recommended.

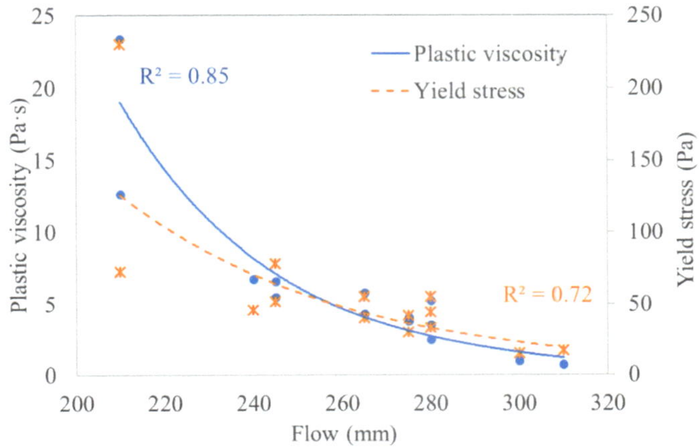

Figure 11. Flow to yield stress and flow to viscosity relationship.

4. Conclusions

Based on the experimental results presented in this study, the following conclusions can be drawn:

- Both yield stress and plastic viscosity decrease with the increase of binder content and water to binder ratio. Binder content has a more substantial effect on both rheology parameters.
- An increase in sodium silicate to sodium hydroxide ratio increases the yield stress. This is due to the formation of early C-S-H products from the reaction of silicate ions from the alkaline solution and calcium ions from the slag.
- Borax is an effective admixture for reducing yield stress. A 2% borax relative to the total binder reduced the yield stress by 49%. This is due to the additional calcium-based borate layer that borax forms at the early stage of the reaction.
- The effect of silica fume on rheology depends on its dosage. Up to 15% replacement of the binder, silica fume decreased the yield stress of the mix due to its lubrication effect; however, further increase resulted in higher water demand and hence higher yield stress.
- Plastic viscosity decreases with the increase of silica fume. This was observed despite the decrease in the flow value when silica fume increased, showing that addition of silica fume is an effective way to reduce viscosity.
- Thixotropic area significantly increases with the decrease in binder content, and decreases with the increase in water content.

Author Contributions: Conceptualization, B.H.T.; methodology, B.H.T.; formal analysis, B.H.T. and L.H.; investigation, B.H.T. and L.H.; writing—original draft preparation, B.H.T.; writing—review and editing, B.H.T. and K.H.; visualization, B.H.T.; supervision, K.H.; project administration, B.H.T. and K.H.; funding acquisition, B.H.T. and K.H. All authors have read and agreed to the published version of the manuscript.

Funding: This research was funded by Alexander von Humboldt Foundation (1206836-AUS-HFST-P) and the APC was funded by HTWK Leipzig.

Institutional Review Board Statement: Not applicable.

Informed Consent Statement: Not applicable.

Data Availability Statement: The data presented in this study are available on request from the corresponding author.

Conflicts of Interest: The authors declare no conflict of interest.

References

1. Herrmann, A.; Koenig, A.; Dehn, F. Structural concrete based on alkali-activated binders: Terminology, reaction mechanisms, mix designs and performance. *Struct. Concr.* **2018**, *19*, 918–929. [CrossRef]
2. Duxson, P.; Provis, J.L.; Lukey, G.C.; van Deventer, J.S. The role of inorganic polymer technology in the development of 'green concrete'. *Cem. Concr. Res.* **2007**, *37*, 1590–1597. [CrossRef]
3. Juenger, M.; Winnefeld, F.; Provis, J.L.; Ideker, J.H. Advances in alternative cementitious binders. *Cem. Concr. Res.* **2011**, *41*, 1232–1243. [CrossRef]
4. Bakharev, T.; Sanjayan, J.; Cheng, Y.-B. Resistance of alkali-activated slag concrete to acid attack. *Cem. Concr. Res.* **2003**, *33*, 1607–1611. [CrossRef]
5. Xie, T.; Visintin, P.; Zhao, X.; Gravina, R. Mix design and mechanical properties of geopolymer and alkali activated concrete: Review of the state-of-the-art and the development of a new unified approach. *Constr. Build. Mater.* **2020**, *256*, 119380. [CrossRef]
6. Tekle, B.H.; Khennane, A.; Kayali, O. Bond behaviour of GFRP reinforcement in alkali activated cement concrete. *Constr. Build. Mater.* **2017**, *154*, 972–982. [CrossRef]
7. Wu, Y.; Lu, B.; Bai, T.; Wang, H.; Du, F.; Zhang, Y.; Cai, L.; Jiang, C.; Wang, W. Geopolymer, green alkali activated cementitious material: Synthesis, applications and challenges. *Constr. Build. Mater.* **2019**, *224*, 930–949. [CrossRef]
8. Panda, B.; Unluer, C.; Tan, M.J. Investigation of the rheology and strength of geopolymer mixtures for extrusion-based 3D printing. *Cem. Concr. Res.* **2018**, *94*, 307–314. [CrossRef]
9. Moeini, M.A.; Hosseinpoor, M.; Yahia, A. Effectiveness of the rheometric methods to evaluate the build-up of cementitious mortars used for 3D printing. *Constr. Build. Mater.* **2020**, *257*, 119551. [CrossRef]
10. Li, L.; Lu, J.-X.; Zhang, B.; Poon, C.-S. Rheology behavior of one-part alkali activated slag/glass powder (AASG) pastes. *Constr. Build. Mater.* **2020**, *258*, 120381. [CrossRef]
11. Criado, M.; Palomo, A.; Fernández-Jiménez, A.; Banfill, P.F.G. Alkali activated fly ash: Effect of admixtures on paste rheology. *Rheol. Acta* **2009**, *48*, 447–455. [CrossRef]
12. Sitarz, M.; Urban, M.; Hager, I. Rheology and Mechanical Properties of Fly Ash-Based Geopolymer Mortars with Ground Granulated Blast Furnace Slag Addition. *Energies* **2020**, *13*, 2639. [CrossRef]
13. Vance, K.; Dakhane, A.; Sant, G.; Neithalath, N. Observations on the rheological response of alkali activated fly ash suspensions: The role of activator type and concentration. *Rheol. Acta* **2014**, *53*, 843–855. [CrossRef]
14. Romagnoli, M.; Sassatelli, P.; Lassinantti Gualtieri, M.; Tari, G. Rheological characterization of fly ash-based suspensions. *Constr. Build. Mater.* **2014**, *65*, 526–534. [CrossRef]
15. Zhang, D.-W.; Wang, D.; Liu, Z.; Xie, F. Rheology, agglomerate structure, and particle shape of fresh geopolymer pastes with different NaOH activators content. *Constr. Build. Mater.* **2018**, *187*, 674–680. [CrossRef]
16. Puertas, F.; Varga, C.; Alonso, M.M. Rheology of alkali-activated slag pastes. Effect of the nature and concentration of the activating solution. *Cem. Concr. Res.* **2014**, *53*, 279–288. [CrossRef]
17. Awoyera, P.; Adesina, A. A critical review on application of alkali activated slag as a sustainable composite binder. *Case Stud. Constr. Mater.* **2019**, *11*, e00268. [CrossRef]
18. Fang, G.; Ho, W.K.; Tu, W.; Zhang, M. Workability and mechanical properties of alkali-activated fly ash-slag concrete cured at ambient temperature. *Constr. Build. Mater.* **2018**, *172*, 476–487. [CrossRef]
19. Nath, P.; Sarker, P.K. Effect of GGBFS on setting, workability and early strength properties of fly ash geopolymer concrete cured in ambient condition. *Constr. Build. Mater.* **2014**, *66*, 163–171. [CrossRef]
20. Lee, N.K.; Lee, H.K. Setting and mechanical properties of alkali-activated fly ash/slag concrete manufactured at room temperature. *Constr. Build. Mater.* **2013**, *47*, 1201–1209. [CrossRef]
21. Palacios, M.; Banfill, P.F.G.; Puertas, F. Rheology and Setting of Alkali-Activated Slag Pastes and Mortars: Effect of Organic Admixture. *ACI Mater. J.* **2008**, *105*, 140–148.
22. Palacios, M.; Puertas, F. Stability of superplasticizer and shrinkage-reducing admixtures Stability of superplasticizer and shrinkage-reducing admixtures in high basic media. *Mater. Constr.* **2004**, *54*, 65–86. [CrossRef]
23. Tong, S.; Yuqi, Z.; Qiang, W. Recent advances in chemical admixtures for improving the workability of alkali-activated slag-based material systems. *Constr. Build. Mater.* **2021**, *272*, 121647. [CrossRef]
24. EN 450-1. *Fly Ash for Concrete—Part 1: Definition, Specifications and Conformity Criteria*; German version EN 450-1:2012; CEN: Brussels, Belgium, 2012.

25. EN 15167-1. *Ground Granulated Blast Furnace Slag for Use in Concrete, Mortar and Grout—Part 1: Definitions, Specifications and Conformity Criteria*; CEN: Brussels, Belgium, 2006.
26. EN 13263-1. *Silica Fume for Concrete—Part 1: Definitions, Requirements and Conformity Criteria*; CEN: Brussels, Belgium, 2005.
27. Tekle, B.H.; Holschemacher, K.; Löber, P.; Heiden, B. Mechanical Behavior and Frost-Resistance of Alkali-Activated Cement Concrete with Blended Binder at Ambient Curing Condition. *Buildings* **2021**, *11*, 52. [CrossRef]
28. Tekle, B.H.; Hertwig, L.; Holschemacher, K. Setting Time and Strength Monitoring of Alkali-Activated Cement Mixtures by Ultrasonic Testing. *Materials* **2021**, *14*, 1889. [CrossRef]
29. EN 1015-3. *Methods of Test for Mortar for Masonry—Part 3: Determination of Consistence of Fresh Mortar (by Flow Table)*. CEN: Brussels, Belgium, 1999.
30. Puertas, F.; Alonso, M.; Gismera, S.; Lanzón, M.; Blanco-Varela, M.T. Rheology of Cementitious Materials: Alkali-Activated Materials or Geopolymers. *MATEC Web Conf.* **2018**, *149*, 1002. [CrossRef]
31. Haist, M.; Link, J.; Nicia, D.; Leinitz, S.; Baumert, C.; von Bronk, T.; Cotardo, D.; Eslami Pirharati, M.; Fataei, S.; Garrecht, H.; et al. Interlaboratory study on rheological properties of cement pastes and reference substances: Comparability of measurements performed with different rheometers and measurement geometries. *Mater. Struct.* **2020**, *53*, 92. [CrossRef]
32. Gugulothu, V.; Gunneswara Rao, T.D. Effect of Binder Content and Solution/Binder Ratio on Alkali-Activated Slag Concrete Activated with Neutral Grade Water Glass. *Arab. J. Sci. Eng.* **2020**, *45*, 8187–8197. [CrossRef]
33. Deb, P.S.; Nath, P.; Sarker, P.K. The effects of ground granulated blast-furnace slag blending with fly ash and activator content on the workability and strength properties of geopolymer concrete cured at ambient temperature. *Mater. Des. (1980–2015)* **2014**, *62*, 32–39. [CrossRef]
34. Correa-Yepes, J.A.; Rojas-Reyes, N.R.; Tobón, J.I. Effect of fly ash and silica fume on rheology, compressive strength and self-compacting in cement mixtures. *DYNA* **2018**, *85*, 59–68. [CrossRef]
35. Memon, F.A.; Nuruddin, M.F.; Shafiq, N. Effect of silica fume on the fresh and hardened properties of fly ash-based self-compacting geopolymer concrete. *Int. J. Miner. Metall. Mater.* **2013**, *20*, 205–213. [CrossRef]
36. Bignozzi, M.C.; Manzi, S.; Natali, M.E.; Rickard, W.D.; van Riessen, A. Room temperature alkali activation of fly ash: The effect of Na_2O/SiO_2 ratio. *Constr. Build. Mater.* **2014**, *69*, 262–270. [CrossRef]
37. Roussel, N.; Lemaître, A.; Flatt, R.J.; Coussot, P. Steady state flow of cement suspensions: A micromechanical state of the art. *Cem. Concr. Res.* **2010**, *40*, 77–84. [CrossRef]
38. Xie, J.; Cui, X.; Guo, N.; Liu, G. Influence of Mix Proportions on Rheological Properties, Air Content of Wet Shotcrete—A Case Study. *Appl. Sci.* **2021**, *11*, 3550. [CrossRef]
39. Dai, X.; Aydin, S.; Yardimci, M.Y.; Lesage, K.; de Schutter, G. Influence of water to binder ratio on the rheology and structural Build-up of Alkali-Activated Slag/Fly ash mixtures. *Constr. Build. Mater.* **2020**, *264*, 120253. [CrossRef]
40. Koehler, E.P.; Fowler, D.W. *Development of a Portable Rheometer for Fresh Portland Cement Concrete*; Research report ICAR–105-3F; International Center for Aggregates Research, The University of Texas at Austin: Austin, TX, USA, 2004.
41. Holschemacher, K.; Tekle, B.H. Review and Experimental Investigation of Retarder for Alkali-Activated Cement. In Proceedings of the 75th Rilem Annual Week and International Conference on Advances in Sustainable Construction Materials and Structures, Mérida, Mexico, 29 August–3 September 2021.
42. Oderji, S.Y.; Chen, B.; Shakya, C.; Ahmad, M.R.; Shah, S.F.A. Influence of superplasticizers and retarders on the workability and strength of one-part alkali-activated fly ash/slag binders cured at room temperature. *Constr. Build. Mater.* **2019**, *229*, 116891. [CrossRef]
43. Revathi, T.; Jeyalakshmi, R. XPS, ^{29}Si, ^{27}Al, ^{11}B MAS -NMR, ATR-IR and FESEM characterization of geopolymer based on borax modified water glass activated Fly ash-GGBS blend. *Mater. Res. Express* **2019**, *6*, 85337. [CrossRef]
44. Jiao, D.; Shi, C.; Yuan, Q.; An, X.; Liu, Y.; Li, H. Effect of constituents on rheological properties of fresh concrete—A review. *Cem. Concr. Compos.* **2017**, *83*, 146–159. [CrossRef]
45. Benaicha, M.; Roguiez, X.; Jalbaud, O.; Burtschell, Y.; Alaoui, A.H. Influence of silica fume and viscosity modifying agent on the mechanical and rheological behavior of self compacting concrete. *Constr. Build. Mater.* **2015**, *84*, 103–110. [CrossRef]
46. Chidiac, S.; Habibbeigi, F.; Chan, D. Slump and Slump Flow for Characterizing Yield Value of Fresh Concrete. *ACI Mater. J.* **2006**, *103*, 413. [CrossRef]
47. Mechtcherine, V.; Schneider, K.; Brameshuber, W. Mineral-based matrices for textile-reinforced concrete. In *Textile Fibre Composites in Civil Engineering*; Elsevier: Amsterdam, The Netherlands, 2016; pp. 25–43. ISBN 9781782424468.
48. Neophytou, M.; Pourgouri, S.; Kanellopoulos, A.; Petrou, M.; Ioannou, I.; Georgiou, G.; Alexandrou, A. Determination of the rheological parameters of self-compacting concrete matrix using slump flow test. *Appl. Rheol.* **2010**, *20*, 62402. [CrossRef]
49. Meng, W.; Khayat, K.H. Improving flexural performance of ultra-high-performance concrete by rheology control of suspending mortar. *Compos. Part B Eng.* **2017**, *117*, 26–34. [CrossRef]
50. Roussel, N. *Understanding the Rheology of Concrete*; Woodhead: Oxford, UK, 2012; ISBN 9780857090287.
51. ACI 238.2T-14. *Concrete Thixotropy: TechNote*; American Concrete Institute: Farmington Hills, MI, USA, 2014.
52. Felekoglu, B. Rheological behaviour of self-compacting micro concrete. *Indian Acad. Sci.* **2014**, *39*, 1471–1495. [CrossRef]
53. Assaad, J.; Khayat, K.H. Assessment of Thixotropy of Self-Consolidating Concrete and Concrete-Equivalent-Mortar—Effect of Binder Composition and Content. *ACI Mater. J.* **2004**, *101*, 400–408. [CrossRef]

54. Ahari, R.S.; Erdem, T.K.; Ramyar, K. Thixotropy and structural breakdown properties of self consolidating concrete containing various supplementary cementitious materials. *Cem. Concr. Compos.* **2015**, *59*, 26–37. [CrossRef]
55. Roussel, N. Correlation between Yield Stress and Slump: Comparison between Numerical Simulations and Concrete Rheometers Results. *Mater. Struct.* **2007**, *39*, 501–509. [CrossRef]
56. Nehdi, M.; Al-Martini, S. Coupled Effects of High Temperature, Prolonged Mixing Time, and Chemical Admixtures on Rheology of Fresh Concrete. *ACI Mater. J.* **2009**, *106*, 231. [CrossRef]

materials

Article

Evaluation of Fly Ash from Co-Combustion of Paper Mill Wastes and Coal as Supplementary Cementitious Materials

Ming-Fu Wu and Wei-Hsing Huang *

Department of Civil Engineering, National Central University, Taoyuan City 320317, Taiwan
* Correspondence: t321655@ncu.edu.tw

Abstract: The applications of waste-derived fuel from paper mills in industrial boilers benefit the reduction of carbon emissions. However, the co-combustion of waste-derived fuel and coal causes significant changes in the characteristics of the ash and brings about the need to find possible means of the utilization of the ash produced. In this work fly, ash samples were collected from circulating fluidized bed (CFB) boilers co-combusting paper mill wastes with coal and analyzed in detail. The chemical, physical, and thermal characteristics of two different co-combustion fly ashes (CCFA) were investigated using X-ray fluorescence (XRF), X-ray diffraction (XRD), thermogravimetry (TG), and scanning electron microscope (SEM). The chemical composition of CCFA is largely affected by the fuel source type. Thermal analyses of CCFA show that the type of desulfurization system used by the boiler influences the form of sulfate present in the fly ash. The presence of calcium sulfite hemihydrate can cause a high loss in the ignition of CCFA. By comparing the physical requirements specified in the ASTM standard for coal fly ash to be used in concrete, the CCFA produced from paper mill wastes was found to show good potential as supplementary cementitious materials.

Keywords: co-combustion; fly ash; solid recovered fuel; paper mill waste; waste-derived fuel; supplementary cementitious materials

Citation: Wu, M.-F.; Huang, W.-H. Evaluation of Fly Ash from Co-Combustion of Paper Mill Wastes and Coal as Supplementary Cementitious Materials. *Materials* **2022**, *15*, 8931. https://doi.org/10.3390/ma15248931

Academic Editor: Yuri Ribakov

Received: 14 November 2022
Accepted: 13 December 2022
Published: 14 December 2022

1. Introduction

With the promotion of carbon neutrality and environmental protection policies, the paper industry has started a net-zero emission action by using combustible wastes in the paper production process as an alternative fuel, including recyclable waste paper, pulp and paper mill sludge (PPMS), and pulper rejects (PR) [1–4]. This improves the utilization efficiency of waste resources and reduces the usage of fossil fuels, consequently achieving the dual goals of sustainable development and carbon reduction. However, while the co-combustion of waste-derived fuel with coal in industrial boilers helps reduce fossil fuel consumption, the co-combustion fly ash (CCFA) experienced extensive alterations in terms of its chemical composition and material properties [4–7]. Variations in the types and sources of waste-derived fuel and the percentage of co-combustion resulted in the diversified chemical composition of the CCFA., which in turn presents difficulties in its recycling and utilization. Moreover, the existing recycling technologies and material quality standards do not apply to CCFA produced by waste-derived fuel, making their recovery and utilization difficult [5,8,9].

In general, industrial boiler type, combustion conditions, the nature of added fuels in the case of co-combustion, and process additives can significantly affect the properties of the resulting ashes. Moreover, coal or waste-derived fuels usually release sulfur oxides when burned and may cause air pollution, thus requiring strict restrictions on emissions from boiler facilities [10,11], especially circulating fluidized bed (CFB) boilers. The characteristics of the combustion ash produced by CFB boilers are significantly different from those of pulverized coal fly ash (PCFA). In addition to the relatively low combustion temperature in CFB boilers [12], in situ desulfurization and flue gas desulfurization (FGD) systems are widely used in CFB boilers to reduce sulfur dioxide emissions [13–16]. However, the captured sulfur oxides will remain in the ash and affect the characteristics of the fly ash significantly [17]. This partly resulted in the exclusion of CFB fly ash from the scope of the general standard specifications for pulverized coal fly ash (PCFA) for use in concrete [5,18–20].

Co-combustion of coal with alternative fuel, such as waste-derived fuel, is a relatively new development for industrial boilers. Effective utilization of the ashes produced from the co-combustion will encourage industrial boilers to adopt waste-derived fuels for co-combustion with coal and thus reducing carbon emissions. This study aimed to address the utilization of CCFAs by analyzing the chemical, physical, and mineralogical characteristics of CCFAs produced from paper mill wastes and evaluating the potential of CCFAs when used as cementitious materials in partial replacement of cement. In this study, two CCFAs were obtained from two different CFB boilers in a paper recycling mill, one was produced from the co-combustion of coal and PPMS, and the other from the co-combustion of coal and pulper rejects.

2. Materials and Experimental Methods

2.1. Waste-Derived Fuel from Paper Mill

The paper production process in a paper recycling mill (Figure 1) produces multiple types of waste, including recyclable waste paper, pulp and paper mill sludge (PPMS), and pulper rejects (PR) [21,22]. The PPMS comes from wastewater treatment sludge with main constituents of fiber, mineral fillers, and organic matter. The PPMS is converted to alternative fuel via a pressurized dehydration process. PR material is the lightweight residue produced from waste paper and paper containers treated with a hydra pulper [2,14]. The PR consists of different fibers and plastic fragments and is converted to solid recovered fuel (SRF) to be co-combusted with coal [4,21–24].

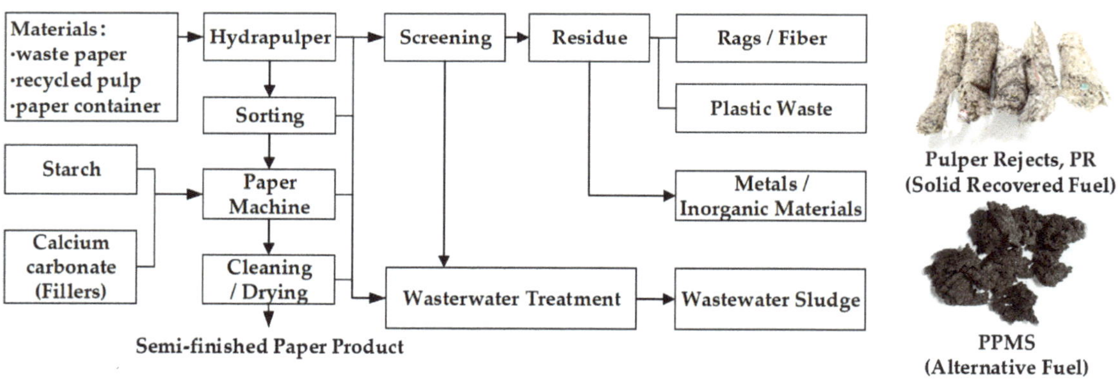

Figure 1. Waste-derived fuel generated during papermaking in a paper recycling mill.

2.2. Co-Combustion Fly Ash from CFB Boilers

Two CCFAs from CFB boilers co-combusting two different auxiliary fuels with coal were studied in this paper. Two samples of co-combustion fly ash (CCFA) were collected separately from the storage hopper of the air pollution control unit in paper recycling mills. The first CCFA is called PPMS-CCFA and is produced from burning 60% coal and 40% PPMS, while the second is PR-CCFA, produced from burning 60% coal and 40% PR.

2.3. Research Methods

ASTM C1709 [25] standard guide was used as a reference for material testing in this study. In addition to testing the basic properties, various tests were used to assess the potential of CCFA as supplementary cementitious materials.

2.3.1. CCFA Characteristics

A chemical composition analysis with XRF and XRD was performed on CCFA to determine its chemical and crystalline constituents. The loss on ignition (LOI) was determined for CCFA at 550 °C and 750 °C. A laser diffraction particle size analysis was used to determine the particle size distribution of CCFA. The appearance of the fly ash was examined by using a Scanning Electron Microscope (Hitachi, FlexSEM 1000 II, Taoyuan, Taiwan).

2.3.2. Evaluation as Cementitious Materials

To evaluate the suitability of CCFA as supplementary cementitious materials for use in concrete, tests were conducted for comparison with the physical and chemical requirements specified in ASTM C618 [26]. These tests included Strength Activity Index (SAI), water requirement, and soundness of cement paste or mortar prepared by replacing 20% cement with CCFA. In addition, the time of setting of cement paste was determined in accordance with ASTM C191 [27].

3. Results

3.1. CCFA Characteristics

The chemical composition of CCFAs was determined using XRF techniques, and the results are shown in Table 1. The chemical composition indicates that the main components of PPMS-CCFA are oxygen compounds of calcium and silicon. The supplement of calcium carbonate in the paper production process provides the source of these components. The excess calcium carbonate enters the wastewater treatment system in the pulper paper production process to form PPMS, causing a high calcium carbonate content in PPMS [2,28]. Because the PPMS is rich in calcium carbonate, it is decomposed into calcium oxide and carbon dioxide during the combustion process in the boiler and undergoes an in-situ desulfurization reaction in the CFB boiler combustion chamber. As a result, the desulfurization reaction product (calcium sulfate) enters the dust collection equipment and is mixed into the fly ash [18]. While the CFB boiler using PPMS as fuel also adopts a semi-dry flue gas desulfurization plant, the use of a desulfurizer caused the PPMS-CCFA to contain calcium hydroxide (CH) and calcium sulfite hemihydrate [19,29].

Table 1. Chemical composition of fly ashes by XRF analysis.

Sample	Chemical Composition (%)							
	SiO_2	Al_2O_3	Fe_2O_3	CaO	MgO	TiO_2	SO_3	P_2O_5
PCFA	55.7	29.3	5.42	3.86	1.34	0.63	0.57	0.51
PPMS-CCFA	29.8	14.5	3.93	43.1	2.34	1.04	3.93	0.59
PR-CCFA	50.0	29.2	6.06	9.47	0.93	2.19	1.38	0.25

PR-CCFA consists primarily of oxides of silicon, aluminum, and calcium. Unlike PPMS having a rich calcium carbonate content, PR is composed of plastic fragments and some fibers. The plastic fragments mainly come from the polyethylene (PE) or polypropylene (PP) film on paper containers [2,21]. Therefore, PR-CCFA's chemical composition is mainly silicon and aluminum minerals. Because the CFB boiler burning pulper rejects as fuel adopts wet flue gas desulfurization, the SO_x in the flue gas is discharged along with the FGD wastewater, leading to a low sulfate composition in PR-CCFA.

As shown in Figure 2, the *crystalline composition* was identified using an X-ray diffraction analyzer (XRD). The results of the diffraction patterns obtained from PPMS-CCFA and PR-CCFA indicate that the fuel type has a significant impact on CCFA properties. If the waste-derived fuel contains a large amount of calcium carbonate or calcium oxide, the fuel can achieve an in situ desulfurization effect in the CFB boiler. However, the FGD gypsum is discharged to the wastewater treatment system in wet flue gas desulfurization. If the CFB boiler uses in situ desulfurization or semi-dry flue gas desulfurization, the calcium sulfite hemihydrate stays in the fly ash. Excessively high contents of calcium sulfite hemihydrates can cause an abnormal setting time when used as cementitious materials in blended cement [16,30].

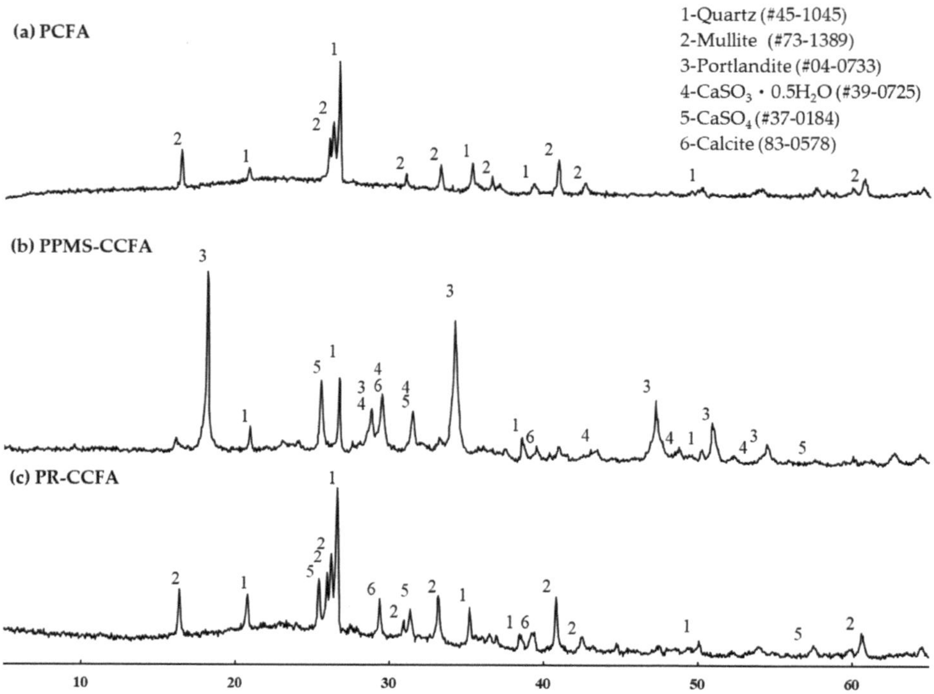

Figure 2. Fly ash X-Ray diffraction patterns.

Table 2 summarizes the test results on the physical properties of CCFAs. The specific gravity test result shows that PPMS-CCFA's specific gravity is greater than that of PCFA and PR-CCFA. This is mainly due to PPMS-CCFA's high calcium content. The particle size of fly ashes is affected by the design parameters of the dust-collecting equipment of the boiler. The fineness of the fly ash plays an important role in the performance of supplementary cementitious materials in concrete [25]. The fineness and laser particle size analysis of the fly ashes show that the particle sizes of the three fly ash sources are not very different, with the PPMS-CCFA having the lowest amount of particles retained on the #325 sieve.

Table 2. Physical Properties of PCFA and CCFA.

	PCFA	PPMS-CCFA	PR-CCFA
Relative Density	2.27	2.48	2.30
Fineness (retained on #325 sieve), %	20	18	24
Particle Size, μm (d_{10})	3.35	3.67	4.05
(d_{50})	21.08	21.04	26.12
(d_{90})	72.81	79.62	101.48
LOI (550 °C), %	0.3	7.2	2.7
LOI (750 °C), %	0.4	13.2	4.7
pH	10.9	8.3	11.2

According to ASTM C618, the loss on ignition (LOI) test of PCFA is conducted at a test temperature of 750 °C. Since PPMS-CCFA contains a large amount of calcium carbonate, which has a decomposition temperature of 600–800 °C [6,29,31,32], a complementary LOI test temperature of 550 °C was added to avoid the impact of weight loss due to the thermal decomposition of calcium carbonate on LOI values.

The test results showed that the weight loss of PPMS-CCFA increased from 7.2% at 550 °C to 13.2% at 750 °C. The weight loss of PR-CCFA increased from 2.7% at 550 °C to 4.7% at 750 °C. The decomposition of carbonate and the loss of the volatile inorganic compounds explain such a difference in LOI.

The XRD analysis showed that PPMS-CCFA's main components include calcium carbonate and calcium hydroxide. TGA heating curves in Figure 3 show that the major weight losses of PPMS-CCFA are observed at 350–450 °C and 600–900 °C. The dehydration of calcium sulfite hemihydrate in PPMS-CCFA occurred at 370–400 °C [16,29]. The thermal decomposition of calcium hydroxide (CH) was observed at 400–460 °C [19,31]. These two events resulted in a weight loss of approximately 4.7%. With a weight loss of approximately 6.9%, the decomposition of calcium carbonate occurred at 600–800 °C. The TGA curve (Figure 3b) shows that the presence of calcium hydroxide and calcium carbonate leads to the difference in loss on ignition of PPMS-CCFA at different temperatures and the high amount of loss on ignition. Because PPMS-CCFA contains calcium sulfite and the decomposition reaction stage of calcium sulfite mainly occurs at 800–950 °C [16,19,33,34], calcium sulfite may change to calcium oxide at these temperatures, resulting in weight loss. The major weight loss of PR-CCFA was observed at 600–800 °C (Figure 3c). XRD analysis showed that PR-CCFA contains calcium carbonate compounds, which changed to calcium oxide, consequently causing weight loss at this temperature range [16,29].

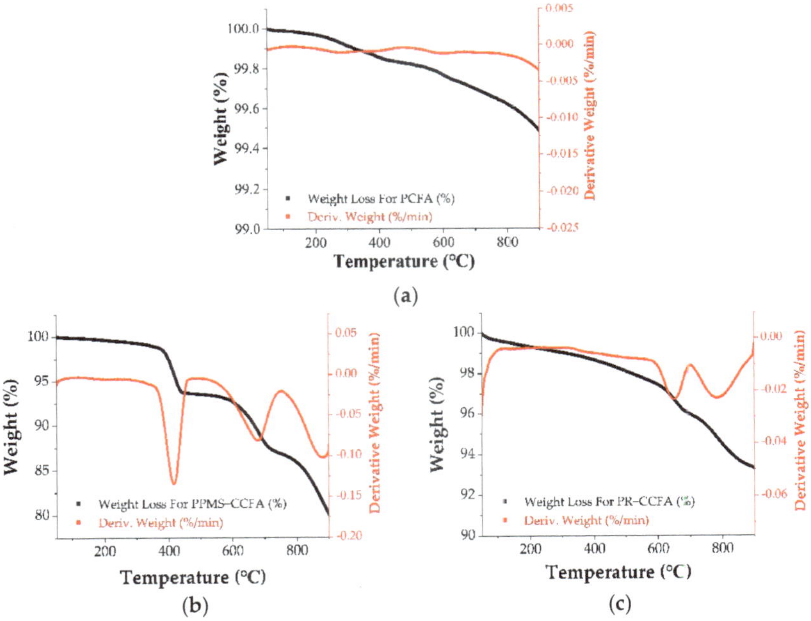

Figure 3. TGA and DTG heating curves of (**a**) PCFA, (**b**) PFMS-CCFA, (**c**) PR-CCFA.

Therefore, CCFA's thermogravimetric loss is affected by multiple compounds, among which calcium hydroxide, calcium carbonate, and calcium sulfite hemihydrate are usually produced by in situ desulfurization or semi-dry flue gas desulfurization in CFB boilers [33]. After being collected by the electrostatic dust collection equipment and mixed into the fly ash, these compounds result in large amounts of loss on ignition of PPMS-CCFA and PR-CCFA.

SEM has been widely used to characterize fly ash's physical or chemical properties. In this study, the SEM technique was used to examine the appearance of the samples of PCFA and CCFA produced by CFB boilers. Figure 4 shows a series of PCFA and CCFA appearance micrographs. The morphology of fly ash particles is controlled by combustion temperature and cooling rate. Because PCFA is formed after cooling in a high-temperature environment, it mainly exhibits a hollow spherical shape with a smooth surface. In contrast, the grains of CCFA have a rough surface texture and irregular shape. This is because in the operating temperature range of 850–900 °C for CFB boilers, the mineral substances in the waste-derive fuel were not subject to partial melting. The dominant particles of CCFA are granular, irregularly shaped, and mingled with some popcorn-like particles. The rough and loose surface texture of CCFA particles generates a large specific surface area that affects the water requirement adversely when used as cementitious materials [5,8,35].

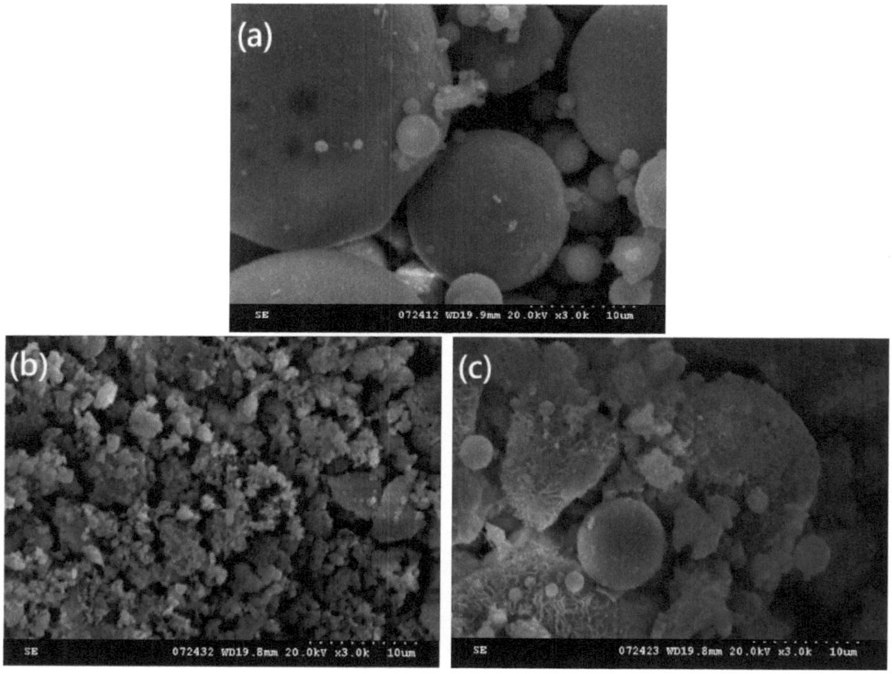

Figure 4. SEM micrographs of (**a**) PCFA, (**b**) PPMS-CCFA, (**c**) PR-CCFA.

Figure 4b,c shows the microscopic images of PPMS-CCFA and PR-CCFA. Due to the presence of calcium hydroxide (CH) and light calcium carbonate, the appearance of the PPMS-CCFA mainly presents typical cubic crystals and spindle-shaped crystals. However, due to different operating temperatures in the boiler, irregular particles and rough appearances can be observed in the PR-CCFA's microcosmic images.

3.2. *Performance Evaluation as Cementitious Materials*

3.2.1. Performance Testing of Cement Paste

The water consumption required for achieving a cement paste consistency and the setting time is often used to indicate the performance of fly ash used as supplementary cementitious materials. The mixed proportion of cement pastes tested for water requirement, time of setting, and autoclave expansion is shown in Table 3, along with the test results. With a replacement level of 20% of the mass of cement by CCFA, the water requirement exhibited increases of 3–5% relative to the control mixture, which is due to the surface roughness and irregular grain shape of CCFA. The initial and final setting times of the

PPMS-CCFA were significantly delayed by 115 and 150 min, respectively, compared to those of the control. The initial and final setting times for PR-CCFA are similar to those of PCFA. The results show that the addition of PPMS-CCFA can seriously delay the setting time of cement paste. This is due to the presence of calcium sulfite components in PPMS-CCFA [36,37]. The increased water requirement also causes delays in the time of setting [16,30,36–38].

Table 3. Water demands and setting time of fly ash substituted cement paste.

Mixture ID	Mix Proportion (%)		Water Requirement for Normal Consistency (%)	Time of Setting (Minute)		Autoclave Expansion (%)
	Cement	FA		Initial Setting	Final Setting	
OPC (control)	100	0	25	165	220	0.10
PCFA	80	20	26	185	235	0.04
PPMS-CCFA	80	20	30	280	375	0.07
PR-CCFA	80	20	28	205	245	0.08

Several studies have confirmed that the presence of calcium sulfate compounds can significantly affect the hydration of cement and setting time [30,33,37,39]. Calcium sulfite has the most significant impact on the setting time of cement due to the low solubility of calcium sulfite hemihydrate in water [30,40]. When the soluble sulfate reacts with C_3A in the cement, calcium monosulfite-aluminate (hemihydrite) is formed [16,33]. If calcium hydroxide exists simultaneously in the environment, the crystallization form of the reaction product will be affected. It covers the cement particles' surface to take the gel form and prevents the entry of water. As a result, it prevents the C_3A from reacting and forming calcium aluminate hydrate, leading to a prolonged setting time [16,33,39]. Compared to the setting retardation effect of calcium sulfite on cement, calcium sulfate dissolves rapidly in water. Therefore sufficient soluble sulfate can react with C_3A in the cement to form ettringite (AFt) and fill the pores of the cement paste [16,30,41,42]. In other words, the setting retardation effect of different types of calcium sulfate on cement is different and particularly dependent upon different characteristics of the solubility and dissolution rate of each calcium sulfate [30].

The autoclave expansion of the cement paste prepared using 20% CCFAs in replacement of cement was determined after exposure to a saturated steam pressure of 20.8 ± 0.7 kgf/cm^2 and a temperature of 215.7 ± 1.7 °C. The results are all much below the 0.8% specified by ASTM C618.

3.2.2. Compressive Strength Test of Cement Mortar

According to the strength activity index test method in the ASTM C311/C311M [43], cubic mortar samples with a cube side length of 5 cm were prepared. Table 4 shows the proportions of sand and water in the cement mortar mixed with CCFAs. Under the controlled condition of a fixed fluidity value, it was found that the use of the two types of CCFA increased the water requirement to 111% and 109%. The difference in the water requirement may be caused by the irregular grain appearance and surface roughness of CCFA particles.

Table 4. Mixture proportions of cement mortars.

Mixture ID	Weight (g)				Water Requirement (%)
	Cement	FA	Sand	Water	
OPC (control)	1000	0	2750	485	100
PCFA	800	200	2750	485	102
PPMS-CCFA	800	200	2750	540	111
PR-CCFA	800	200	2750	530	109

Figure 5 shows the variation in the compressive strength of different mortar samples. To evaluate whether CCFA can provide the pozzolanic activity or supplementary cementitious activity, the test was conducted to provide the compressive strength and the corresponding Strength Activity Indices (SAI) of the mortar samples from 3 to 90 days.

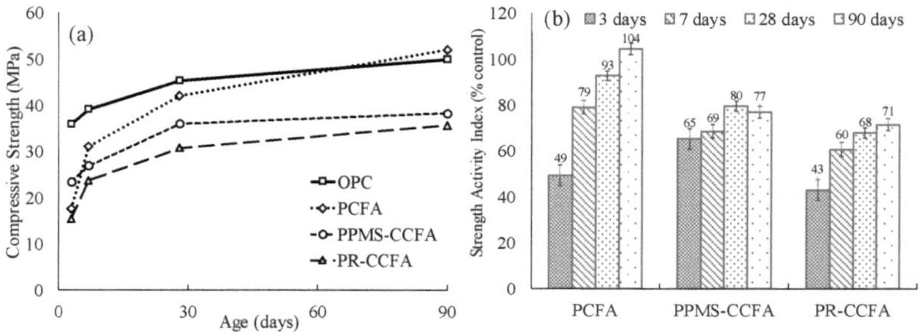

Figure 5. Strength development and strength activity index of different mortar mixtures (**a**) Compressive Strength, (**b**) Strength Activity Index.

The results show that the mortar mixed with PCFA had a lower early compressive strength than the OPC. Additionally, it was found that the PCFA mortar's late compressive strength was greater than that of the OPC in the duration between 56 and 90 days under the pozzolanic activity. According to ASTM C618, the strength activity indices of PCFA or natural pozzolanic material should be greater than 75% at 7 and 28 days. However, because CCFA is not included in the standard category of supplementary cementitious materials, reference values according to ASTM C618 can be used to judge the potential of CCFA as supplementary cementitious materials for partial replacement of cement in making concrete [25].

The test results show that PPMS-CCFA had a higher initial compressive strength than other fly ashes at three days due to the presence of calcium hydroxide (CH) in the PPMS-CCFA sample. As the age of specimens increases, the compressive strength becomes flat in the growth trend and reaches the 75% requirement at 28 days.

Although the chemical composition of PR-CCFA is similar to that of PCFA, due to the low operating temperature of the CFB boiler, the glassy phase of the material is low, inevitably resulting in a deteriorated cementitious activity. As expected, when PR-CCFA is mixed with cement, a low strength activity index of 68% was obtained, indicating that PR-CCFA has a relatively low strength activity when used in cement mixtures. However, it should be noted that the strength activity index of PR-CCFA is still increasing in the later age up to 90 days, indicating a slow pozzolanic reaction.

4. Conclusions

This study characterized the material properties of CCFA produced from CFB boilers in a paper recycling mill, and the potential of the CCFA as cementitious materials was evaluated by comparing the test results with the standard specifications for PCFA. The main conclusions are summarized as follows.

1. The material characteristics of CCFA are largely affected by the fuel source type. The high calcium content of PPMS-CCFA comes from the calcium carbonate filler in PPMS, while the chemical compositions of PR-CCFA are mainly silicon and aluminum minerals, as the fuel sources are mainly plastics and fibers.
2. The mineralogical and thermal analyses of CCFA highlighted the differences in the type of desulfurization system in the form of sulfate present in the fly ash. The presence of calcium sulfite hemihydrate in PPMS-CCFA causes a high loss on ignition and significant increases in setting time when the ash is used in replacement of cement.

3. The particle size of CCFA falls in the range of about 4 to 100 μm. The dominant particles of CCFA are granular and irregularly shaped, with a rough and porous surface texture. The porous surface of CCFA particles tends to increase the water requirement as the ash is used in cement mixtures.
4. The compressive strength of mortars containing 20% CCFA shows a relatively slow development with curing time. The strength activity index of the two CCFAs studied is either satisfactory according to the requirement for coal fly ash or increasing with time and exhibiting a characteristic of pozzolanic reaction. These results confirmed that fly ash from the co-combustion of paper mill wastes and coal has the potential to be used as cementitious materials in partial replacement of cement.

This study characterized the fly ash produced from CFB boilers co-combusting paper mill wastes and coal and assessed possible means for utilization of CCFA of such origin. Further studies on the CCFA obtained from different fuel sources are highly recommended, such that feasible utilization of CCFA can be developed to facilitate the use of waste-derived fuels in industrial boilers.

Author Contributions: Conceptualization, M.-F.W. and W.-H.H.; methodology, M.-F.W. and W.-H.H.; data curation, M.-F.W.; writing—original draft preparation, M.-F.W.; writing—review and editing, W.-H.H.; supervision, W.-H.H. All authors have read and agreed to the published version of the manuscript.

Funding: This research was funded by the Ministry of Science and Technology of Taiwan (MOST-110-2221-E-008-018 and MOST-111-2221-E-008-022).

Institutional Review Board Statement: Not applicable.

Informed Consent Statement: Not applicable.

Data Availability Statement: Not applicable.

Conflicts of Interest: The authors declare no conflict of interest.

References

1. Haile, A.; Gelebo, G.G.; Tesfaye, T.; Mengie, W.; Mebrate, M.A.; Abuhay, A.; Limeneh, D.Y. Pulp and paper mill wastes: Utilizations and prospects for high value-added biomaterials. *Bioresour. Bioprocess.* **2021**, *8*, 35. [CrossRef]
2. Monte, M.C.; Fuente, E.; Blanco, A.; Negro, C. Waste management from pulp and paper production in the European Union. *Waste Manag.* **2009**, *29*, 293–308. [CrossRef] [PubMed]
3. Hilber, T.; Maier, J.; Scheffknecht, G.; Agraniotis, M.; Grammelis, P.; Kakaras, E.; Glorius, T.; Becker, U.; Derichs, W.; Schiffer, H.P.; et al. Advantages and possibilities of solid recovered fuel cocombustion in the European energy sector. *J. Air Waste Manag. Assoc.* **2007**, *57*, 1178–1189. [CrossRef]
4. Tsai, M.-Y.; Wu, K.-T.; Huang, C.-C.; Lee, H.-T. Co-firing of paper mill sludge and coal in an industrial circulating fluidized bed boiler. *Waste Manag.* **2002**, *22*, 439–442. [CrossRef] [PubMed]
5. Sarabèr, A.J. Co-combustion and its impact on fly ash quality; full-scale experiments. *Fuel Process. Technol.* **2014**, *128*, 68–82. [CrossRef]
6. Rajamma, R.; Ball, R.J.; Tarelho, L.A.C.; Allen, G.C.; Labrincha, J.A.; Ferreira, V.M. Characterisation and use of biomass fly ash in cement-based materials. *J. Hazard. Mater.* **2009**, *172*, 1049–1060. [CrossRef]
7. Tian, Y.; Themelis, N.J.; Zhao, D.; Bourtsalas, A.C.T.; Kawashima, S. Stabilization of Waste-to-Energy (WTE) fly ash for disposal in landfills or use as cement substitute. *Waste Manag.* **2022**, *150*, 227–243. [CrossRef]
8. Faleschini, F.; Zanini, M.A.; Brunelli, K.; Pellegrino, C. Valorization of co-combustion fly ash in concrete production. *Mater. Des.* **2015**, *85*, 687–694. [CrossRef]
9. Azam, M.; Jahromy, S.S.; Raza, W.; Wesenauer, F.; Schwendtner, K.; Winter, F. Comparison of the Characteristics of Fly Ash Generated from Bio and Municipal Waste: Fluidized Bed Incinerators. *Materials* **2019**, *12*, 2664. [CrossRef]
10. Lee, J.; Lee, T. Effects of High CaO Fly Ash and Sulfate Activator as a Finer Binder for Cementless Grouting Material. *Materials* **2019**, *12*, 3664. [CrossRef] [PubMed]
11. Du, X.; Huang, Z.; Ding, Y.; Xu, W.; Zhang, M.; Wei, L.; Yang, H. Feasibility Study of Grinding Circulating Fluidized Bed Ash as Cement Admixture. *Materials* **2022**, *15*, 5610. [CrossRef] [PubMed]
12. Lin, W.-T.; Lin, K.-L.; Chen, K.; Korniejenko, K.; Hebda, M.; Łach, M. Circulation Fluidized Bed Combustion Fly Ash as Partial Replacement of Fine Aggregates in Roller Compacted Concrete. *Materials* **2019**, *12*, 4204. [CrossRef] [PubMed]
13. Córdoba, P. Status of Flue Gas Desulphurisation (FGD) systems from coal-fired power plants: Overview of the physic-chemical control processes of wet limestone FGDs. *Fuel* **2015**, *144*, 274–286. [CrossRef]

14. Kim, Y.B.; Gwak, Y.R.; Keel, S.I.; Yun, J.H.; Lee, S.H. Direct desulfurization of limestones under oxy-circulating fluidized bed combustion conditions. *Chem. Eng. J.* **2019**, *377*, 119650. [CrossRef]
15. Li, D.; Ke, X.; Kim, M.; Cai, R.; Yang, H.; Zhang, M.; Jeon, C.-H. Attrition and product layer development of limestone during simultaneous calcination and sulfation in a fluidized bed reactor. *Fuel* **2021**, *293*, 120280. [CrossRef]
16. Navarrete, I.; Vargas, F.; Martinez, P.; Paul, A.; Lopez, M. Flue gas desulfurization (FGD) fly ash as a sustainable, safe alternative for cement-based materials. *J. Clean. Prod.* **2021**, *283*, 124646. [CrossRef]
17. Sun, H.; Qian, J.; Peng, S.; Xiong, M.; Fan, C.; Fan, Y. Utilization of circulating fluidized bed combustion ash to prepare supersulfated cement. *Constr. Build. Mater.* **2022**, *318*, 125861. [CrossRef]
18. Brylicki, W.; Łagosz, A. Utilization of Flue GAS Desulphurization by-Products in the Cellular Concrete Technology. In *Studies in Environmental Science*; Goumans, J.J.J.M., Senden, G.J., van der Sloot, H.A., Eds.; Elsevier: Amsterdam, The Netherlands, 1997; pp. 571–580.
19. Bigham, J.M.; Kost, D.A.; Stehouwer, R.C.; Beeghly, J.H.; Fowler, R.; Traina, S.J.; Wolfe, W.E.; Dick, W.A. Mineralogical and engineering characteristics of dry flue gas desulfurization products. *Fuel* **2005**, *84*, 1839–1848. [CrossRef]
20. Al-Shmaisani, S.; Kalina, R.D.; Ferron, R.D.; Juenger, M.C.G. Critical assessment of rapid methods to qualify supplementary cementitious materials for use in concrete. *Cem. Concr. Res.* **2022**, *153*, 106709. [CrossRef]
21. Brown, L.J.; Collard, F.-X.; Görgens, J. Pyrolysis of fibre residues with plastic contamination from a paper recycling mill: Energy recoveries. *Energy Convers. Manag.* **2017**, *133*, 110–117. [CrossRef]
22. Brown, L.J.; Collard, F.X.; Gottumukkala, L.D.; Görgens, J. Fermentation-pyrolysis of fibre waste from a paper recycling mill for the production of fuel products. *Waste Manag.* **2021**, *120*, 364–372. [CrossRef]
23. Arenales Rivera, J.; López, V.P.; Casado, R.R.; Hervás, J.-M.S. Thermal degradation of paper industry wastes from a recovered paper mill using TGA. Characterization and gasification test. *Waste Manag.* **2016**, *47*, 225–235. [CrossRef] [PubMed]
24. Ouadi, M.; Greenhalf, C.; Jaeger, N.; Speranza, L.G.; Hornung, A. Thermo-catalytic reforming of pulper rejects from a secondary fibre mill. *Renew. Energy Focus* **2018**, *26*, 39–45. [CrossRef]
25. *ASTM Standard C1709*; Standard Guide for Evaluation of Alternative Supplementary Cementitious Materials (ASCM) for Use in Concrete. ASTM International: West Conshohocken, PA, USA, 2018.
26. *ASTM Standard C618*; Standard Specification for Coal Fly Ash and Raw or Calcined Natural Pozzolan for Use in Concrete. ASTM International: West Conshohocken, PA, USA, 2019.
27. *ASTM Standard C191*; Standard Test Methods for Time of Setting of Hydraulic Cement by Vicat Needle. ASTM International: West Conshohocken, PA, USA, 2021.
28. Wan, H.-P.; Chang, Y.-H.; Chien, W.-C.; Lee, H.-T.; Huang, C.C. Emissions during co-firing of RDF-5 with bituminous coal, paper sludge and waste tires in a commercial circulating fluidized bed co-generation boiler. *Fuel* **2008**, *87*, 761–767. [CrossRef]
29. de Castro, R.d.P.V.; de Medeiros, J.L.; Araújo, O.d.Q.F.; Cruz, M.d.; Ribeiro, G.T.; de Oliveira, V.R. Fluidized bed treatment of residues of semi-dry flue gas desulfurization units of coal-fired power plants for conversion of sulfites to sulfates. *Energy Convers. Manag.* **2017**, *143*, 173–187. [CrossRef]
30. Shen, Y.; Qian, J.; Zhang, Z. Investigations of anhydrite in CFBC fly ash as cement retarders. *Constr. Build. Mater.* **2013**, *40*, 672–678. [CrossRef]
31. Kumar, P.; Malhotra, B.; Phalswal, P.; Khanna, P.K.; Salim, A.; Singhal, R.; Mukhopadhyay, A.K. Effect of reaction rate on the properties of chemically synthesized calcium hydroxide nanoparticles. *Mater. Today Proc.* **2020**, *28*, 2305–2310.
32. Modolo, R.C.E.; Tarelho, L.A.C.; Teixeira, E.R.; Ferreira, V.M.; Labrincha, J.A. Treatment and use of bottom bed waste in biomass fluidized bed combustors. *Fuel Process. Technol.* **2014**, *125*, 170–181. [CrossRef]
33. Lagosz, A.; Malolepszy, J. Tricalcium aluminate hydration in the presence of calcium sulfite hemihydrate. *Cem. Concr. Res.* **2003**, *33*, 333–339. [CrossRef]
34. Zheng, D.; Lu, H.; Sun, X.; Liu, X.; Han, W.; Wang, L. Reaction mechanism of reductive decomposition of FGD gypsum with anthracite. *Thermochim. Acta* **2013**, *559*, 23–31. [CrossRef]
35. Mlonka-Mędrala, A.; Dziok, T.; Magdziarz, A.; Nowak, W. Composition and properties of fly ash collected from a multifuel fluidized bed boiler co-firing refuse derived fuel (RDF) and hard coal. *Energy* **2021**, *234*, 121229. [CrossRef]
36. Li, H.; Zhang, H.; Li, L.; Ren, Q.; Yang, X.; Jiang, Z.; Zhang, Z. Utilization of low-quality desulfurized ash from semi-dry flue gas desulfurization by mixing with hemihydrate gypsum. *Fuel* **2019**, *255*, 115783. [CrossRef]
37. Zahedi, M.; Jafari, K.; Rajabipour, F. Properties and durability of concrete containing fluidized bed combustion (FBC) fly ash. *Constr. Build. Mater.* **2020**, *258*, 119663. [CrossRef]
38. Huang, W.-H. Improving the properties of cement–fly ash grout using fiber and superplasticizer. *Cem. Concr. Res.* **2001**, *31*, 1033–1041. [CrossRef]
39. Sypek, M.; Latawiec, R.; Pichór, W. Gypsum dehydration in cement and its impact on air-void structure in air-entrained concrete. *Constr. Build. Mater.* **2019**, *220*, 396–402. [CrossRef]
40. Lagosz, A.; Malolepszy, J.; Garrault, S. Hydration of tricalcium aluminate in the presence of various amounts of calcium sulphite hemihydrate: Conductivity tests. *Cem. Concr. Res.* **2006**, *36*, 1016–1022. [CrossRef]
41. Dung, N.T.; Chang, T.-P.; Chen, C.-T. Engineering and sulfate resistance properties of slag-CFBC fly ash paste and mortar. *Constr. Build. Mater.* **2014**, *63*, 40–48. [CrossRef]

42. Ragipani, R.; Escobar, E.; Prentice, D.; Bustillos, S.; Simonetti, D.; Sant, G.; Wang, B. Selective sulfur removal from semi-dry flue gas desulfurization coal fly ash for concrete and carbon dioxide capture applications. *Waste Manag.* **2021**, *121*, 117–126. [CrossRef]
43. *ASTM Standard C311/C311M*; Standard Test Methods for Sampling and Testing Fly Ash or Natural Pozzolans for Use in Portland-Cement Concrete. ASTM International: West Conshohocken, PA, USA, 2022.

materials

MDPI

Article

Using Experimental Statistical Models for Predicting Strength and Deformability of Self-Compacting Concrete with Ground Blast-Furnace Slag

Vadim Zhitkovsky [1], Leonid Dvorkin [1], Dmyrto Kochkarev [1] and Yuri Ribakov [2,*]

1 Department of Constructing Products Technology and Material Science, National University of Water and Environmental Engineering, 33028 Rivne, Ukraine; v.v.zhitkovsky@nuwm.edu.ua (V.Z.); l.i.dvorkin@nuwm.edu.ua (L.D.); d.v.kochkarev@nuwm.edu.ua (D.K.)
2 Department of Civil Engineering, Ariel University, Ariel 40700, Israel
* Correspondence: ribakov@ariel.ac.il; Tel.: +972-3-9066-327

Abstract: Ground blast-furnace slag is one of the waste products available in Ukraine and other countries. It is obtained at metallurgical enterprises in huge quantities and can be efficiently used for concrete production. The article is devoted to obtaining experimental-statistical models of the influence of technological factors that determine the composition of self-compacting concrete (SCC) based on ground blast-furnace slag and polycarboxylate superplasticizer on compressive strength, tensile strength, prismatic strength, elastic modulus and crack resistance. Analysis of the investigated factors' influence on the specified SCC properties is carried out and positive influence of blast-furnace slag and superplasticizer simultaneous action on durability and deformation characteristics is studied. A design method of SCC composition design using the obtained mathematical models is developed. It allows for the consideration of a set of necessary parameters simultaneously. A numerical example is given.

Keywords: self-compacting concrete; blast-furnace slag; superplasticizer; strength; modulus of elasticity; experimental-statistical model; concrete composition design

Citation: Zhitkovsky, V.; Dvorkin, L.; Kochkarev, D.; Ribakov, Y. Using Experimental Statistical Models for Predicting Strength and Deformability of Self-Compacting Concrete with Ground Blast-Furnace Slag. *Materials* **2022**, *15*, 4110. https://doi.org/10.3390/ma15124110

Academic Editor: Alessandro P. Fantilli

Received: 4 May 2022
Accepted: 8 June 2022
Published: 9 June 2022

1. Introduction

Applying effective plasticizers and superplasticizers in concrete compositions enable to obtain concrete mixtures with high workability (cast mixtures) and hardened concrete with high physical, mechanical and operational characteristics based on ordinary Portland cement and aggregates [1,2]. At the same time, along with obvious advantages, concrete mixtures with high workability have a number of disadvantages, such as loss of fluidity, delamination, slow hardening, etc. The effect of those disadvantages can be significantly reduced by using third-generation superplasticizers and highly dispersed silica-containing materials of man-made origin, especially silica fume [2,3].

Wide use of self-compacting concrete (SCC) in construction is quite attractive, but it is limited by the need to use highly active mineral admixtures and the lack of sufficient experimental basis for predicting its properties [3]. In turn, for proper structural design, data on deformability and crack resistance of such concrete are required, and the effect of its structural properties caused by high content of binders, active mineral admixtures and superplasticizers on structural behavior and durability should be investigated [4,5].

One of the widespread active mineral additives for concrete is ground domain granular slag [6]. Of all types of metallurgical slags used for producing construction materials, blast-furnace slag is the most widely used one. This is because blast-furnace slag has a chemical and mineralogical composition that is close to that of cement and has the ability to acquire hydraulic activity at rapid cooling [7].

Blast-furnace slag is formed in large quantities at metallurgical enterprises. It is an artificial silicate material formed during smelting metals from ore due to the high-temperature interaction of ore, fuel, floodplains and gaseous medium. For many years, a significant amount of blast-furnace slag has accumulated in dumps, occupied agricultural land and polluted the environment. Therefore, disposal of metallurgical slag is important for environmental protection and resource reuse worldwide.

SCC is usually characterized by high cement consumption [8], so the issue of efficient use of blast-furnace slag to replace part of cement, which is addressed in this study, is very important to improve the environmental friendliness of construction.

In order to use SCC in structural elements, the influence of technological factors on concrete deformability, crack resistance and behavior under loading is important. Development of proper methodology for predicting the SCC deformability at the concrete composition design stage is also relevant.

One of the most important and well-predictable concrete deformation characteristics is its modulus of elasticity. It is known [9] that ordinary and high-strength concrete modulus of elasticity closely correlates with its prismatic strength. Empirical dependences that interrelate those parameters have been proposed [10]. These dependences are included in various design codes. The empirical equations are often significantly different each from other by the equation form and also by coefficients, which vary in a rather wide range.

As known, due to the lower content of coarse aggregate, which creates a rigid frame, SCC has lower modulus-of-elasticity values [11]. For low-class concrete, the decrease in SCC modulus of elasticity is about 40% [12]. As the concrete class becomes higher, the difference between SCC modulus of elasticity and that of ordinary concrete decreases to 5%. It is obvious that various technological and structural factors cause significant changes in modulus of elasticity; therefore, it is difficult to find a close relationship, which could predict the SCC modulus of elasticity.

2. Research Aim, Scope and Novelty

In order to design composition of SCC with blast-furnace slag, taking into account a set of specified properties, it is important to obtain mathematical dependences allowing evaluation of technological and concrete composition factors' impact on the required properties. The method of mathematical experiments planning [13,14] is effective to obtain such dependences. The method enables the receipt of regression equations by statistical processing of data obtained from the "active" experiment, i.e., the experiments are conducted according to a mathematically based scheme that helps to minimize the number of experimental points [15].

In order to identify the complex effect of active mineral additives and polycarboxylate-type superplasticizers on SCC strength and deformable properties in a wide range of strength classes, as well as to establish mathematical relationships between properties and technological factors, a planned experiment was conducted.

The novelty of this study is that based on the experimental results, mathematical models of strength and deformation characteristics of SCC based on blast-furnace slag are obtained. These models allow for a design of concrete composition that provides the required concrete characteristics at different hardening stages with minimum possible cement consumption.

3. Experimental Program

As variable factors were selected Portland cement consumption C, kg/m^3 (X_1), superplasticizer content SP, % (X_2) and consumption of mineral additives (blast-furnace granulated slag) Sl, kg/m^3 (X_3). The concrete-mixture workability at the points of the experiment was maintained at the level of S5 (Slump—250–270 mm, which corresponds to the cone spread of 560–620 mm (F5 according to [16]). 204 cubic specimens of 100 mm × 100 mm × 100 mm and 48 prisms of 100 mm × 100 mm × 400 mm were prepared from the concrete mixture. Compressive strength (f_{cm}, MPa), splitting tensile strength

(f_{ctm}, MPa) and modulus of elasticity (E_{cm} were determined at 1 and 28 days. Mechanical properties of concrete (strengths of cubes and prisms) under a single short-term load were determined by standard methods [17].

For determining the strength of concrete cubic specimens, the experiment was performed according to a three-level three-factor plan B_3 [15] to obtain mathematical polynomial models of the form:

$$Y = b_0 + b_1X_1 + b_2X_2 + b_3X_3 + b_{11}X_1^2 + b_{22}X_2^2 + b_{33}X_3^2 + b_{12}X_1X_2 + b_{13}X_1X_3 + b_{23}X_2X_3, \quad (1)$$

where Y is the output parameter, $X_1 \ldots X_3$ are variables, $b_0 \ldots b_{23}$ are the equation coefficients.

Investigation of deformation characteristics was carried out in accordance with the two-level plan of the three-factor plan [15]. The type of mathematical model is:

$$Y = b_0 + b_1X_1 + b_2X_2 + b_3X_3 + b_{12}X_1X_2 + b_{13}X_1X_3 + b_{23}X_2X_3, \quad (2)$$

The planning conditions are given in Table 1.

Table 1. Experiment-planning conditions.

No.	Factors		Variation Levels			Variation Interval
	Natural	Coded	−1	0	+1	
1	Cement consumption (C, kg/m^3)	X_1	200	400	600	200
2	Superplasticizer content (SP, %)	X_2	0	0.5	0.1	0.5
3	Consumption of blast-furnace granulated slag (Sl, kg/m^3)	X_3	0	100	200	100

The following materials were used in the research: Portland cement CEM I 42.5 (Table 2), sand with M_f = 2.0, locally available granite crushed-stone fraction 5–20 mm, blast-furnace granular slag, polycarboxylate-type superplasticizer PCE50 (LLC UA-Chemical, Ukraine). The experiment-planning matrix and the concrete composition at the experimental points are given in Table 3.

Table 2. Characteristics of Portland cement and blast-furnace granular slag.

Material	Chemical Composition, %								Specific Surface Area, m^2/kg
	SiO_2	Al_2O_3	Fe_2O_3	CaO	MgO	SO_3	R_2O *	L.O.I **	
Portland cement	21.80	5.12	4.11	65.80	0.95	0.63	0.85	0.81	340
Blast-furnace granular slag	39.52	6.49	0.12	47.13	3.10	1.74	-	-	270

Notes: * $R_2O = Na_2O + K_2O$, ** L.O.I—loss on ignition.

By statistical analysis of the experimental results (Table 3), mathematical models of the output parameters, as well as correlations between individual initial parameters were obtained. The coefficients of mathematical models are given in Table 4.

Table 3. The results of determining the experiment output parameters.

No.	Coded Factors			W/C	Compressive Strength		Splitting Tensile Strength		Prismatic Strength		Modulus of Elasticity, MPa	
	X_1	X_2	X_3		f_{cm1}, MPa	f_{cm28}, MPa	f_{ctm1}, MPa	f_{ctm28}, MPa	f_{prm1}, MPa	f_{prm28}, MPa	E_{cm1}	E_{cm28}
1	1	1	1	0.26	25.2	81.1	3.4	8.4	22.7	45.0	25,529	47,563
2	1	1	−1	0.25	53.8	103.9	5.4	8.9	31.8	85.0	27,335	53,612
3	1	−1	1	0.56	9.9	40.4	1.4	5.3	9.2	35.0	12,946	27,102
4	1	−1	−1	0.44	13.3	44.7	1.8	4.3	9	32.5	20,022	35,270
5	−1	1	1	0.63	9	51.8	1.1	6.65	7.5	50.0	16,545	43,033
6	−1	1	−1	0.73	4	20	0.4	2.3	2.25	10.7	9165	9947
7	−1	−1	1	1.08	2.8	14.3	0.4	1.95	2.25	11.7	14,709	16,633
8	−1	−1	−1	0.86	2.7	14.4	0.6	1.35	2.5	13.6	10,697	28,078
9	1	0	0	0.29	43.3	92.3	5.2	8.3	-	-	-	-
10	−1	0	0	0.56	10.3	46.7	1.3	5.35	-	-	-	-
11	0	1	0	0.29	23	90.8	2.7	6.6	-	-	-	-
12	0	−1	0	0.55	9.3	43.7	1.2	4.15	-	-	-	-
13	0	0	1	0.36	17.5	53.5	2.2	6	-	-	-	-
14	0	0	−1	0.30	29	76	3.6	6.4	-	-	-	-
15	0	0	0	0.31	22	71.3	2.5	6.1	-	-	-	-
16	0	0	0	0.31	22.4	70.9	2.5	6.3	-	-	-	-
17	0	0	0	0.31	21.8	70.5	2.4	6.0	-	-	-	-

Table 4. Coefficients of experimental-statistical models of output parameters.

	b_0	b_1	b_2	b_3	b_{11}	b_{22}	b_{33}	b_{12}	b_{13}	b_{23}
			Experimental statistical models according to Equation (1)							
W/C	0.30	−0.21	−0.13	0.03	0.13	0.13	0.04	0.01	0	−0.05
f_{cm1}	23.6	11.67	7.71	−3.85	1.98	−8.67	−1.54	6.06	−4.63	−2.53
f_{cm7}	58.5	20.85	19.22	−2.12	−4.01	−10.96	−2.80	10.61	−5.89	1.04
f_{cm28}	74.3	21.51	19.00	−1.79	−6.81	−9.01	−11.56	7.11	−7.34	1.69
f_{ctm1}	2.8	1.34	0.76	−0.34	0.23	−1.01	−0.12	0.63	−0.38	−0.09
f_{ctm28}	6.4	1.76	1.58	0.51	0.21	−1.24	−0.42	0.26	−0.56	0.28
f_{cm}/f_{ctm}	11.68	0.72	0.49	−1.20	−1.71	0.51	−1.25	0.61	−0.07	0.40
			Experimental statistical models according to Equation (2)							
f_{prm1}	10.9	7.3	5.2	−0.5	-	-	-	3.9	−0.5	−0.5
f_{prm28}	35.5	13.9	12.2	−0.03	-	-	-	3.5	−9.3	−0.144
E_{pr1}	17,119	4340	2525	314	-	-	-	2449	−2534	1080
E_{pr28}	32,655	8232	5884	928	-	-	-	3817	−4482	5831

4. Results and Discussion

SCC with organic-mineral modifier, which includes polycarboxylate SP and blast-furnace granulated slag, had compressive strength in the range from 3.6 to 51 MPa after 1 day of hardening. As expected, factors X_1 (cement consumption) and X_2 (SP content), which cause a significant decrease in W/C, contribute to a significant increase in strength at 1 day.

The optimal content of the SP additive is significantly related to the cement consumption—at a minimum content of C just 0.5% of SP is enough to maximize the strength; at a maximum cement consumption the optimal SP content is 0.3–1.1. With the increase in the slag content, the optimal content of the additive and its effectiveness slightly decreases (Figure 1).

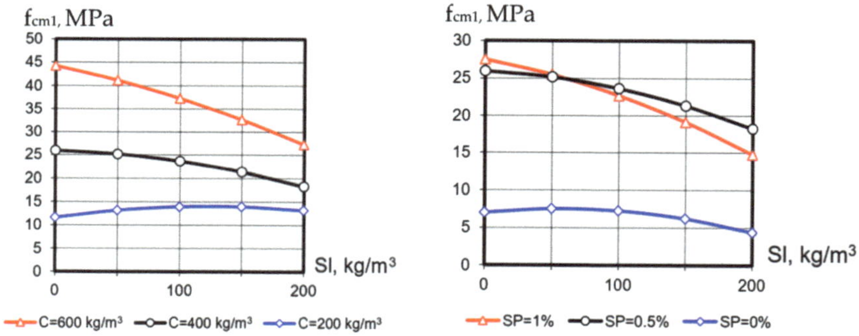

Figure 1. Dependence of compressive strength at 1 day on the blast-furnace-slag content.

Increasing the blast-furnace-slag content (X_3) mostly reduces the strength at 1 day of SCC hardening by 40–70%. As known [18], blast-furnace slag reacts rather slowly with cement-hydration products (primarily with portlandite), so a greater effect should be expected at longer hardening duration. At low cement consumptions, blast-furnace slag plays a positive role; there is an increase in f_{cm1} by 20–25%.

The concrete classes by strength at 28 days that were investigated in the frame of the present research ranged from C8/10 to C 70/85. In general, the tendencies of the factors affect, detected for strength at 1 day, are similar to those at 28 days (Table 4, Figure 2). It should be mentioned that the efficiency of factor X_3 (blast-furnace-slag consumption) becomes higher. The extreme influence of blast-furnace-slag content becomes more noticeable, which makes it possible to determine its optimal amount from the position of achieving maximum strength: at a cement consumption of 600 kg/m^3 the optimal slag content from the strength viewpoint is 40–50 kg/m^3 and it is 120–150 kg/m^3 at cement consumption of 200 kg/m^3.

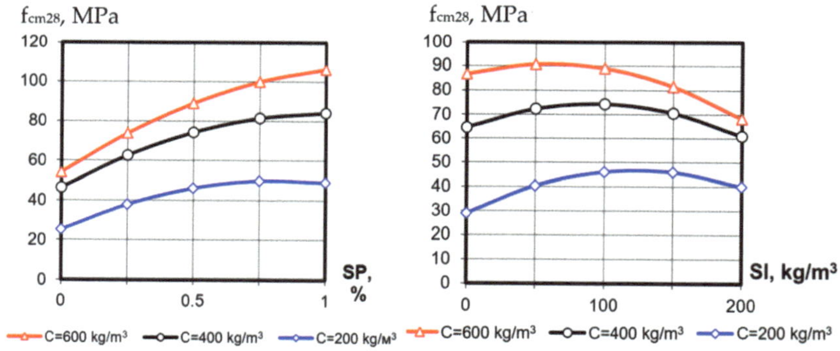

Figure 2. The influence of SP and blast-furnace-slag content at different levels of cement consumption on the SCC compression strength at 28 days (the value of the third factor in the graphs is at the average level (Table 1)).

The values of splitting tensile strength at the 1st day (f_{ctm1}) varied from 0.4 to 5.44 MPa and at 28 days (f_{ctm28}), from 1.35 to 8.9 MPa. Following the mathematical models, the nature of the variable factors' influence on splitting tensile strength is almost the same as for compressive one (Figures 3 and 4). Only the more noticeable positive effect of blast-furnace-slag content on f_{ctm} at 28 days is noteworthy, especially at low cement consumption. As known [19], splitting tensile strength is mainly determined by the cement stone strength and its amount in concrete, so at low cement consumption, such an effect of increasing the total amount of binder with a long interaction of active additives and cement is expected.

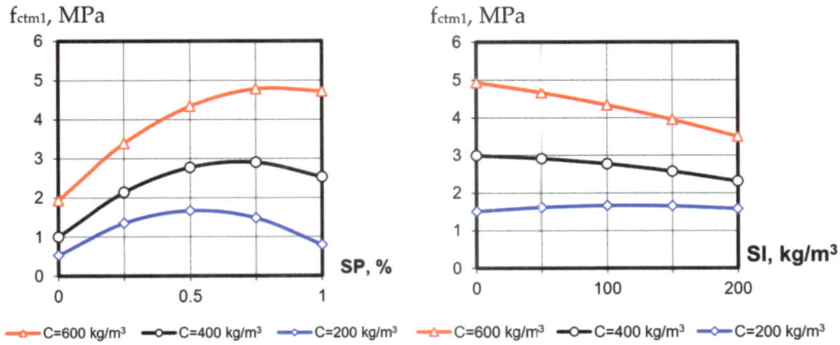

Figure 3. The influence of SP content and blast-furnace-slag content at different levels of cement consumption on SCC splitting tensile strength at 1 day (the value of the third factor on the graphs is at the average level (Table 1)).

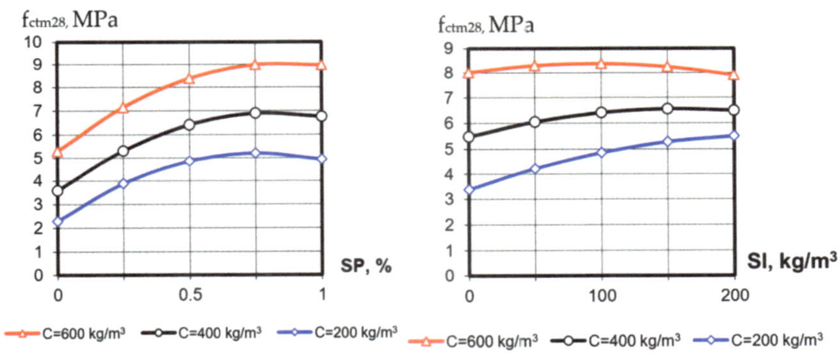

Figure 4. The influence of SP and blast-furnace-slag content at different levels of cement consumption on SCC splitting tensile strength at 28 days (the value of the third factor in the graphs is at the average level (Table 1)).

One of the criteria for crack resistance of concrete is the ratio between compressive and tensile strengths f_{cm}/f_{ctm}. A decrease in this ratio indicates an increase in concrete resistance to cracking [9]. Using the calculated values from Table 3, a mathematical model of the ratio f_{cm}/f_{ctm} at 28th days is obtained (Table 4). The main graphical dependences of the crack resistance index on the investigated factors are shown in Figure 5. The value of this criterion according to experimental data is in the range from 7.4 to 13.8. Analysis of the model shows that the achievement of the minimum criterion values is greatly facilitated by the increase in the blast-furnace-slag content (factor X_3) at a minimum cement consumption (X_1) (Figure 5). According to the available results [9], during hardening of cements with high slag content, a significant amount of low-basic calcium hydrosilicates is formed. As known [20], due to their nature these hydrosilicates are more deformable compared to highly basic calcium hydrosilicates, formed by hydration of Portland cement clinker.

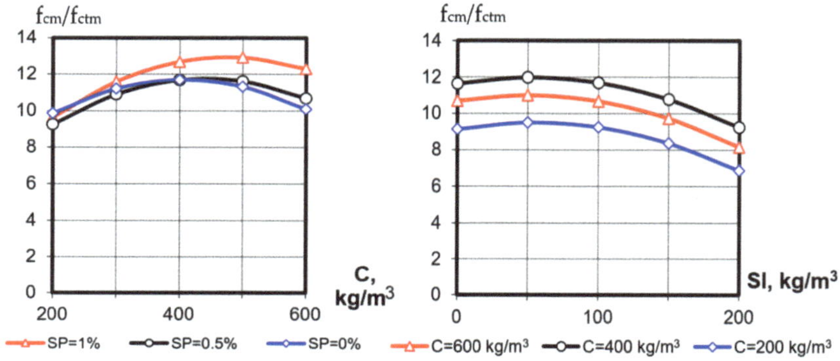

Figure 5. Influence of varied factors on the crack-resistance criterion f_{cm}/f_{ctm}.

Prismatic strength and modulus of deformation of SCC with organic-mineral modifier were studied by testing 100 mm × 100 mm × 400 mm prisms. The influence of the same three factors (cement consumption (X_1) C, kg/m^3, superplasticizer content (X_2) SP, %, blast-furnace granular-slag consumption (X_3) Sl, %) was studied according to the linear plan of the full factorial experiment [15]. The coefficients of the mathematical model are given in Table 4.

The prismatic strength after one day of hardening ranged from 2 to 32 MPa, which corresponded to 78% of concrete cubes' strength at the same age. The highest influence on the prismatic strength is caused by factors X_1 (C) and X_2 (SP), which first of all cause a decrease in W/C in transition from the lower to the upper level (Figure 6). Factor X_3 (content of blast-furnace granulated slag) has little effect on the prismatic strength after one day of hardening.

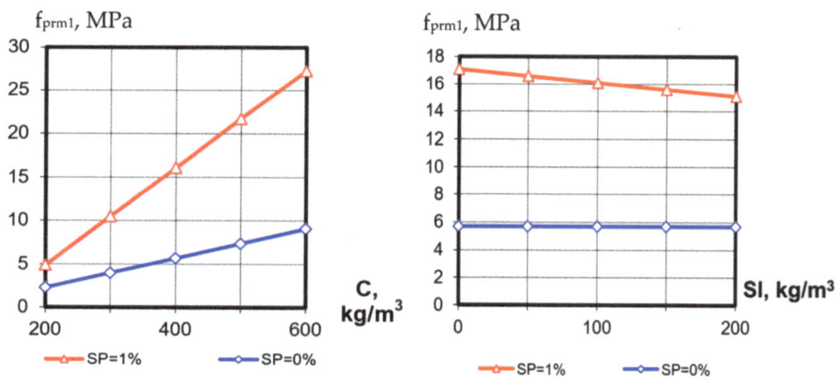

Figure 6. Influence of varied factors on SCC prismatic strength at 1 day (the value of the third factor on the graphs is at the average level (Table 1)).

After 28 days of hardening (Figure 7) interaction of factors X_3 (slag content) with X_1 (cement consumption) becomes more noticeable: for low cement consumption, an increase in slag content has a positive effect on concrete prismatic strength, while at high cement consumption the same increase in slag consumption causes a decrease in strength. As known [9], concrete deformations play a greater role in concrete prisms' destruction (compared to the test of cubes), so the prismatic strength can be considered more as a concrete-deformation characteristic. The positive effect of the cement slag component at high values of W/C also confirms the structural role of low-basic calcium hydrosilicates, which are formed to a higher extent than in clinker cements.

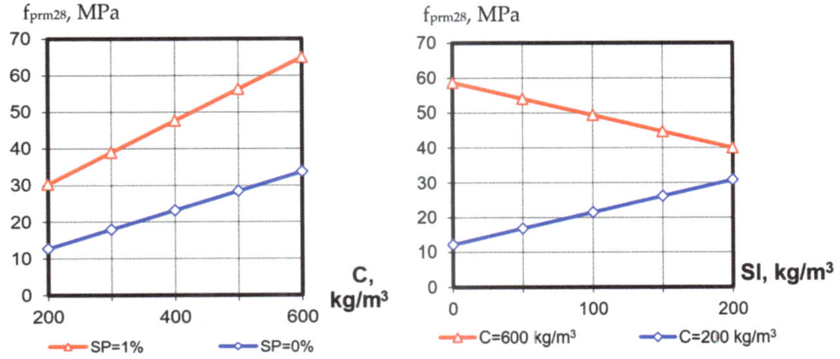

Figure 7. Influence of varied factors on SCC prismatic strength at 28 days (the value of the third factor on the graphs is at the average level (Table 1)).

The influence of other factors (cement consumption and SP content) on the prismatic strength at 28 days is similar to that after 1 day of hardening. The strength varies from 11 to 85 MPa. The mean value of the ratio between cubic and prismatic strength at 28 days is like after one day of hardening: 78%. The correlation between these concrete characteristics is shown in Figure 8.

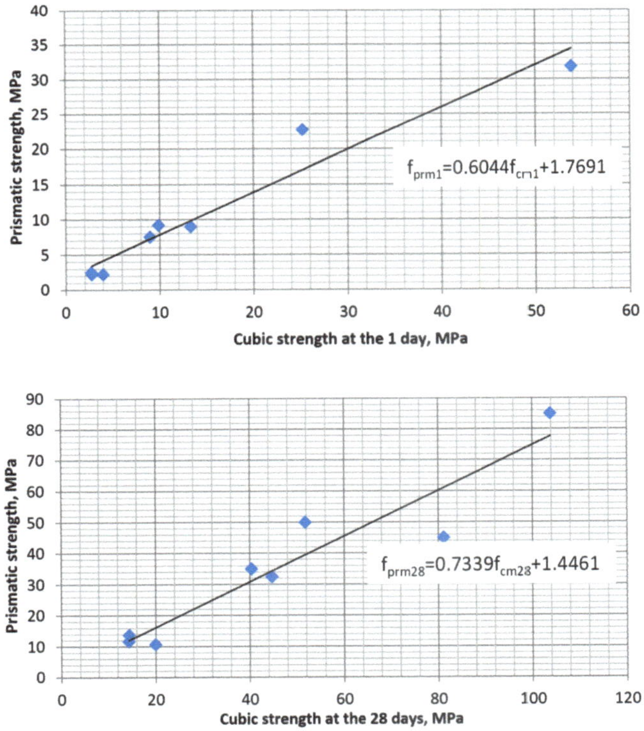

Figure 8. Correlation between the SCC cubic and prismatic strength.

The values of concrete modulus of elasticity at the first day were in the range from 10,000 to 27,000 MPa, at the 28th day—from 19,000 to 50,000 MPa. Analyzing the mathematical models, it should be noted that all three investigated factors significantly affect the SCC modulus of elasticity. The interactions between these factors are also significant.

At transition of factors X_1 (C, kg/m^3) and X_2 (SP, %) from the lower level to the upper one the SCC modulus of elasticity increases on average from 10,000–12,000 to 20,000–26,000 MPa at 1 day and from 20,000–30,000 to 40,000–50,000 MPa at 28 days. The positive interaction coefficient of these factors (b_{12}, Table 4) indicates a significant increase in the modulus of elasticity values with a simultaneous increase in cement consumption and superplasticizer content, both after 1 day and after 28 days of hardening. Factor X_3 (slag consumption) on average has less effect on the modulus of elasticity than X_1 and X_2, but its effect is significantly enhanced by the simultaneous influence of other factors. At low cement consumption (C = 200 kg/m^3) the increase in slag consumption increases E_{pr} by 1.5–1.7 times both at early hardening stages and at 28 days; however, at high cement consumption the same factor causes a significant reduction in SCC modulus of elasticity (Figures 9 and 10).

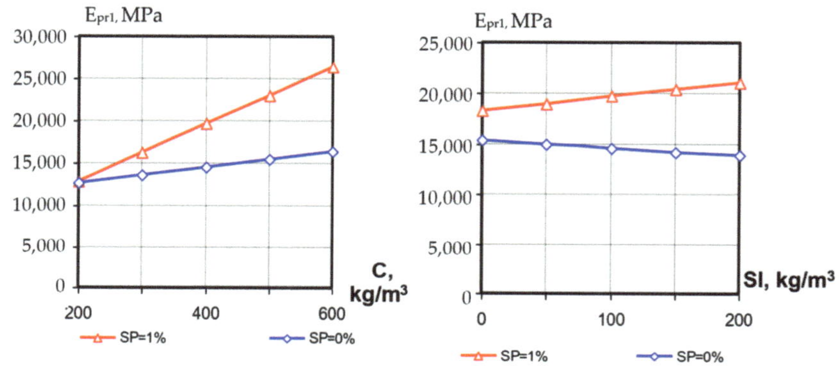

Figure 9. Influence of cement, slag and SP contents on SCC modulus of elasticity at 1 day.

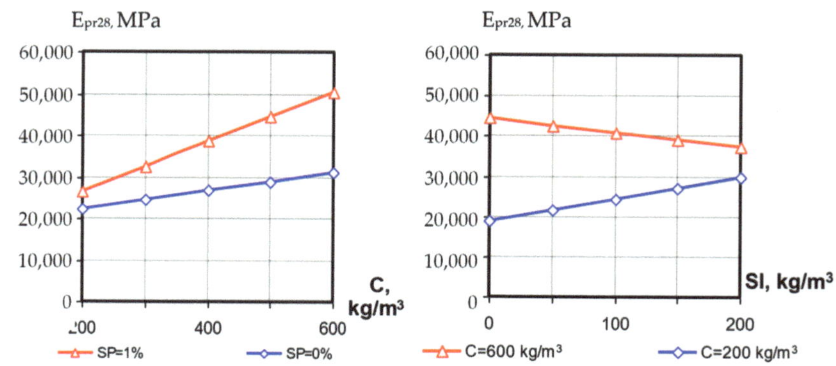

Figure 10. Influence of cement, slag and SP contents on SCC modulus of elasticity at 28 days.

The action of superplasticizer has a significant influence on the change of SCC modulus of elasticity vs. the blast-furnace-slag content; if in compositions without SP the positive effect of factor X_3 is manifested only at low cement consumption, then the diluting and water-reducing action of polycarboxylate SP contributes to the positive effect of the slag component, and at higher cement costs it is probably leveling the significant increase in concrete-mixture water demand and causing formation of a denser concrete structure.

For predicting the concrete modulus of elasticity under loading at the age of τ days the most used dependencies have the following form:

$$E_c = \frac{E_m R_\tau}{S + R_\tau}, \quad (3)$$

where R_τ is compressive strength of concrete cubes after a certain curing duration (τ); E_m and S are empirical constants. Modern codes for structural design recommend the following values: E_m = 52,000; S = 23 [9].

According to the European Concrete Committee recommendation [9], the following dependence should be used:

$$E = C(R_\tau)^\gamma, \quad (4)$$

where C = 1900, γ = 0.5.

Different modifications of Equation (4) and coefficient values were proposed [10]. According to ACI-318-83 [21], concrete modulus of elasticity can be calculated as follows:

$$E = 4540\sqrt{f_C}, \quad (5)$$

For high-strength concrete, the American Concrete Association has proposed a different equation [11]:

$$E = 3320\sqrt{f_C} + 6900, \quad (6)$$

Deformability characteristics of concrete with modifying additives differ from the usual one. Elastic properties of modified concrete, including SCC, increase 1.4 times, and plastic ones decrease twice [10]. For modified concrete, the following dependence was obtained:

$$E = 5370.8\sqrt{f_C}, \quad (7)$$

This dependence is fundamentally similar to Equations (3) and (4) and differs by a refined coefficient for modified high-strength concrete. The difference between the values of E calculated by Equations (3)–(7) increases (up to 35%) as the concrete strength increases [9].

Based on the obtained data, the dependence between the modulus of elasticity of self-compacting concrete with a multifunctional modifier containing blast-furnace slag and superplasticizer was approximated (Figures 11 and 12). The obtained equations have the following form:

- after 1 day of hardening:

$$Epr1 = 8334{f_{pr1}}^{0.34}, \quad (8)$$

- after 28 days of hardening:

$$Epr28 = 18319 ln(f_{pr28}) - 28342 \quad (9)$$

Table 5 shows the modulus of elasticity values, calculated by empirical Equations (1), (3)–(5) and (9) by strength indicators, as well as experimental values obtained for SCC. As it can be seen from Table 5, the average value of errors for calculations by all equations varies within a fairly narrow range: 18–35%. Significant differences in the errors obtained for calculations according to each of the equations (see Table 5 vertically) allows one to suggest that along with close correlation between strength and modulus of elasticity, which is noted by many authors [9,11,22,23], this indicator is significantly dependent on concrete qualitative composition, and especially the content of mineral and chemical additives that modify the cement stone structure. Comparing the data in Table 5, it becomes obvious that using correlations between strength and modulus of elasticity is appropriate only for approximate establishment of the required deformation characteristics, for example, for the design of reinforced concrete structures.

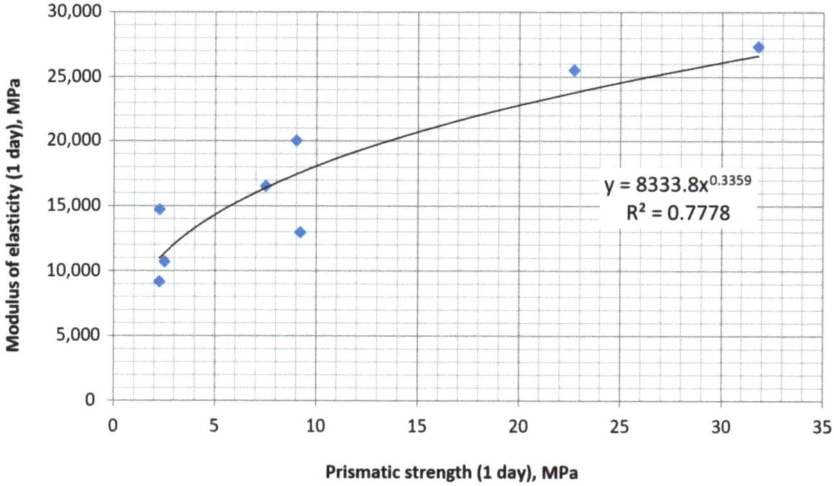

Figure 11. Correlation between the modulus of elasticity and prismatic strength at 1 day.

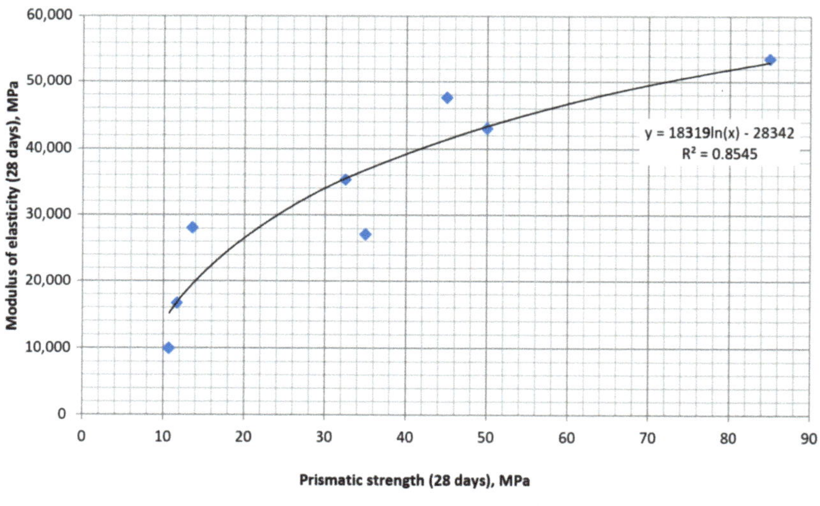

Figure 12. Correlation between the modulus of elasticity and prismatic strength at 28 days.

A method of creating and using multifactor dependences obtained, for example, by mathematical experiments planning has a higher resolution to perform tasks of predicting concrete properties at the composition design stage. The obtained experimental-statistical models with coefficients given in Table 4 enable us to predict a set of strength and deformation properties at certain values of the investigated factors: consumptions of cement, blast-furnace slag and superplasticizer. The use of mathematical optimization methods [16] and the corresponding software allows one to find the values of the above-mentioned factors that provide the necessary properties with minimal resource consumption [24].

Table 5. Comparison of SCC modulus of elasticity values calculated using different empirical dependences with experimental data.

No.	Experimental Data		Calculated Data, according to Equations E_{pr28}, MPa Error (Δ, %)				
	f_{pr28}, MPa	E_{pr28}, MPa	(1)	(3)	(4)	(5)	(9)
1	45	47,563	30,455 / 36	53,182 / 12	29,171 / 39	36,030 / 24	41,392 / 13
2	85	53,612	41,857 / 22	100,455 / 87	37,509 / 30	49,518 / 8	53,043 / 1
3	35	27,102	26,859 / 1	41,364 / 53	26,541 / 2	31,775 / 17	36,788 / 36
4	32.5	35,270	25,882 / 27	38,409 / 9	25,827 / 27	30,619 / 13	35,431 / 0,5
5	50	43,033	32,103 / 25	59,091 / 37	30,376 / 29	37,979 / 12	43,322 / 1
6	11	9,947	15,057 / 51	13,000 / 31	17,911 / 80	17,814 / 79	15,585 / 57
7	12.3	16,633	15,890 / 4	14,477 / 13	18,520 / 11	18,799 / 13	17,557 / 6
8	13.8	28,078	16,835 / 40	16,250 / 42	19,211 / 32	19,916 / 29	19,673 / 30
	Average error, %		26	35	31	24	18

Numerical example. Calculate composition of high-strength SCC (Slump—250–270 mm) using ground blast-furnace slag with a specific surface area of 270 m^2/kg and polycarboxylate superplasticizer. The required indicators are that concrete class C40/50 (cubic strength at 28 days at least 64.3 MPa), cubic strength and modulus of elasticity at 1 day must meet the normative indicators for class C12/15: at least 19.3 MPa and 27,000 MPa, respectively.

Characteristics of raw materials: true densities of cement, slag, sand and crushed stone are 3100 kg/m^3, 2900 kg/m^3, 2650 kg/m^3, 2700 kg/m^3. The part of crushed stone in the SCC should not exceed 0.35.

5. Solution

1. According to experimental statistical models of f_{cm1}, f_{cm28}, E_{pr1} (see Table 4) using the methods of mathematical optimization [25], find the values of factors (consumptions of cement and slag, and superplasticizer content) that provide required indicators and satisfying the condition that the total cost of these components is minimal (Table 6). For calculations purposes, the following cost of components was assumed: cement—88 EUR/t, ground blast-furnace slag—27 EUR/t, superplasticizer—2.85 EUR/t. These cost values were taken on the basis of market prices in Ukraine on 10 April 2021.

2. According to the mathematical model of W/C (see Table 4), calculate the value of W/C at the optimal factors values and concrete-mix water demand:

$$W/C = 0.26,\ W = C\cdot(W/C) = 492\cdot 0.26 = 126\ l/m^3$$

3. Find the crushed stone consumption from the condition that its volume should be 0.35 of concrete volume:

$$CS = 0.35\cdot\rho_{CS} = 0.35\cdot 2700 = 945\ \text{kg/m}^3$$

4. Find the sand consumption:

$$S = \left(0.65 - \frac{C}{\rho_C} - \frac{W}{\rho_W} - \frac{Sl}{\rho_{Sl}}\right)\cdot\rho_{CS} = \left(0.65 - \frac{492}{3100} - \frac{126}{1000} - \frac{200}{2900}\right)\cdot 2650 = 786\text{kg/m}^3$$

Table 6. Optimization results for mathematical models.

Indicator	Units	Value
Optimal values of factors		
Cement consumption per 1 m^3	kg	492
Slag consumption per 1 m^3	kg	200
SP content	%	1
Concrete quality indicators		
Normative compression strength at 1 day	MPa	19.3
Calculated * compression strength at 1 day	MPa	21.2
Normative value of modulus of elasticity at 1 day	MPa	23,000
Calculated * value of modulus of elasticity at 1 day	MPa	23,000
Normative compression strength at 28 days	MPa	64.3
Calculated * compression strength at 28 days	MPa	81.0

* The values of indicators calculated according to mathematical models for corresponding optimal values of factors (cement, slag and SP contents).

5. Find the superplasticizer content:

$$Б_{SP} = \frac{C \cdot SP}{100} = \frac{492 \cdot 1}{100} = 4.92 kg$$

Concrete composition per 1 m^3 is:

Cement—492 kg; water—128 *l*; SP—4.92 kg; slag—200 kg; sand—780 kg; crushed stone—945 kg. A spreadsheet with the calculation performed can be viewed in Supplementary Materials (Table S1: Numerical example).

When using raw materials that differ from those used in the obtaining models (Table 4), it is effective to use the method of experimental-calculation adaptation [26], allowing to adjust the values of models' coefficients considering laboratory control data of the required indicators.

6. Conclusions

The properties of self-compacting concrete with organic-mineral modifier, including blast-furnace granulated slag and polycarboxylate-type superplasticizer, have been studied. Mathematical models of cubic and prismatic strength, as well as modulus of elasticity at different concrete ages were obtained.

The combined action of blast-furnace slag and polycarboxylate superplasticizer contributes to an increase in the self-compacting concrete elasticity modulus by 1.5–1.7 times. A method for composition design of self-compacting concrete using blast-furnace slag and polycarboxylate superplasticizer is proposed. The method allows for the consideration of the strength of prisms and also the influence of technological factors that characterize the SCC composition when determining its modulus of elasticity.

The use of the obtained mathematical models enables the optimization of the SCC composition in order to provide the necessary strength and deformability characteristics at minimum cost of the final product.

Supplementary Materials: The following supporting information can be downloaded at: https://www.mdpi.com/article/10.3390/ma15124110/s1, Table S1: Numerical example.

Author Contributions: Conceptualization, V.Z. and L.D.; methodology, V.Z. and D.K.; software, D.K. and Y.R.; validation, V.Z., L.D. and Y.R.; formal analysis, V.Z. and D.K.; investigation, V.Z.; resources, Y.R.; data curation, V.Z.; writing—original draft preparation, V.Z. and L.D.; writing—review and editing, Y.R.; visualization, V.Z.; supervision, L.D.; project administration, Y.R.; funding acquisition, Y.R. All authors have read and agreed to the published version of the manuscript.

Funding: This research received no external funding.

Conflicts of Interest: The authors declare no conflict of interest.

References

1. Okamura, H.; Ozawa, K.; Ouchi, M. Self-compacting concrete. *Struct. Concr.* **2000**, *1*, 3–17. [CrossRef]
2. Ohta, A.; Sugiyama, T.; Tanaka, Y. Fluidizing Mechanism and Applications of Polycarboxylate-Based Superplasticizers. In Proceedings of the SP-173: The Fifth CANMET/ACI International Conference, Rome, Italy, 7–10 October 1997; pp. 359–378.
3. Long, G.; Gao, Y.; Xie, Y. Designing more sustainable and greener self-compacting concrete. *Constr. Build. Mater.* **2015**, *84*, 301–306. [CrossRef]
4. Khudhair, J.; Chkheiwer, A. Structural Behavior of Reinforced Self Compacting Concrete. *Al-Muthanna J. Eng. Sci.* **2016**, *3*, 1–11.
5. Kolanjinathan, P.; Perumal, P. An Experimental Investigation and Comparison of Structural Behavior of RCC Columns and Reinforced Fly Ash Added Self Compacting Concrete Columns. *Eur. J. Sci. Res.* **2011**, *61*, 131–143.
6. Esping, O.; Löfgren, I. Investigation of early age deformation in self-compacting concrete. In Proceedings of the 2nd International Symposium on Advances in Concrete through Science and Engineering, Quebec City, QC, Canada, 11–13 September 2006; pp. 1–16. [CrossRef]
7. Yu, J. Research on the Mechanical Properties of Self-compacting Waste Rubberised Aggregate Concrete. In Proceedings of the 2016 International Conference on Civil, Transportation and Environment, Guangzhou, China, 30–31 January 2016; Atlantis Press: Amsterdam, The Netherlands, 2016; pp. 88–92. [CrossRef]
8. Awal, A.; Rahman, M.M. Deformation behaviour of self-compacting concrete containing high volume palm oil fuel ash. In Proceedings of the 9th Asia Pacific Structural Engineering and Construction Conference (APSEC 2015) & 8th Asean Civil Engineering Conference (ACEC 2015), Kuala Lumpur, Malaysia, 3–5 November 2015; pp. 1–6.
9. Dvorkin, L.I.; Dvorkin, O.L. *Bases of Concrete Science*; Stroybeton: Saint Petersburg, Russia, 2006; 690. (In Russian)
10. Yu, M.; Bazhenov. *Modified High Quality Concrete*; Association of Civil Engineering Education: Moscow, Russia, 2006; 368. (In Russian)
11. Safiuddin, M.; Jumaat, M. Fresh properties of self-consolidating concrete incorporating palm oil fuel ash as a supplementary cementing material. *Chiang Mai J. Sci.* **2011**, *38*, 389–404.
12. Domone, P. A review of the hardened mechanical properties of self-compacting concrete. *Cem. Concr. Compos.* **2007**, *29*, 1–12. [CrossRef]
13. Montgomery, D.C. *Design and Analysis of Experiments*, 5th ed.; Wiley: Hoboken, NJ, USA, 2000; 688p.
14. Box, G.E.P.; Hunter, J.S.; Hunter, W.G. *Statistics for Experimenters: Design, Discovery, and Innovation*, 2nd ed.; Wiley: Hoboken, NJ, USA, 2005; 672p.
15. Dvorkin, L.; Dvorkin, O.; Ribakov, Y. *Mathematical Experiments Planning in Concrete Technology*; Nova Science Publishers: Hauppauge, NY, USA, 2012; 175p.
16. *EN 206-1:2000*; Concrete—Part 1: Specification, Performance, Production and Conformity. CEN: Brussels, Belgium.
17. *EN 12390-1:2021*; Testing Hardened Concrete—Part 1: Shape, Dimensions and Other Requirements for Specimens and Moulds. CEN: Brussels, Belgium.
18. *Lea's Chemistry of Cement and Concrete*, 4th ed.; Hewlett, P. (Ed.) Butterworth Heinemann: Oxford, UK, 2003; 1092p, p. 1092.
19. Neville, A.M. *Properties of Concrete*, 4th ed.; Wiley & Sons: New York, NY, USA, 1996; 844p.
20. Topçu, I.B. High-volume ground granulated blast furnace slag (GGBFS) concrete. *Eco-Effic. Concr.* **2013**, 218–240. [CrossRef]
21. *ACI 318-19*; Building Code Requirements for Structural Concrete. American Concrete Institute: Farmington Hills, MI, USA, 2019.
22. Batrakov, B.G. *Modified Concrete*; Stroyizdat: Moscow, Russia, 1998; 768p. (In Russian)
23. Sviridov, N.V.; Kovalenko, M.G. Concrete based on ordinary cement with strength of 150 MPa. *Beton Zhelezobeton* **1990**, *2*, 7–9. (In Russian)
24. Dvorkin, L.; Bordiuzhenko, O.; Zhitkovsky, V.; Marchuk, V. Mathematical modeling of steel fiber reinforced concrete properties and selecting its effective composition. *IOP Conf. Ser. Mater. Sci. Eng.* **2019**, *708*, 012085. [CrossRef]
25. Kuneva, V.; Milev, M.; Gocheva, M. Modeling the transportation assesment with MS excel solver. *AIP Conf. Proc.* **2021**, *2333*, 150005. [CrossRef]
26. Dvorkin, L.; Dvorkin, O.; Ribakov, Y. *Multi-Parametric Concrete Compositions Design*; Nova Science Pub Inc.: New York, NY, USA, 2013; 233p.

MDPI AG
Grosspeteranlage 5
4052 Basel
Switzerland
Tel.: +41 61 683 77 34

Materials Editorial Office
E-mail: materials@mdpi.com
www.mdpi.com/journal/materials

www.ingramcontent.com/pod-product-compliance
Lightning Source LLC
Chambersburg PA
CBHW041402210926
51850CB00013B/3154